Maintenance Management and Engineering

Maintenance Management and Engineering

[For B. Tech., M. Tech. and MBA Courses]

Tanmoy Deb

Head, Department of Electrical Engg.
Apeejay College of Engineering, Gurgaon.
Ex-plant Manager, Whirlpool of India.

Ane Books Pvt. Ltd.

New Delhi ♦ Chennai

Maintenance Management and Engineering
Tanmoy Deb

First Edition 2011
Reprint : 2015, 2018, 2020, 2023

Published by

Ane Books Pvt. Ltd.
4821, Parwana Bhawan, 1st Floor, 24 Ansari Road, Darya Ganj, **New Delhi** - 110 002, India
Tel.: +91(011) 23276843-44, Fax: +91(011) 23276863
e-mail: kapoor@anebooks.com, anebooks@vsnl.net,
Website: www.anebooks.com

Branch Office :

- Avantika Niwas, 1st Floor, 19 Doraiswamy Road, T. Nagar,
  Chennai - 600 017, Tel.: +91(044) 28141554, 28141209
  e-mail: anebookschennai@gmail.com, rathinam@anebooks.com

ISBN : 978-93-8061-858-6

Printed at: Repro Knowledgecast Limited, Thane

Dedicated

to

my Beloved Parents

Late Sh. B. B. Deb and Smt. Maya Deb

Preface

When I started my professional career in maintenance at Tata Chemicals, I could not locate any book on this subject. So I had to learn the basics on the job, from seniors and from numerous training programs that I attended later on. In fact, I did not known where to start! So I understand the difficulties faced by fresh engineering graduates entering this field. Now-a-days, this subject is taught at B. Tech. (Mechanical), B. Tech. (Production), B. Tech. (Industrial), B. Tech. (Mining Machinery), M. Tech. (Industrial), M. Tech. (Maintenance Engg.) and MBA (Operations Management). However, very few books are available in the market written by India authors. Some of the books contain rigorous mathematical treatment or do not cover relevant topics. Keeping this problem in view, this book has been written with a wide variety of related topics and covers syllabus prescribed by various universities including associateship examination of Institution of Engineers (India), Indian Institution of Industrial Engineers *etc*.

The subject matter has been presented in a plain and simple language devoid of rigorous mathematical treatment. Sufficient number of real life case studies have been included at the end of each chapter to supplement theoretical input. Practice questions and caselets (wherever necessary) have been added at the end of each chapter. The content of this book are an ideal mix of theory and practice. This book will be useful to students as well as to fresh engineering graduates and practising engineers.

All relevant topics are grouped under a common heading. There are five different common headings or "themes" designated as parts I to V. All topics logically interlinked are grouped under these heads. Hence, reader is advised to go through the book chapterwise for better understanding of the subject. Parts I to IV are related to maintenance management while Part V is dedicated to maintenance engineering.

I take this opportunity to thank my father late Sh. B. B. Deb and mother Shrimati Maya Deb for being constant source of inspiration. I thank my brother Dr. Tapomoy Deb, Dy. G. M. (J. K. Lakshmi cement) for continuous encouragement and support. He had been a pillar of strength and had provided critical inputs during various stages of preparation of manuscript. I thank my sister inlaw Mrs Swagata Deb for constant encouragement.

I am also thankful to Sh. Amod Bhatnagar, Manager, Indian Institution of Plant Engineers for allowing unrestricted access to the library.

I also thank my publisher for publishing this book in a short time. Ultimately I thank the almighty who made it possible.

Any suggestion for the improvement of this book would be gratefully acknowledged.

—Author

Contents

Detailed Contents

Part-III

OPERATIONAL ASPECTS OF MAINTENANCE MANAGEMENT

Part-V

MAINTENANCE ENGINEERING

PART I

Fundamentals of Maintenance Management

Strategic Outline

This part deals with the basic concept of maintenance. It familiarizes the reader with complexities of maintenance function. The objectives, functions and importance of maintenance department have been explained in detail. Interlinking and difference with respect to other departments are discussed. This part gives a bird's eye view of maintenance function and sets the ball rolling for in-depth study.

Chapters in this section

1. Concept of Maintenance
2. Maintenance Strategy and Organisation

1

Concept of Maintenance

Learning Objectives

After reading this chapter, you will be able to:

- Understand the concept of maintenance.
- Understand the complexities and challenges in maintenance.
- Apply basic management tools for analysis of maintenance problems.
- Get an overview of maintenance function.

1.1 OPENING CASE

Once upon a time, technocrats' meeting was held wherein engineering professionals from diverse field *viz*. Production, Quality control, Research and development, Project, Planning, Vendor development *etc.*, participated.

A challenge was thrown before everyone to make a cow to nod its head (to say no). A huge sum of money was offered as a reward. Engineers from every discipline tried their luck but the cow would nod in affirmative (nodding its head up and down).

At the end, there was no hope and reward was lying unclaimed. Suddenly, a shabbily dressed man with a harassed look appeared on the scene. Seeing him approaching, the crowd burst into peel of laughter. However, the person unmindful of surrounding, whispered something into cow's ear. Lo and behold! The cow vehemently nodded its head implying 'no'. The crowd was taken-in for a surprise and there was a pin-drop-silence.

Somebody from the crowd mustered courage and asked the man, "Sir, what magic did you do"? The man smiled at last and said, "I just whispered—*I am a maintenance engineer, would you like to take up my job for a day as I am getting married*"?

This essentially brings out the role and responsibility of a maintenance engineer in an organisation. The maintenance job is extremely challenging, stressful and hazardous and involves frequent running-up. Inspite of putting up his best efforts, at the end of the day, he ends up getting blame for problems. Today, maintenance is not only a science but an art. Given the globalisation of economy, competition has become complex, uncertain and cut-throat. Efficiency and cost of production has assumed unparalled importance. No organisation can afford to compete without improving plant efficiency and effectiveness, both continuously and consistently. Plant efficiency and effectiveness is dependent upon the individual efficiency, their maintenance and operation. Due to this, maintenance is a technical and a management function. A maintenance engineer has to handle machinery, manpower, money and technology.

Fig.1.1 What ? maintenance job !!!

In the global village of 21st century, India and China have emerged as preferred destination as production bases. North American and European multi-national companies are shifting their production bases to Asia.

There are two major reasons for this global shift—cost advantage and availability of large pool of scientific and technical manpower. This has resulted in unprecedented growth of technology of production. Maintenance function is no more confined to back-room rather with infusion of latest technology, it requires induction of technocrats with a fresh mind-set. The discipline of maintenance has become an independent field of specialization both in engineering and in management. Various universities and Institutes are offering postgraduate level courses *viz.*M. Tech. (Maintenance Engineering), MBA (Operations Management) with maintenance management specialization.

In the above context, maintenance discipline is explored both from engineering and management aspect.

1.2 HISTORICAL BACKGROUND OF MAINTENANCE

Machines were invented by human beings to make life more comfortable and to perform those tasks which were extremely difficult, hazardous and monotonous.

Out of several inventions, some noteworthy are Iron furnace (10th century), Spinning wheel (13th century), Steam engine (18th century) etc. The concept of integrated manufacturing was developed by Henry Ford in 1913. Thereafter, MIT (Massachusetts Institute of Technology), USA, developed numerically controlled machines in 1952. These machines were used to manufacture complex profiles with extremely good accuracy. Later on, industrial robots were developed in 1954 (Cyril Walter Kenward, UK).

In the utility domain, dedicated band of persons were running and maintaining space heating-cooling systems, water distribution system, electricity generation and distribution, sewage disposal systems etc. These people worked as 'operations man' and took-up the job of "maintenance man" when the system failed to operate. Hence, the maintenance philosophy was "to fix when system fails". This period was 1940s and 50s (Ist generation of maintenance). Then in 1960s onward upto 1980s or so, a need was felt for higher plant availability and longer equipment life. Hence, maintenance system changed into second gear involving planning and scheduling of maintenance work and periodic overhauling concept. From 1980s and onwards—computers made in-roads into maintenance profession.

Introduction of computers totally changed the scenario. The maintenance system was no more limited to scheduled maintenance alone rather condition based maintenance, system reliability studies, hazard studies and expert systems were used. This is depicted in Fig. 1.2.

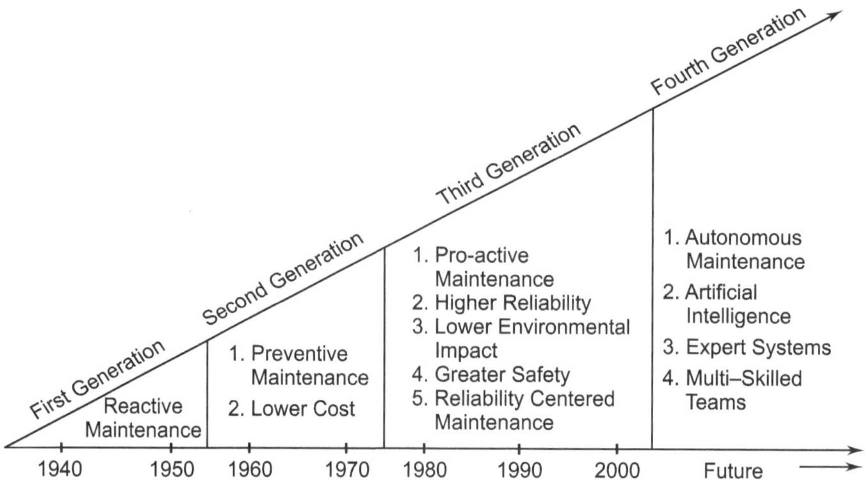

Fig. 1.2 History of Maintenace Practices.

The first generation of maintenance saw a reactive maintenance approach *i.e.*, fix it when it broke. The second generation ushered-in preventive maintenance concepts *i.e.*, fix it before it breaks. The third generation went on to pro-active maintenance *i.e.*, monitoring condition of machines before doing maintenance with higher stress on reliability and safety and causing lower environmental impact on the eco-system. The fourth generation is going to be autonomous maintenance (self-driven maintenance) stage employing self diagnostic and highly automated machines, based on artificial intelligence and expert systems. The maintenance will be carried out by multi-skilled teams.

To sum up, in the words of Henry Ford (father of mass production) "*Put all machinery in the best possible conditions, keep it that way and insist on absolute cleanliness everywhere in order that a man may learn to respect his tools, his surroundings and himself*".

1.3 BASICS OF MAINTENANCE

Each one of us is involved in some kind of maintenance in our day-to-day activities. Have you not ever replaced a broken heater element? Or a fuse in distribution board? Or a flat tyre? Or lubricated wheels of a kid's tri-cycle? All these activities are examples of simple maintenance.

But when a complex and big automated plants are concerned, the maintenance is more than adjustment or lubrication. It involves good deal of management and engineering skills of very high order.

The focus of maintenance function is to ensure that all organisational assets are so maintained that these assets perform designated functions efficiently and are available on-demand.

A maintenance system requires input such as Manpower, Material, Technology and Energy. The output is functionality of machine so that it performs its intended function effectively. This is depicted in Fig. 1.3.

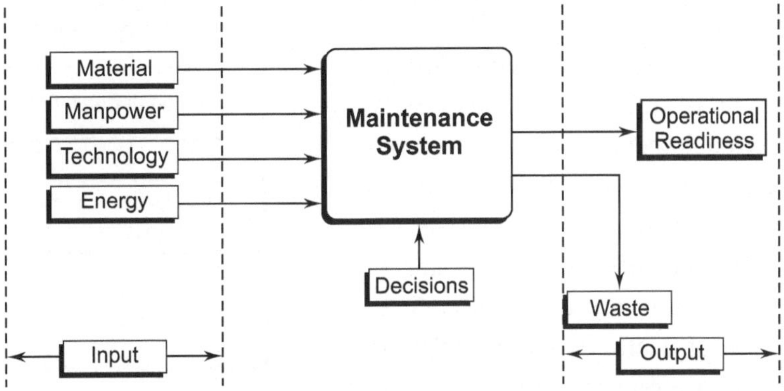

Fig. 1.3 A Typical Maintenance System.

All machines undergo wear and tear due to continuous running and may sometimes fail to perform due to various reasons. The maintenance function restores operation of these machines back to running condition. Also, it involves enhancement of life of running machines through overhauling, modifications, retrofitting, refurbishment etc.

So maintenance can be defined as—Process of keeping equipments/ Machineries/systems/plants in operational state at designed efficiency and capacity.

British standard (BS 3811:1984) defines Maintenance as “A combination of any action carried to retain an item in or restore it to an acceptable condition”.

European Federation of National Maintenance Society defines maintenance as “All actions which have an objective to retain an item or restore it to a state in which it can perform the required function. The action include the combination of all technical and corresponding administrative, managerial and supervision control”.

To summarize, maintenance is scientific upkeep of equipments and associated systems so that equipments are available to users at minimum cost, optimum efficiency and without interruption.

Hence, maintenance involves whole gamut of upkeep of physical assets from equipment birth (purchase) to death (de-commissioning or disposal).

Figure 1.4 shows life cycle of equipment. First step is the decision to purchase an equipment. After this, specification are decided and finally equipment is purchased. The newly purchased equipment is commissioned and put to use. After several years of use, the equipment shows deterioration in performance then life extension strategies are applied so as restore its condition. After completion of useful economic life, the equipment is decommissioned.

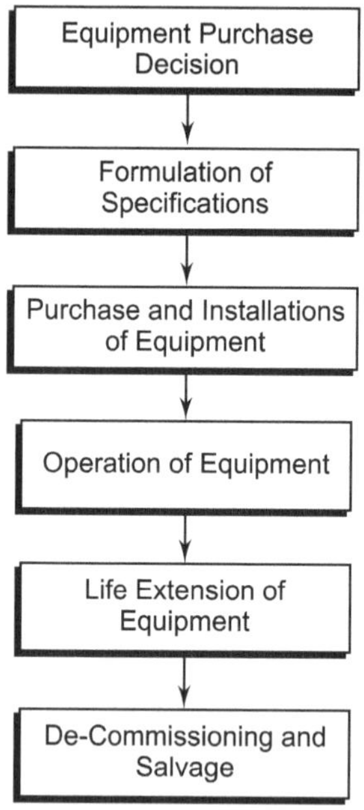

Fig. 1.4 Life Cycle of an Equipment.

1.4 CHALLENGES IN MAINTENANCE

Maintenance function is characteristically different from other disciplines such as Production, Quality Control, Project, Planning, Research and Development etc. The differences are due to the following factors:

1. **Complex Nature of Job:** No two maintenance jobs are similar. Every maintenance job is unique due to varieties, complexities and risks involved.
2. **Changing Technology:** Today, technology is changing very fast. For example, a new wrist watch purchased three months ago becomes 'out of fashion' or latest models of gizmos *viz.*, laptops, i-Pods *etc.*, become obsolete within a year. Even car manufacturers bring-in new models every 1–2 years. This change in technology requires change in maintenance practices and approach. Hence, maintenance function is continuously evolving keeping in pace with the technological changes.
3. **Tight Schedule:** Maintenance activities are always under pressure due to production need. This compresses the 'normal' available time required for maintenance.

4. **Multifunctional Team Approach:** Maintenance jobs are multi-disciplinary in nature and hence require a multifunctional team approach. Maintenance crew needs to be multiskilled to perform complex maintenance job. For example, job may require knowledge of mechanical, electrical, electronics or even software disciplines.
5. **Limited Resources:** Maintenance resources *viz.*, spares, skilled manpower, tools and tackles etc. are limited and a maintenance engineer requires optimum use of available resources. At times, he has to be resourceful and ingenuous in the absence of required resources.
6. **Uncertainty and Risks:** All maintenance jobs involves a certain degree of uncertainty and risks due to complexity of problem, availability of limited resources and time.
7. **Obsolescence:** Obsolescence of technology requires frequent and costly updation of equipments, system and manpower skills.

   Hence, maintenance function is a complex and heavily dependent upon technology and skilled manpower.

1.5 OBJECTIVES OF MAINTENANCE

Objective of maintenance is not merely to fix a "breakdown" but much more. Followings are the major objectives of maintenance:

(*i*) To maximize equipment availability at minimum cost and time.

(*ii*) To ensure safe operation of equipments.

(*iii*) To monitor and improve energy consumption.

(*iv*) To optimize inventory level of spares parts and consumables.

(*v*) To enhance life of capital equipments.

(*vi*) To provide input to management for capacity expansion, renovation and modernization.

(*vii*) To comply with the latest laws of the land.

(*viii*) To improve designed productivity of the system.

(*ix*) To maintain harmonious industrial relations.

(*x*) To improve skills set of maintenance personnel.

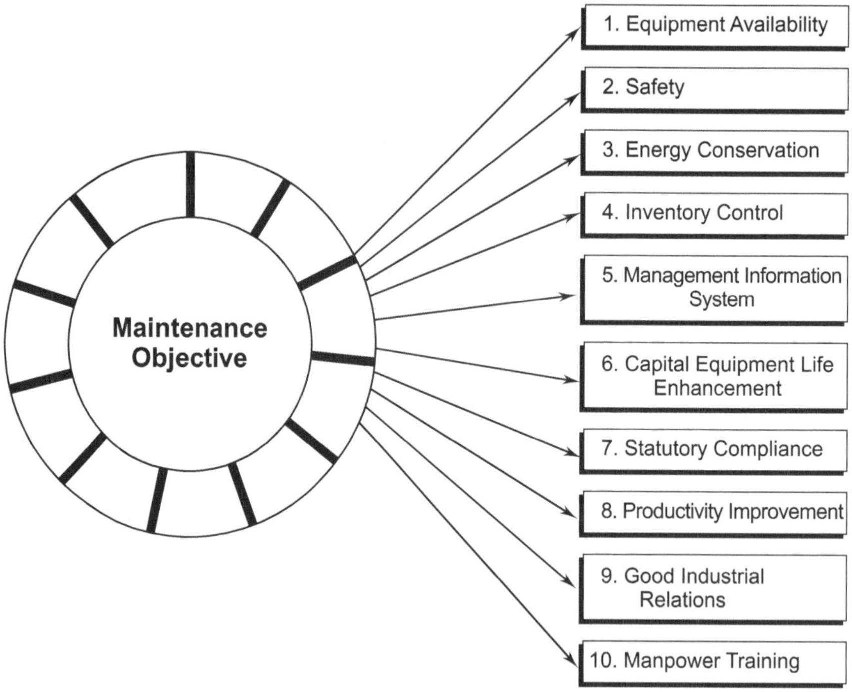

Fig. 1.5 Maintenance Objectives.

1.6 BENEFITS OF MAINTENANCE

Followings are the benefits of Maintenance:

(*i*) Improved reliability and availability of equipments.

(*ii*) Safety of operating personnel.

(*iii*) Higher return on investment.

(*iv*) Extended useful life of equipments.

(*v*) Reduced pollution and waste generation.

(*vi*) Satisfied users leading to industrial peace.

(*vii*) Reduced material and man-hours requirement.

(*viii*) Improved productivity.

(*ix*) Optimum equipment life cycle cost.

(*x*) Reduced energy cost.

1.7 FUNCTIONS OF MAINTENANCE

Followings are the functions of Maintenance:

(*i*) To perform scheduled maintenance on plant and machinery.

(*ii*) To plan and execute capital and revenue job budgets.

(*iii*) To procure and control materials (spares, consumables *etc.*) within budgeted norm.

(*iv*) To monitor and improve equipment performance.

(*v*) To meet statutory requirement as per law of the land.

(*vi*) To practice TPM, Six-Sigma, GEMBA-KAIZON, 5-S *etc.*

(*vii*) To plan and execute capacity expansion and modernization.

(*viii*) To consistently and continuously improve knowledge and skill level of maintenance manpower.

(*ix*) To develop standard operating procedures, maintenance policies and maintenance manual.

(*x*) To periodically generate MIS (Maintenance Information System).

(*xi*) To maintain proper record (*e.g.*, equipment history, modifications, relevant drawings and specifications, department correspondence, all licenses/permits/No Objection Certificate as required by law, various test reports, breakdown and preventive maintenance record *etc.*).

(*xii*) To ensure compliance of quality systems such as ISO, QS, TS, OHSAS *etc.*, as adopted by the organisation.

1.8 BASIC MAINTENANCE MANAGEMENT TOOLS AND TECHNIQUES

Before delving into details of maintenance management, it is necessary for the reader to acquaint himself with widely used analytical tools.

These basic tools are used to define and analyse a problem. There are number of such tools starting from simple bar chart to force-field analysis.

The numerous management techniques used for analysis are described below:

(*i*) **Bar and Pie Chart:** These are extensively used for general purpose analysis. Consider the following bar and pie charts applications:

(*a*) ***Bar Charts*:** Bar Chart on LHS shows monthly breakdown hours of three plant *A*, *B* and *C*. It clearly shows that breakdown at plant *B* is highest (22 hours) while it is lowest at plant *C* (7.2 hours). On Y-axis, it could be other operational parameter such as maintenance cost, overtime hours, plant power factor *etc.*, to suit the analytical need.

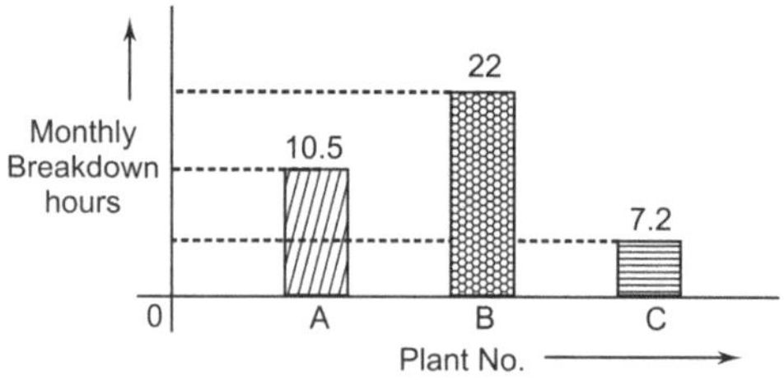

Fig. 1.6 Plants *vs* Breakdown Hours.

Bar Chart in Fig. 1.7 shows the duration of maintenance of various equipments during shutdown maintenance. This chart can be used for estimation of duration of overhaul and for progress review. Similar charts can be used for projects/installation and commissioning of equipments.

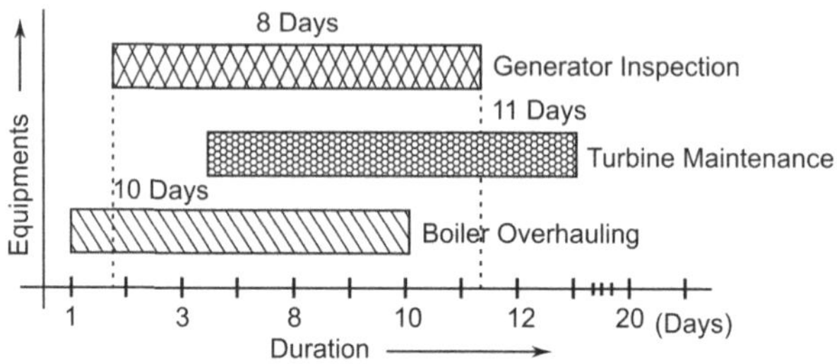

Fig. 1.7 Equipment Maintenance *vs* Duration Bar Chart.

(*b*) ***Pie Chart:*** Pie chart can be used to depict fuel cost of different fuels used in a plant, preventive maintenance cost as part of overall maintenance cost *etc.*

Figure 1.8 shows annual plant maintenance expenditure of *A*, *B*, *C*, *D* and *E* plants.

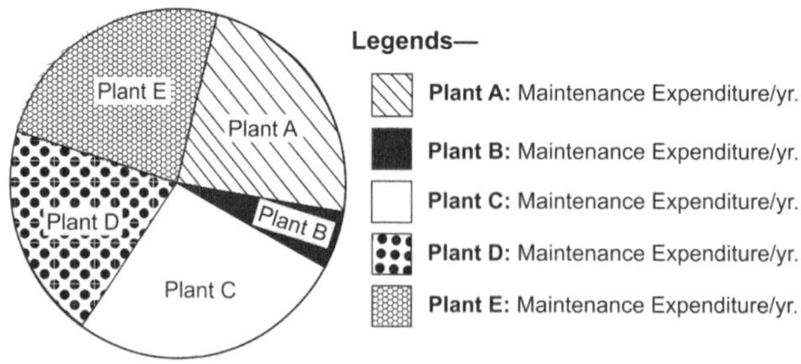

Pie charts shows that plant B has lowest maintenance expenditure while plant C has highest expenditure. This visual presentation helps to grasp the facts at the first book.

Fig. 1.8 Annual Plant Maintenance Expenditure Pie Chart.

(*ii*) **Pareto Analysis:** Pareto analysis or 80–20 rule was founded by Italian economist Wilfredo Pareto. He discovered that 80% of the wealth in society is held by 20% of the people.

This analysis helps in finding "vital few and trivial many", For example, in the analysis of monthly household expenses, a large chunk of money is required for meeting food, education and travel expenses, while other expenses require comparatively less money.

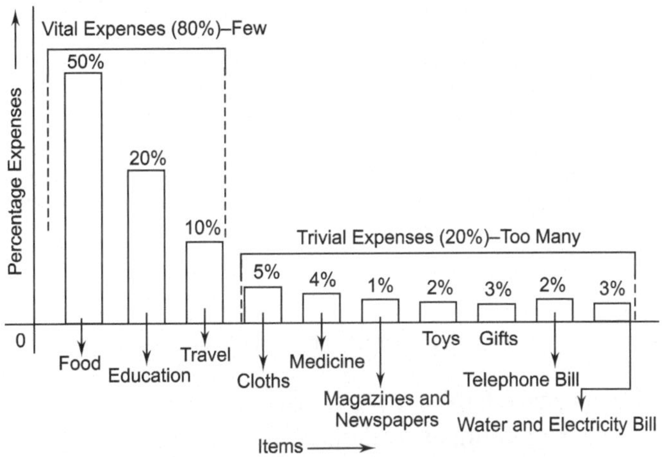

Fig. 1.9 Pareto Chart for Household Expenses.

As can be seen from Fig. 1.9, depicting household expenses. Major expenses are incurred on three items—food, education and travel which constitute 80% of the total expenses. While remaining 20% expenses are incurred on several items. This analysis helps to concentrate on few vital problems.

(*iii*) **Ishikawa or Fish-bone Diagram:** This is a "Cause and effect" diagram used extensively in failure analysis. The pictorial representation looks like a fish-bone and hence the name. Let us use this technique to analyse cause of defeat in a football match.

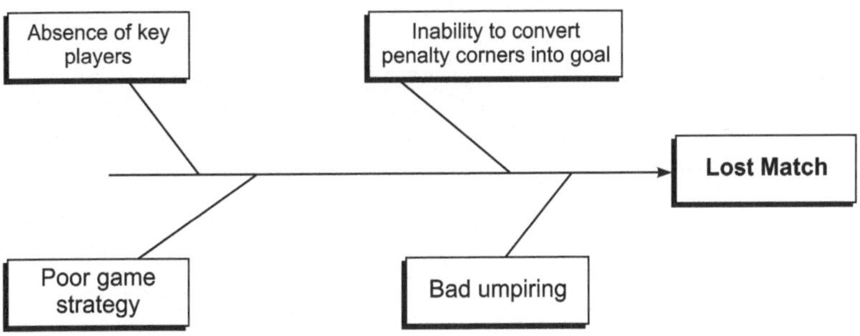

Fig. 1.10 Fish-bone Diagram of a Football Match.

This shows various causes leading to loss of a football match.

(*iv*) **Trend Chart:** Trend chart is used for monitoring a rising or a falling trend of a parameter over a period.

Consider monthly expenses of a student staying in a hostel.

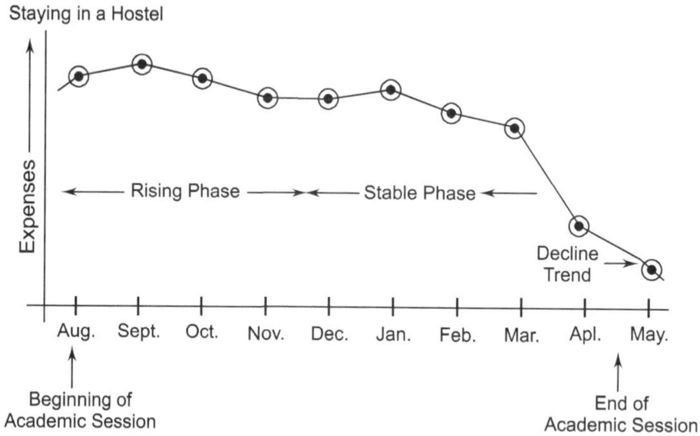

Fig. 1.11 Trend Chart for Students Expenses.

Let us analyse the expenses of a student over one academic calendar. At the beginning of academic session, the expenses shoots up for first three months due to purchase of books, uniforms, fees *etc*. Thereafter, it stabilizes for next four months due to fixed routine expenses. The expenses fall rapidly at the end of the term.

Trend analysis helps us to find "peaks" and "valleys" or unusual sharp changes. This helps us in finding reasons for such peaks and valleys and the probable solutions for the same.

(*v*) **Decision Tree:** A decision tree helps us to take right decision when several options are available. Consider hosting a marriage party. We can construct a decision tree as given below:

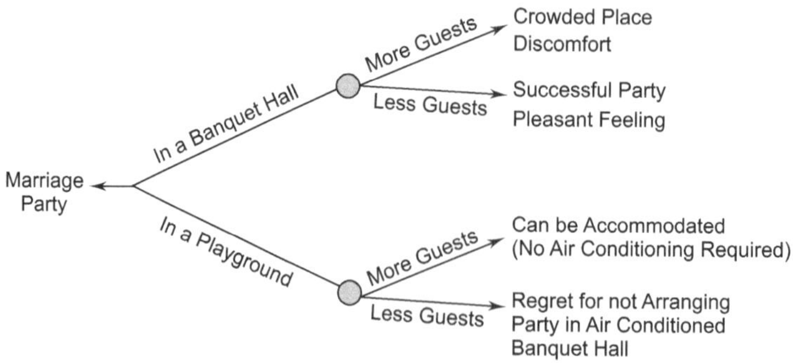

Fig. 1.12 Decision Tree for holding a Marriage Party.

Decision tree depicts a situation when celebration is held in a banquet hall or on a playground. This is first decision stage. The next is, arrival of guests. What happens when more than or less than invited guests turn-up? As we move away from one decision, we are faced with still more alternatives or choices. The process verifies all possible alternatives and their outcomes. This helps us to choose right decision with desired outcome.

(*vi*) **Force-field Analysis:** Before initiating an action, force-field analysis is carried out to find out barriers or hindrances (negative forces) and facilitating or helping influences (positive forces) that will help to achieve goal.

Goal: *Achievement of 80% marks in term-end exam.*

Table 1.1 Force-field analysis

Positive forces (Acting towards achievement of goal)	Negative forces (Acting against achievement of goal)
1. Scholarship (for future studies)	1. Absence of strategy—Devise appropriate strategy
2. Positive image (family and friends)	2. Non-availability of funds for coaching classes—Manage loans from banks parents/friends
3. Good chance of placement	3. Low grades in previous internal tests—Find the reason and eliminate it

The purpose of force-field analysis is to estimate relative strengths and devise ways and means to overcome negative influences.

(*vii*) **Process Flow Diagram:** It shows step-by-step transformation *e.g.*, it shows how a raw material undergoes transformations when it enters input terminal and appears as final product at the output terminal. Consider an example—where a wooden study table is to be fabricated. The process will consist of the following steps:

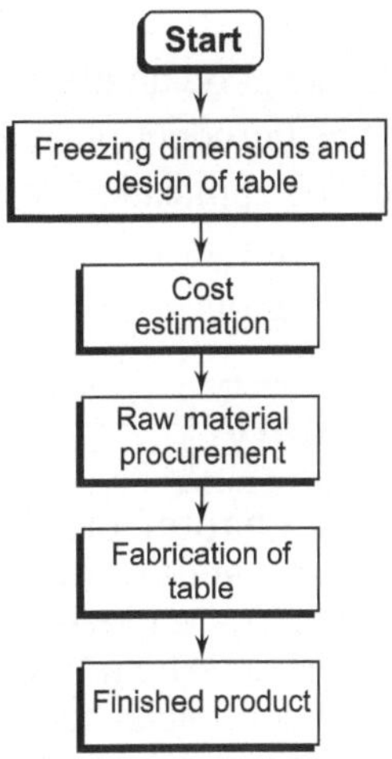

Fig. 1.13 Process Flow Diagram for Fabrication of Study Table.

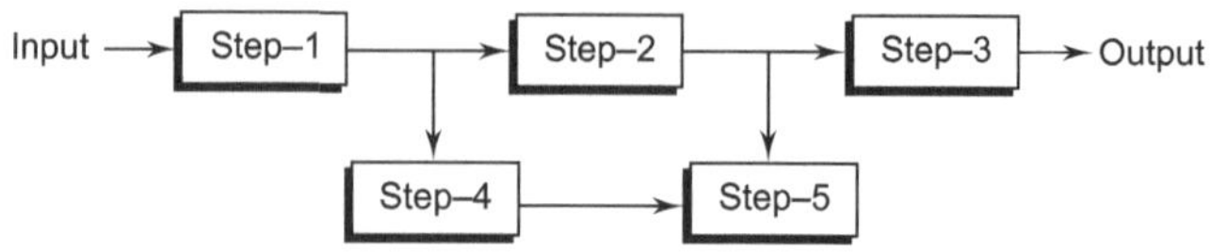

Fig. 1.14 Generalised Process Flow chart.

In general, there can be multiple series or parallel paths depending upon the complexity of process.

(viii) **Scatter Diagram:** It tries to establish relationship between two variables.

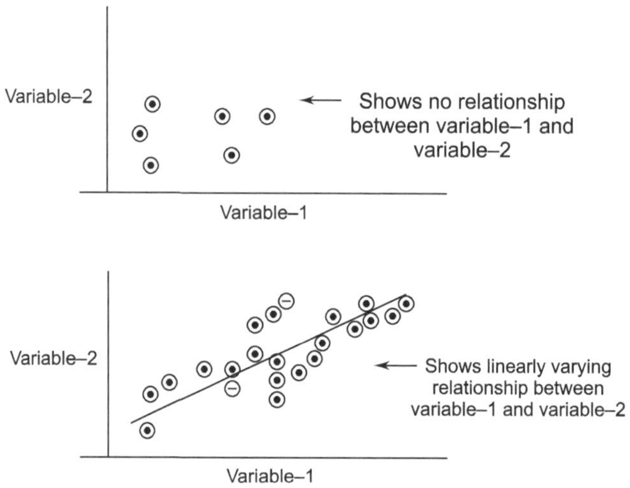

Fig. 1.15 Scatter Diagram.

There are good number of problem solving or analytical tools available but at this stage, introduction to these basic eight tools will suffice. Other analytical tools and their applications will be discussed in subsequent chapters. The purpose of this chapter is to introduce readers to multiple uses of such simple management tools.

Summary

- This chapter provides bird's eye-view of maintenance field. Maintenance grew from a "back-room operation" to the present day sophisticated level. The maintenance philosophy has undergone sea-change from 'reactive maintenance' to "autonomous maintenance". First generation maintenance was reactive maintenance followed by second generation maintenance which laid the foundation for preventive maintenance. Third generation maintenance was characterized by "pro-active" maintenance which laid stress on

Contd...

equipment reliability enhancement. While fourth generation maintenance relies on expert system and artificial intelligence.

- Maintenance has a system concept which consists of inputs and outputs. Man, material, energy, technology and decisions form the input to the system. The system output is "operational readiness". The system also generates waste products such as worn out spare parts, used oils and chemicals *etc.*
- Maintenance is a unique field and thus has its own challenges. This is due to the fact that each maintenance situation is "unique". It is characterized by fast changing technology, multi-team approach and limited resources.
- Basic management tools are discussed at the end of the chapter. These are bar chart, pie chart, pareto analysis, trend analysis, Nishikawa diagram etc. These tools help to identify and analyse a problem which leads to a solution.
- Maintenance as a function is complex and requires knowledge of many fields *viz.* engineering and management. Maintenance engineering principally deals with technical aspects while management part deals with commercial and operational aspects.

Review Questions

1. Define Maintenance. What are the various challenges in a maintenance function?
2. Discuss in detail, benefits of maintenance in an industrial set up.
3. What are functions and objectives of maintenance?
4. How is maintenance function different from production and project?
5. Discuss system concept of maintenance.
6. What is meant by life cycle of equipment?
7. Why does maintenance require multi-functional team approach?
8. Give examples of various management tools used in maintenance function.
9. What changes in technology affect maintenance function?
10. How do you foresee role of maintenance function, ten years from hence?

Discussion Questions

1. Discuss—"Maintenance function require multi-disciplinary approach".
2. Elaborate—"Maintenance and production interests are always conflicting".
3. "Technology is the biggest driver in the maintenance field". Do you agree?
4. "Reliability of system can be improved by using self diagnostic sub-systems". Comment.
5. "Higher the system complexity, lesser the reliability", Do you agree?
6. "Maintenance is not merely lubrication or adjustment rather optimization of available resources to the fullest extent". Discuss.

7. Discuss—"Skilled manpower is the most important asset of a maintenance department".
8. Assess—"Automation is the key to reduced maintenance and higher productivity".
9. "Out of man, material, technology and energy, it is most difficult to manage man". Do you agree?
10. Discuss—"Maintenance is a profit and not an expense centre".

Multiple Choice Questions

1. Output of maintenance system is
   (*a*) Production (*b*) Operational readiness
   (*c*) Work done (*d*) None of these.
2. Function of maintenance is
   (*a*) Cleaning and lubrication (*b*) Reliability improvement
   (*c*) Reduction of energy cost (*d*) All of these.
3. Maintenance is a challenging field due to
   (*a*) Job complexity (*b*) Limited resources
   (*c*) Tight schedule (*d*) All of these.
4. Use of equipment begins with
   (*a*) Purchase of equipment (*b*) Drafting of specifications
   (*c*) Commissioning of equipment (*d*) None of these.
5. Objective of maintenance is
   (*a*) To fix breakdowns (*b*) To improve product quality
   (*c*) To optimize equipment availability
   (*d*) None of these.

True/False Questions

1. Maintenance is cleaning, lubrication and adjustment only.
2. Maintenance function does not produce any waste.
3. Maintenance is highly complex and technology driven.
4. Maintenance requires multi-disciplinary team approach.
5. Maintenance helps to reduce energy consumption and pollution.

ANSWERS

Multiple Choice Questions

1. (*b*) **2.** (*d*) **3.** (*d*) **4.** (*c*) **5.** (*c*)

True/False Questions

1. False **2.** False **3.** True **4.** True **5.** True

❑❑❑

2

Maintenance Strategy and Organisation

Learning Objectives

After reading this chapter, you will be able to:

- Comprehend maintenance strategies and their selection.
- Explore application of strategies to specific situations.
- Examine components and structure of maintenance organisation.
- Design an organisation structure for specific industry.

2.1 INTRODUCTION

Maintenance strategy is a long-term plan covering every single aspect of maintenance management. The strategy contains action plans for achieving future goals. Every organisation decides strategies and is broadly mentioned in a maintenance manual of the organisation. The strategies change broadly due to change in technology, skill level of manpower, criticality and type of equipments, available financial resources *etc.*

2.2 MAINTENANCE STRATEGIES

Maintenance strategies are guided by many factors such as cost of downtime, cost of equipment, availability of skilled manpower *etc.* The maintenance strategies got evolved over a period of time. Initially, it was "breakdown maintenance" or fix the equipment when it breaksdown which no longer works. This type of strategy works when cost of equipment is low and cost of down time is minimal. However, with the development of technology, equipments became more complex and expensive. Hence, a new strategy called "preventive maintenance" was developed. Preventive maintenance means fixing the equipment before it breaks down. This strategy helped to reduce unwanted and costly breakdowns. The next in line was "predictive maintenance" *i.e.* anticipating trouble and taking corrective action before actual failure occurs. Hence, maintenance strategies changed over a period of time due to high cost of breakdown and growing equipment complexity. Routine maintenance refers to fixed time based adjustment, cleaning, lubrication etc.

Let us illustrate various maintenance strategies through an example of a car. A car can have breakdowns such as flat tyre, engine failure, battery failure etc. which leaves us stranded in the middle of a road. To prevent occurrence of such incidents in future, following maintenance strategies can be adopted.

Table 2.1

S. No.	Problem	Maintenance Solution	Maintenance Strategy
1.	Flat Tyre.	Repair when required.	Breakdown maintenance.
2.	Low level of coolant.	Regular checking and topping up.	Preventive maintenance.
3.	Low level of electrolyte in battery.	Regular checking and topping up.	Preventive maintenance.
4.	Vibration.	Wheel alignment and balancing, bolt tightening *etc.*	Routine maintenance.
5.	Non-uniform wear of tyre.	Tyre rotation, correcting tilting *etc.*	Predictive maintenance.
6.	Poor mileage.	Carburettor tuning.	Preventive maintenance.
7.	Engine problems.	Periodic checking.	Preventive maintenance.

The above example of car maintenance illustrates various maintenance strategies adopted for different sections of car. These maintenance approaches evolved over a period of time and are illustrate ahead:

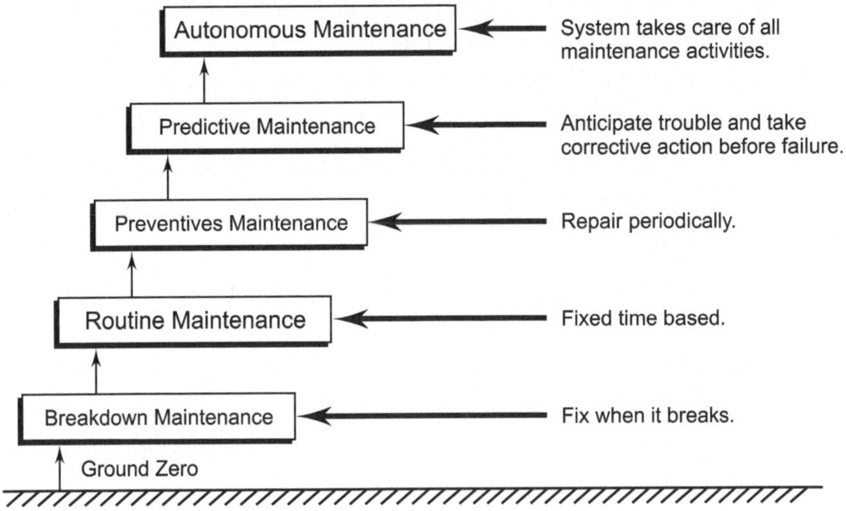

Fig. 2.1 Types of Maintenance Approaches.

It is seen that for same car, different maintenance strategies are adopted for different sections of the car. Here, the choice of strategy is dictated by economics, convenience and consequences.

Similarly, in an organisation, selection of maintenance strategy is complex and depends on the following factors:

(*i*) Criticality of operation (from production point of view).

(*ii*) Operational safety.

(*iii*) Cost of machine *Vs* maintenance cost.

(*iv*) Availability of skilled manpower and spares.

(*v*) Energy consumption and environmental pollution.

(*vi*) Start up and shutdown cost.

(*vii*) Availability of financial resources.

(*viii*) Complexity of equipment.

Let us understand the effect of above mentioned factors on selection of maintenance strategy:

(*i*) **Criticality of Operation:** In any production process, the equipment which is most critical is looked after well. It is like extra care for 'strikers' and 'goal keeper' in a game of football. Critical equipments are those whose stoppage leads to complete disruption of production process, huge down time cost and extremely high cost of repair.

For example, reactor or autoclave in a Urea manufacturing plant is most critical as complete chemical reaction takes place in the reactor. For such critical machines, combination of maintenance policies are used *i.e.*, preventive, predictive and routine maintenance.

(*ii*) **Operational Safety:** Operation of certain equipments may endanger safety of personnel and surroundings due to very nature of their operation.

For example, in a petro-chemical plant, oxidizer is a equipment where the main chemical reaction takes place (oxidiser for "Caprolactum" – a raw material for Nylon). A slight leakage from oxidiser may result in complete destruction of the entire plant.

Hence, periodic inspection, testing and on-line monitoring plays a vital role in the safe operation of the equipment.

(*iii*) **Cost of Machine *Vs* Maintenance Cost:** Sometimes cost of maintenance may turn out to be more than the cost of equipment itself. In such situations, it is prudent to operate the equipment till failure. These are "use and discard" type. For example, in gas-turbine operation, atmospheric air is sucked through number of filters. The air is required for combustion and cooling purposes. The differential pressure across filter bank is monitored and if exceeds a set value, the existing filters are replaced by new ones.

(*iv*) **Availability of Skilled Manpower and Spares:** Maintenance of major capital equipments *e.g.* blast furnace, boilers, turbines, *etc.* are planned by taking into account availability of skilled manpower (in house as well as outsourced) and critical spares among other factors. Maintenance of such equipments are carried out by proper planning during annual or bi-annual maintenance plant shutdowns. Many of the critical spares needed may require long lead time for procurement and hence, proper planning takes into account lead time required for procurement of spares as well as tie-up with original equipment manufactures for expert service.

(*v*) **Energy Consumption and Environment Pollution:** With rising energy cost and concern for environmental pollution, energy consumption is one of the major factors that decides the type of maintenance philosophy adopted.

For example, a proper and timely maintenance of a car results in more mileage and less pollution. This is applicable in case of diesel engines of captive power generation, fired boilers, furnaces *etc*. A timely periodic maintenance not only reduces energy bill but also lowers environmental pollution.

(*vi*) **Start-up and Shutdown Cost:** Some equipments require considerable start up and shutdown time. This results in significant amount of energy being consumed during startup and shut down. Hence, this factor is taken into account for deciding maintenance policy.

For example, a large melting furnace in steel plant requires considerable amount of time and energy in start up and shut down. In case of boilers too, a significant amount of energy is required for start up and shutdown. Hence, when the plant load is very low (*e.g.*, due to major plant breakdown, national holiday *etc.*) instead of stopping the boiler, it is run and kept at low energy input. So, as and when situation improves/ changes, the boiler can brought in normal operation within little time thereby avoiding expensive start up and shut down time and cost. So maintenance of such equipment is planned accordingly.

(*vii*) **Availability of Financial Resources:** It is one of the most important factor that decides the type of maintenance practices adopted for an organisation. Sometimes due to financial compulsions, a major overhauling of a critical equipment may be deferred.

(*viii*) **Complexity of Equipment:** Maintenance policy is influenced by the technology of the equipment. A highly complex and sophisticated equipment requires well trained manpower, expensive spares, special tools and tackles *etc.*

Due to non-availability of well trained in-house maintenance team, the job may be carried out by original equipment manufacturer (OEM). The time of maintenance, in such situation, is dictated by the availability of OEM engineers.

Now, it is clear that maintenance policy is dictated by several factors. Also, different policy may be adopted for different equipments.

2.3 DUTIES AND RESPONSIBILITIES OF MAINTENANCE DEPARTMENT

The entire maintenance operation that includes whole gamut of activities is carried out by a dedicated maintenance department. Maintenance department like any other department is well structured with specialists from various fields *viz.*, Mechanical, electrical, instrumentation, electronic and civil engineering disciplines. The entire department though diverse in constitution works for focused departmental objectives.

Duties and responsibilities of Maintenance department are:

(*i*) Establishing departmental policies, procedures and manual.

(*ii*) Upgradation of maintenance technology and methods.

(*iii*) Upgradation of skills of manpower through training.

(*iv*) Proper maintenance of optimum level of spares inventory.

(*v*) Providing timely service to internal customers.

(*vi*) Proper upkeep of all physical assets of the organisation at optimum efficiency and cost.

(*vii*) Co-ordination with external agencies (contractors, vendors, suppliers *etc.*) and internal customers for proper planning and scheduling of maintenance activities

(*viii*) Maximisation of equipment availability at minimum cost.

(*ix*) Minimisation of cost of spares, energy, manpower and technology.

(*x*) Collection and analysis of maintenance data—technical and financial – for betterment.

(*xi*) Maintenance of harmonious industrial relations.

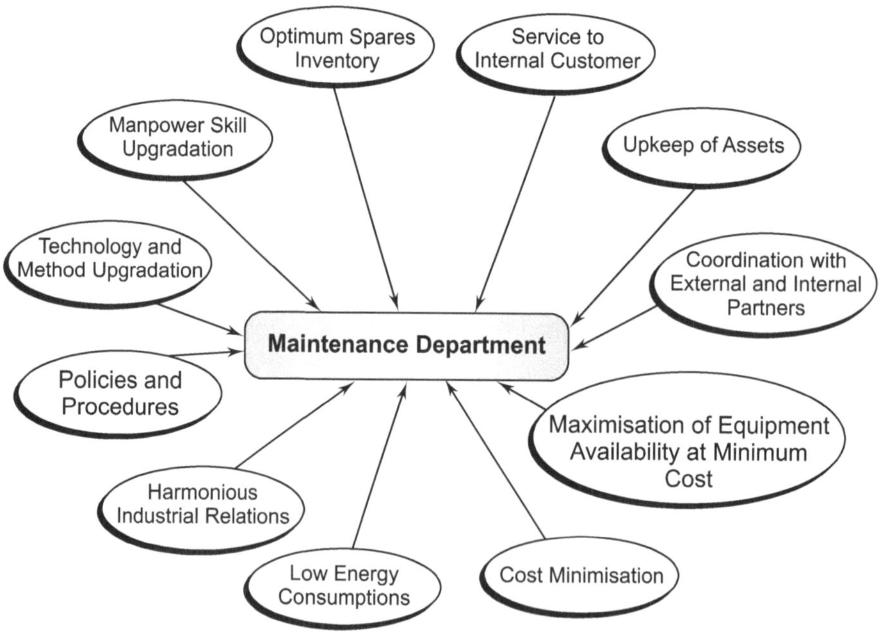

Fig 2.2 Duties and Responsibilities of Maintenance Department.

2.4 DESIGN OF MAINTENANCE ORGANISATION

The purpose of maintenance organisation is to serve internal customers in an efficient and effective manner. However, there are numerous facilitating and hindering factors that decide the type of maintenance organisation. Following key factors influence the structure and type of maintenance organisation:

(*i*) Size and type of organisation.

(*ii*) Maintenance workload.

(*iii*) Cost of breakdown.

(*iv*) Criticality of process operation.

(*v*) Secondary maintenance workload.

(*vi*) Geographic location of plant.

(*vii*) Complexity of equipment/system technology.

(*i*) **Size and Type of Plant:** Generally, organisations are of three categories: large, medium and small. Examples of large organisations are—Petrochemical plants, Refineries, Cement plants, Fertilizer plants, Iron and steel, Pulp and paper *etc.* Examples of medium scale organisations are—Refrigerator manufacturing plant, colour picture tube manufacturing plant, pharmaceutical units *etc.* Examples of small scale organisations are—automotive ancillary, consumer electronics assembly units, telephone assembly units *etc.* The size is defined on the basis of capital invested [according to first schedule of industries (Development and Regulation) Act, 1951].

All large and medium scale organisations maintain their own dedicated maintenance staff. However, in case of small scale, a skeleton maintenance structure is maintained or it is completely outsourced.

Moreover, in case of continuous process plant such as petrochemical fertilizer, cement *etc*, an independent and self sufficient maintenance organisation is maintained.

While in case of batch process plants, organisations may have skeleton maintenance structure or may opt for contract maintenance.

(*ii*) **Maintenance Workload:** If the maintenance workload is heavy and is continuous in nature then dedicated maintenance structure is required.

(*iii*) **Cost of Breakdown:** In case where longer breakdown hours can be tolerated and the plant is situated in the city limits then contract maintenance is one of the preferred option.

(*iv*) **Criticality of Process Operation:** Due to very nature of criticality of process operation, a fully dedicated maintenance structure is preferred.

(*v*) **Secondary Maintenance Workload:** If the organisation requires secondary maintenance workload such as looking after air conditioning system, power generation, water supply, effluent treatment, hot water generation, *etc.* then a well structured and dedicated maintenance organisation is preferred.

(*vi*) **Geographic Location of Plant:** Generally, plants which are located far away from city limits maintain a dedicated self reliant maintenance department. In case of plants located within periphery of city limits, a combination of dedicated maintenance personnel and contract-maintenance strategy is adopted.

(*vii*) **Complexity of Equipment/System Technology:** Plants with state-of-the-art equipments, require dedicated maintenance staff who are well trained in handling these equipments. Since, highly skilled maintenance personnel may not be available on contract.

Hence, whether to have a full-fledged and self-reliant maintenance department or to have part contract maintenance and part own maintenance or completely contract maintenance depends on several factors as discussed above.

2.5 TYPES OF MAINTENANCE ORGANISATION

Once, it is decided to have maintenance organisation, the next step is to decide its structure.

Let us consider the example of a large scale continuous process industry having a completely dedicated and self-reliant maintenance organisation. The proposed structure of such an organisation is depicted in fig. 2.3.

Figure 2.3 shows structure of maintenance organisation. Due to large size of plant, hierarchy is multi-layered. The first structure of organisation is based on functional specialisation *viz.*, operations, finance, HR, sales and marketing *etc.* It is further split into engineering specialisation *e.g.*, electrical maintenance, mechanical maintenance, electronic and instrumentation maintenance and civil maintenance. Each individual engineering discipline is further split into specific tasks such as workshop, plant maintenance, planning, spares and documentation and so on. This layer is further sub-divided into engineers and ultimately to trade specific technicians.

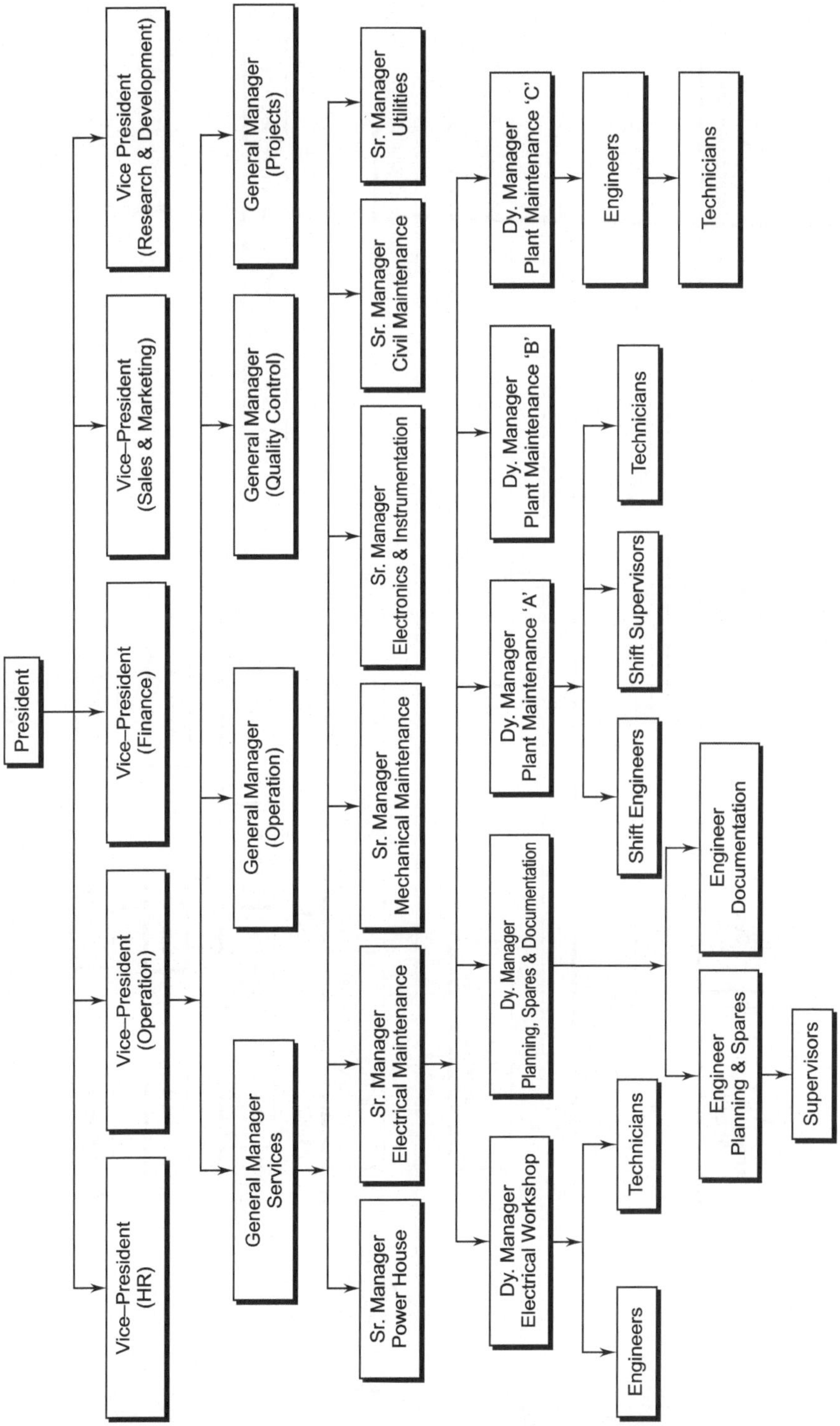

Fig. 2.3 Structure of Maintenance Organisation (Large Plant).

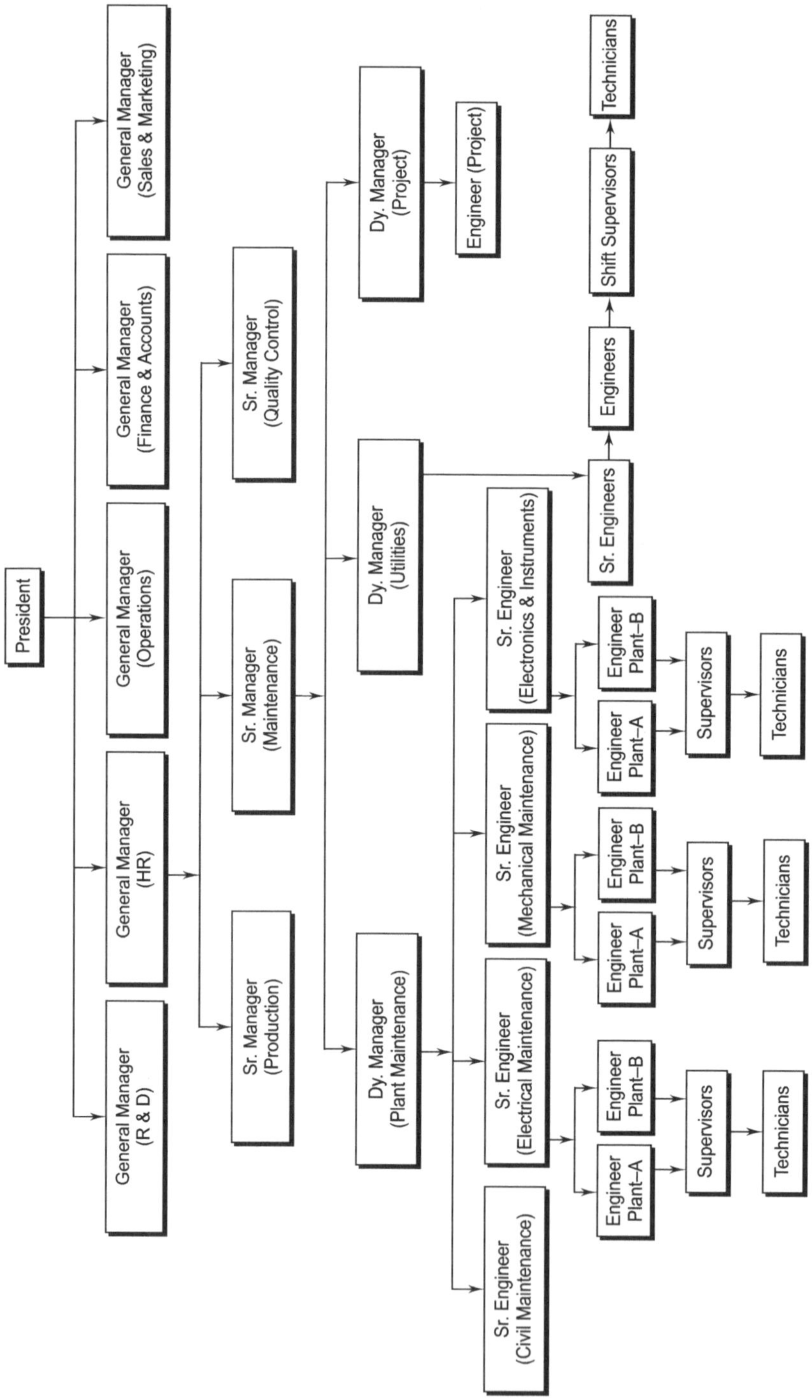

Fig. 2.4. Structure of Maintenance Organisation (Medium scale).

Figure 2.3 shows sub-divisions of electrical maintenance. Such divisions exist for power house, mechanical maintenance, electronics and instrumentation maintenance and utilities.

Figure 2.4 shows the structure of maintenance organisation of medium scale industry. Here, due to smaller size of plant, these are lesser number of layers at maintenance structure. It is broadly divided into civil, electrical, mechanical, electronics and utilities. All plants are looked after by all trades of maintenance in a centralized way.

In decentralized set-up, individual plants have their dedicated maintenance staff covering all trades. Central services such as power generation, water treatment, air conditioning, compressed air, effluent treatment plant remain common to all the plants. Such a plant is shown below—

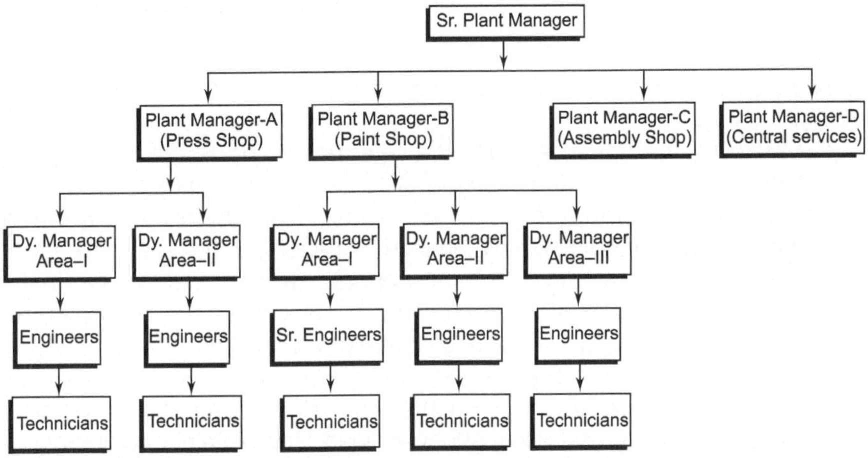

Fig. 2.5 Decentralised Maintenance Organisation.

Now consider the structure of a small scale organisation. As can be seen, number of layers in the organisation are minimum and are discipline specific.

Apart from organisation structure mentioned above, other approaches can be full contract maintenance, past contract part own maintenance *etc*.

It can now be concluded from study of various types of maintenance organisations that each maintenance organisation is unique and is tailor-made for organisations' requirements.

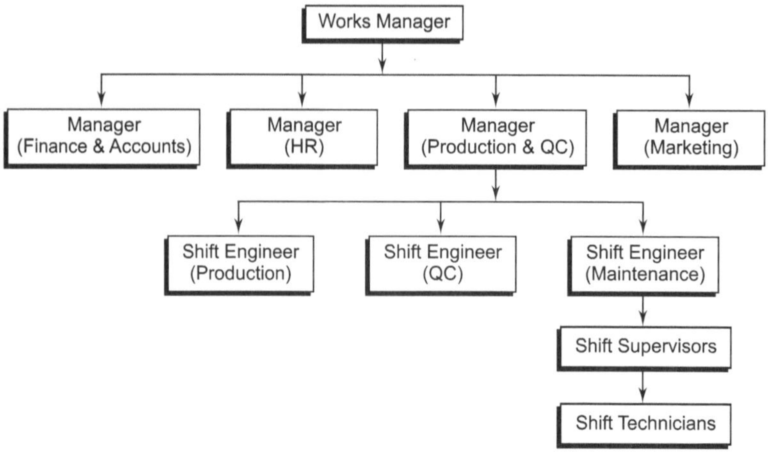

Fig. 2.6 Structure of Maintenance Organisation (Small scale).

Summary

For achieving maintenance target, strategies need to be decided. Strategy is a long-term plan for achievement of a goal. Maintenance strategies are breakdown maintenance, preventive maintenance, routine maintenance and predictive maintenance. Depending upon organisational philosophy and other factors, an appropriate strategy is adopted. For critical equipments, predictive maintenance strategy is adopted while for equipment with lesser importance, breakdown or routine maintenance strategy could be more appropriate. Selection of maintenance strategy depends upon several factors such as criticality of operation, cost of machine, availability of skilled manpower, financial constraints, complexity of equipment *etc.*

Maintenance department discharges a wide range of responsibilities such as upkeep of assets, spares inventory, industrial relations, cost reduction, training of maintenance personnel *etc.* It also maintains all licenses/permits *etc.* required as per law (such as for boilers, electrical systems, fuel storage systems, pollution control *etc.*)

A maintenance organisation is tailored to meet the objective of the organisation. The key factors determine the organisational structure, are size and type of organisation, maintenance workload, criticality of operation, geographic location of plant and equipment complexity. The organisational structure could be centralized or decentralized. It could be based on functional specialisation or geographical location (for multi-location organisation).

Hence, the success of a maintenance department depends upon the selected strategies and structures.

Review Questions

1. How do you choose suitable maintenance strategy for an organisation? Draw hierarchy chart for a car assembly department.
2. Explain the factors responsible for selection of a suitable maintenance policy for an organisation.
3. Why "Cost factor" plays a vital role in selection of suitable maintenance strategy?
4. What are the functions of a maintenance organisation?
5. List and discuss important components of a maintenance organisation.
6. Explain with the help of block diagram, the decision procedure for establishing a maintenance administrative structure.
7. Differentiate among maintenance structures of large, medium and small scale organisations? What differences do you notice?
8. What are various factors that decide the changes in the policies and strategies of a maintenance department?
9. Why are maintenance structures organized as per specific tasks? Is there any other way of organizing maintenance department?
10. What should be the maintenance strategy of a small garment manufacturing concern operating within city limits?

Discussion Questions

1. "Maintenance department requires different kind of organisational structure". Do you agree?
2. "Maintenance manager can improve productivity of an organisation by judiciously selecting appropriate organisation structure". Discuss.
3. Elaborate, "selection of maintenance policies and strategies decide the success or failure of a maintenance department in the long run"
4. Justify, "Overlapping of functional areas result in organisational failure".
5. "Contract-out maintenance could be best suited for small scale organisations.". Defend.
6. "While deciding maintenance structure, a multi-discipline team is preferable compared to discipline specific structure in small scale organisations". Discuss.
7. "Centralized maintenance structure is beneficial for closely located plant clusters". Justify.
8. "For a geographically remote plant, a full fledged and self reliant maintenance department is essential". Discuss.
9. "Economics decides the choice of a particular maintenance system". Elaborate.
10. "Continuous technology upgradation is a pre-requisite to keep away obsolescence". Discuss.

Multiple Choice Questions

1. Preventive maintenance is
   (*a*) Fix when it breaks (*b*) Fix before it breaks
   (*c*) Repair periodically
   (*d*) Anticipate trouble and take corrective action.

2. Selection of maintenance strategy depends on–
   (*a*) Financial resources (*b*) Skilled manpower
   (*c*) Cost of equipment (*d*) All of these.
3. Duty of maintenance department is not
   (*a*) Maintain spares (*b*) Maintain good industrial relations
   (*c*) Reduce energy cost (*d*) Increase production.
4. Contract maintenance is suitable for
   (*a*) Batch process plant (*b*) Large continuous process plant
   (*c*) Medium continuous process plant
   (*d*) All of the above.
5. A Complex and state-of-the-art machinery requires
   (*a*) Preventive maintenance (*b*) Breakdown maintenance
   (*c*) Predictive maintenance (*d*) None of these.

True/False Questions

1. Mere adoption of appropriate maintenance policies and strategies do not result in achievement of organisational goal.
2. A progressive organisation must keep changing policies and strategies for best possible results.
3. Written-down policies are always better in avoiding ambiguities and conflicts.
4. Large organisations that are self reliant do not hire outside services.
5. In a technological complex environment, a multi-discipline approach is the best.

ANSWERS

Multiple Choice Questions

1. (*b*) **2.** (*d*) **3.** (*d*) **4.** (*a*) **5.** (*c*)

True/False Questions

1. True **2.** True **3.** True **4.** False **5.** True

CASELET

Mr. Gurupreet Singh owns fleet of trucks carrying goods all over the country. The trucks keep on experiencing frequent breakdowns due to their age and bad roads. He is in a fix to select right maintenance policy for various sections of truck. Can you decide the type of maintenance strategy to be followed for the following sections:

(*i*) Engine, (*ii*) Tyres, (*iii*) Lighting bulbs, (*iv*) Air and diesel filters.

Give reasons for choice of a particular strategy.

❑❑❑

PART II

Maintenance Management Systems

Strategic Outline

This part deals with familiarisation with various types of maintenance systems, policies and philosophies. It explains basis for choice of a particular policy. It examines the importance and application of planning and scheduling in maintenance. The use of various planning and scheduling techniques have been discussed with examples.

Systems and procedures related to various subsystems of maintenance are explained along with documentation. Various laws as applicable to maintenance function are summarised.

Project management process including tendering has been discussed in detail.

Lastly, salient features of contract management *viz.*, types, element *etc.*, are discussed.

Chapters in this section

3

Maintenance Systems

Learning Objectives

After reading this chapter, you will be able to:

- Understand various types of maintenance systems.
- Explore requirements and elements of preventive maintenance.
- Examine types and philosophies of various maintenance policies.
- Explore policies for other areas of maintenance.

3.1 INTRODUCTION

With rapid industrialisation and advancement of technology, maintenance of industrial assets and facilities is getting more and more complex. With state-of-the-art automated machines, breakdown of even few minutes is not tolerated.

Even a best designed machine requires periodic maintenance which if ignored results in unexpected breakdown and consequent loss of production.

It is easy to carry out maintenance when machines are not in operation, say in a batch process plant or plant operating under single shift. However, in a continuous process plant operating 24×7 and 365 days a year, there is hardly any time available for maintenance. Hence, new techniques are devised to monitor machine performance—both on-line and off-line. Hence selection of maintenance policy becomes utmost important.

In this chapter, attention is focused on various types of maintenance systems, policies and their relative merits and demerits. This knowledge will help us to choose most suitable policy for a particular plant, equipment or a sub-system.

3.2 TYPES OF MAINTENANCE SYSTEMS

There are various types of maintenance systems. The classification of maintenance systems is shown by a flow chart in Fig. 3.1.

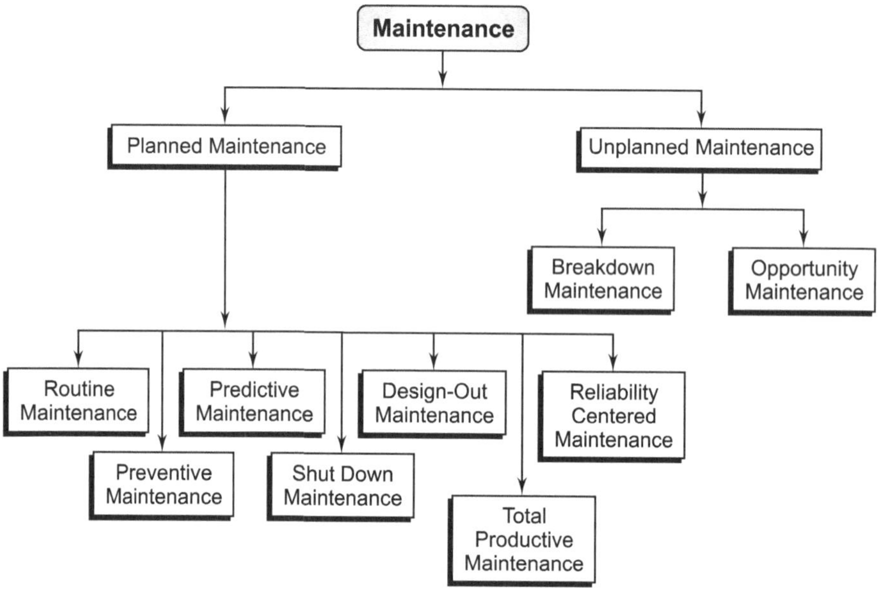

Fig. 3.1 Types of Maintenance.

Maintenance system can be broadly divided into two categories—planned and unplanned.

1. **Planned Maintenance:** Planned maintenance is the essence of modern maintenance system. It refers to maintenance of equipments with advance planning in regard to manpower, material, tools and tackles *etc.* This type of maintenance follows a schedule for time and date of carrying out maintenance. This type of maintenance enhances the life of equipment and prevents unexpected outages. This is further sub-divided into the following:

   (*i*) *Routine Maintenance*: It refers to fixed time based general inspection and minor repairs *e.g.*, lubrication, nut-bolt tightening, general cleaning *etc.* It is carried out as per fixed schedule and helps to minimise equipment outage due to minor problems.

(*ii*) *Preventive Maintenance (PM)*: The purpose of preventive maintenance is to prevent likelihood of failure of equipments.

This type of maintenance is carried out as per fixed schedule. Each equipment has a checklist of activities that need to be carried out on specific equipment. The job may require replacement of worn-out spares. All these activities are properly documented.

(*iii*) *Predictive Maintenance*: Predictive maintenance tells about an impending failure. It employs a number of condition monitoring techniques like vibration measurement, shock-pulse measurement, wear-debris analysis, infra-red thermography, ultrasonic leak detection, magnetic chip detection, blotest *etc.* (these are discussed in sufficient detail in subsequent chapters).

These are some of the diagnostic techniques that forewarns about an impending failure. These techniques are of two types—online and off-line. In on-line system, diagnostic tests are carried out when the equipment is running. In off-line system, diagnostic tests are conducted when the equipment is stopped.

This type of maintenance requires technical expertise and are expensive. But proves to be economical in the long run.

(*iv*) *Shutdown Maintenance (S/D Maintenance)*: In continuous process industries (Fertilizer, refineries, heavy chemicals, cement *etc.*) due to the nature of production process, equipments are run continuously and are not available for maintenance. Hence, a part of the plant or entire plant is stopped for one week to two weeks (or more) to carry out maintenance. The periodicity of the shutdown may be once in a year or two years. Generally, lean season (low market demand for the product) is chosen for shutdown maintenance. For example, car manufacturers choose December for shutdown, similarly refrigerator manufacturers choose July-August for shutdown.

During this period, there is total stoppage of plant and all preventive maintenance as well as major jobs such as overhauling, major modification, equipment replacement, additions and deletions in the existing structures are carried out.

(*v*) *Design-out Maintenance*: When a repetitive failure of an equipment takes place due to design deficiency or change in site conditions or change in mode of operation, the regular maintenance is not the solution. To eliminate such failures for once and all, design-out maintenance is carried out. In this type of maintenance, root cause of failure is analysed, and required modifications are carried out in the system/equipment/component so that the problem is eliminated forever.

(*vi*) *Total Productive Maintenance (TPM)*: It is a complete paradigm shift in maintenance management. In this system of maintenance, both maintenance and production personnel work together as a team. Minor maintenance activities such as lubrication, cleaning, bolt tightening, minor adjustments *etc.*, are carried out by production operator. In case of major problem, maintenance personnels are called.

(*vii*) *Reliability Centered Maintenance (RCM)*: The purpose of RCM is to achieve maximum operational reliability so that the equipment performs in a cost-effective manner during its lifetime. This is achieved by gathering data during lifetime of equipment. This data collected over a period of time is used for improving reliability of the equipment. The purpose is to focus on the most important functions of system and remove unnecessary maintenance activities thereby reducing maintenance cost. Its aim is to provide required reliability and availability at lowest cost.

2. **Unplanned Maintenance:** In this type of maintenance, no maintenance activities are carried till the equipment fails or one gets a chance to carryout inspection/repair/maintenance. Hence, there is no proper planning for periodic inspection/repair *etc.*

(*i*) *Breakdown Maintenance*: As the name suggests, maintenance is carried out when the equipment breaks-down or stops working. Hence, purpose of maintenance is to restore normal operation of the equipment. This type of maintenance is preferred when the component or part of the equipment is low cost and hence "use and throw" principle is followed (*e.g.*, oil filters, air filters etc)

(*ii*) *Opportunity Maintenance*: This type of maintenance is carried out when production is stopped due to various reasons such as raw material shortage, manpower shortage, no-demand of finished product or a particular section is under breakdown and hence unable to feed semi-finished product to next section of production line.

Under such situation, the available opportunity is utilised for maintenance purpose. Let us illustrate this with an example—suppose an overhead conveyor carrying semi-finished car body to paint shop breakdown. Now while the overhead conveyor is being repaired, maintenance of other equipments can be carried out simultaneously. So on one hand breakdown maintenance (emergency repair) is being carried out on overhead conveyor while on the other hand opportunity maintenance is being carried out on equipments lying idle due to conveyor breakdown.

3.3 PLANNED *Vs* UNPLANNED MAINTENANCE

Planned Maintenance	Unplanned Maintenance
1. This type of maintenance requires proper planning & scheduling of resources such as manpower, materials etc.	1. Requires no planning of resources.
2. It is carried out at- periodic intervals.	2. It is carried out when equipment breaks down or is available due to a chance.
3. It proves to be cheaper on a longer run.	3. It proves expensive on a longer run.
4. It helps to improve safety, enhance life of equipment and reliability.	4. Since the maintenance job is carried out hurridly, the life of equipment is reduced and safety is compromised.
5. No unexpected production loss due to equipment failure.	5. Production loss due to sudden and unexpected failures.
6. Lesser number of standby equipments are needed.	6. More number of standby equipments are required.

3.4 PREVENTING MAINTENANCE

Preventive maintenance is carried out on equipments at planned intervals so that unexpected breakdowns can be prevented. Interval for inspection is decided on the basis of load on the machine, site conditions, past experience and manufacturer's recommendations.

Preventive maintenance consists of inspection, adjustment, lubrication, replacement of worn-out parts *etc.*

3.4.1 Objectives of Preventive Maintenance

The objectives of preventive maintenance are:

(*i*) To obtain maximum availability of plant and machinery by avoiding breakdowns.

(*ii*) To keep equipments in proper condition so as to ensure quality of product and reduce waste.

(*iii*) To enhance life of equipments by minimising wear and tear.

(*iv*) To ensure safe operation of plant and machinery.

(*v*) To run plant at optimum efficiency.

(*vi*) To ensure optimum utilisation of maintenance manpower.

(*vii*) To improve equipment and plant asthetics

3.4.2 Requirements for Good Preventive Maintenance Plan

Following are requirements for good preventive maintenance plan:

(*i*) Good supervision and administration.

(*ii*) Proper planning and scheduling of jobs in consultation with production department.

(*iii*) Proper maintenance record keeping and analysis.

(*iv*) Advanced planning and stocking of spares and consumables.

(*v*) Proper training of manpower.

3.4.3 Elements of Preventive Maintenance

Elements of preventive maintenance are:

(*i*) Inspection.

(*ii*) Lubrication.

(*iii*) Planning and scheduling.

(*iv*) Record keeping and analysis.

(*v*) Planning and stocking of spare parts.

(*vi*) Training of technicians and Supervisors.

3.4.4 Economics of Preventive Maintenance

The extent of preventive maintenance should be such that, it does not become uneconomical. If cost of preventive and breakdown maintenance are plotted, the following curve shown in fig. 3.2 is obtained.

From Fig. 3.2, it is observed that as preventive maintenance is increased, breakdown maintenance decreases. So the total cost also decreases. However, with further increase in preventive maintenance, there is no consequent reduction in breakdown and hence total cost tends to increase. The point at which the total cost is minimum is the point of optimum value of maintenance.

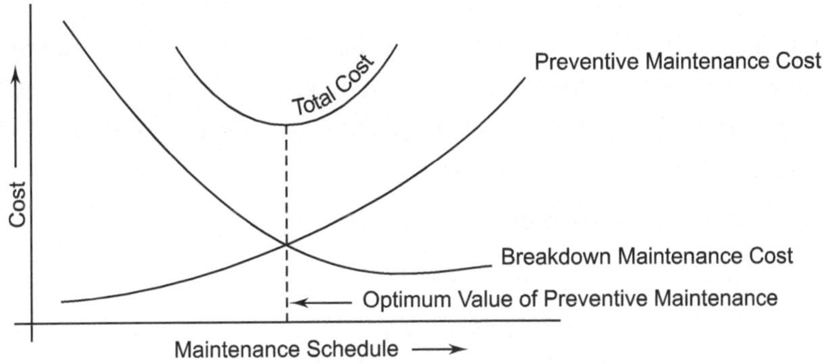

Fig. 3.2 Comparison of Cost of Preventive and Breakdown Maintenance.

Hence, in deciding preventive maintenance, it is necessary to decide interval between two consecutive inspections. Over maintenance is as harmful as no-maintenance.

3.5 BREAKDOWN MAINTENANCE

In breakdown maintenance, repair is carried out when the equipment is no longer able to discharge its intended function. After repair, no maintenance is carried out on the equipment till it breaks down and this cycle is repeated.

An equipment breakdown is due to two major reasons:

(*i*) Catastrophic failure of component.

(*ii*) Gradual wear and tear of component.

Failure due to gradual wear and tear of components can be avoided by preventive maintenance while the chance of catastrophic failure is always present even with preventive maintenance.

3.5.1 Breakdown Analysis

All breakdown are reported in daily log book of maintenance department by shift engineer/supervisor. At the end of a month, breakdown data is collected from the log book and analysis of breakdowns are carried out.

After thorough analysis, cause of breakdown is established and effort is made to remove the cause. Generally, a pareto analysis of breakdowns is carried out in terms of either down time or frequency of breakdowns. Top three breakdowns are chosen for detailed root-cause analysis and corrective action is taken. All the action taken are recorded for future reference.

3.6 TYPES OF MAINTENANCE POLICY

Maintenance policy is a statement of principle used to guide maintenance management decision making. Types of maintenance policy to be adopted is decided by the maintenance department taking several factors into consideration. Various types of maintenance policies are:

(*i*) Operate to failure policy.

(*ii*) Fixed time based maintenance policy

(*iii*) Opportunity based maintenance policy.

(*iv*) Shutdown based maintenance policy.

(*v*) Condition based maintenance policy.

(*vi*) Total productive maintenance policy.

(*vii*) Contract based maintenance policy.

(*i*) **Operate of Failure Policy:** This type of policy requires no planned action for maintenance. Maintenance action is required when the equipment fails to perform. The quantity of spare parts required to be stocked is based on past consumption pattern. This policy may be followed for non-critical equipments.

(*ii*) **Fixed Time Based Maintenance Policy:** Here, maintenance is carried out after a fixed time interval which in turn is based on expected wear out time of components. This type of policy is cost effective compared to operate to failure policy. However, such a policy is not preferable for complex equipments because failure pattern is seldom time dependent.

(*iii*) **Opportunity Based Maintenance Policy:** Opportunity maintenance is carried out as and when there is availability of equipments for maintenance. This type of policy may be used together with other policy in a continuous process plant, where equipment availability for preventive maintenance is almost nil.

(*iv*) **Shutdown Maintenance Policy:** Generally once in a year, entire plant or part thereof are stopped and made available for maintenance. Time chosen for maintenance stoppage is decided by availability of raw material, market demand, health of operating equipment *etc.* This policy is extensively followed in continuous process plants such as chemical, fertilizer, cement *etc.*

(*v*) **Condition Based Maintenance Policy:** The appropriate time for maintenance is determined by condition of the equipment. This system requires extensive use of expensive instrumentation and trained manpower. This policy helps to eliminate probabilistic nature of failure.

Selection of monitoring techniques and periodicity, depends on the complexity of equipment, criticality and cost involved apart from availability of trained manpower. The various condition monitoring techniques are wear-debris analysis, vibration monitoring, ultra sound testing, *etc.* (all these techniques are discussed in detail in subsequent chapters).

(*vi*) **Total Productive Maintenance Policy:** This policy is followed in most progressive industrial organisations. The operator of equipment is responsible for operation and minor maintenance. So production operator and maintenance personnel work jointly. This not only improves equipments uptime but also productivity and quality. Beginning of shift starts with cleaning, oiling and general checking of machine by the operator. Any minor discrepancy or set-up issue is rectified by the operator himself. This frees the maintenance personnel and they can concentrate on modifications, equipment efficiency improvement, planned maintenance activities *etc.*

(*vii*) **Contract Based Maintenance Policy:** Under this policy, entire maintenance activities including spare parts *etc.* are outsourced from a third party. This type of policy is followed where the equipments are complex and sophisticated and in-house maintenance capability is not there or is too expensive to maintain a dedicated team. Whenever maintenance need to be carried out, the maintenance contractor is called for the jobs. For example, maintenance of computers (Hardware and software), fax machine, photo copiers, water filters, telephone repair *etc.* are contract based. Even in small organisations, maintenance of production equipments may also be on contract basis.

3.7 SELECTION OF MAINTENANCE POLICY

Selection of maintenance policy depends mainly on cost, age and failure pattern of equipments. For random failures, with low Mean Time to Failure (MTTF), condition based maintenance policy or total productive maintenance policy is followed. With high MTTF, operate to failure is best suited.

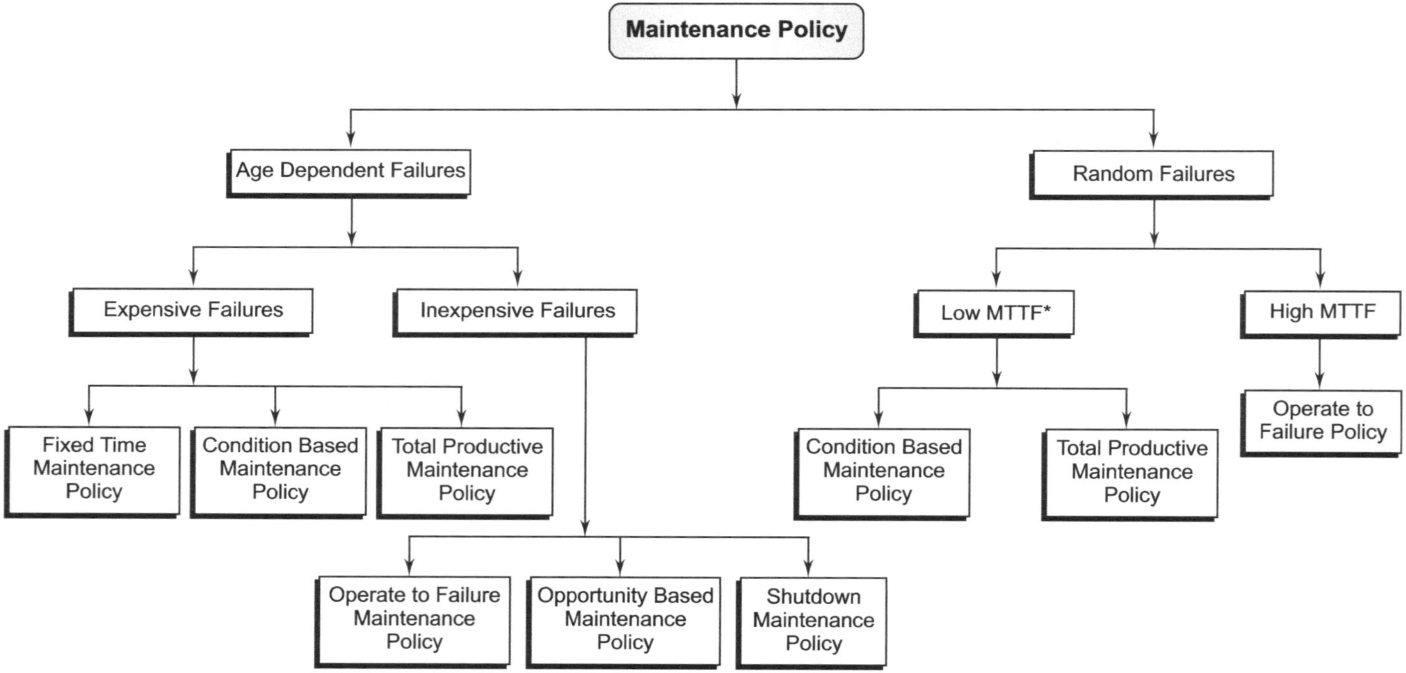

*MTTF—(Mean Time to Failure) Time Interval between Two Failures

Fig. 3.3 Selection of Maintenance Policy.

Equipments exhibiting age dependent failure patterns, a distinction is made between expensive and inexpensive failures. In case of expensive failures, any of the three policies *viz.*, fixed time based maintenance policy or condition based maintenance policy or total productive maintenance policy could be adopted. However, in case of inexpensive failures, suitable policies are operated to failure maintenance policy, opportunity based maintenance policy and shutdown maintenance policy.

3.8 MAINTENANCE POLICIES FOR OTHER AREAS

Apart from selection of type of maintenance to be adopted, the policy is decided for the following key areas:

(*i*) Policies for repair *Vs* replacement decisions

(*ii*) Policies for contract *Vs* in house jobs.

(*iii*) Policies for spares and inventory level.

(*iv*) Policies for refurbishment/retrofitting/major overhauling.

(*v*) Policies for scrapping of old/inefficient equipments.

The policies for above areas are decided by top management based on organizational goal, allocated budget, organizational culture and philosophy, age of plant and machinery, availability of skilled manpower *etc.*

Maintenance policies are stated in maintenance manual and vary from organisation to organisation.

Summary

In today's world of industrial systems, employing state of the art equipments, plant stoppage for even few minutes is not desirable. Hence, the equipments and its peripheral systems are supposed to function non-stop for 365 days a year. For this to happen, the maintenance system needs to be selected accordingly. The various types of maintenance are broadly classified as planned and unplanned maintenance. Under planned maintenance, routine maintenance, preventive maintenance, predictive maintenance, shutdown maintenance, design-out maintenance, total productive maintenance and reliability centered maintenance are there. Under unplanned maintenance comes breakdown and opportunity maintenance.

Preventive maintenance is followed universally. This system requires periodic checking of equipments thereby minimising chance failures. However, it requires dedicated band of personnel for the job.

Contd...

Type of maintenance to be adopted is decided by maintenance policy. The maintenance policy could be operated to failure policy, fixed time based maintenance policy, shutdown based maintenance policy, condition based maintenance policy, total productive maintenance policy and contract based maintenance policy. The choice of appropriate policy is guided by cost, equipment critically, availability of trained manpower and equipment complexity. Different policies are adopted for different equipments. For example, for filters, fixed time based policy is appropriate while for a gas turbine condition based maintenance policy is more appropriate. For small plants with very little skilled maintenance back-up, contract based maintenance policy is appropriate. However, the choice of maintenance policy is decided by failure pattern and cost. For expensive failure and age dependent failures, fixed time based or condition based or total productive maintenance policy is preferable. While for inexpensive and age dependent failures, operate to failure or opportunity based maintenance policy or shutdown maintenance policy is more appropriate. But for random failure with low MTTF (Mean Time to Failure), condition based maintenance policy or total productive maintenance policy is suitable. Similarly for random failures with high MTTF, operate to failure policy is more appropriate.

So in a plant, different sections or equipments may have different maintenance policy. Apart from maintenance policy, appropriate policy may be decided for replacement decisions, outsourcing jobs, spares level, overhauling *etc.* These policy guidelines are mentioned in "Maintenance Manual" of the organisation.

Review Questions

1. Explain three types of maintenance systems giving their applications.
2. Explain the procedure for choosing a suitable maintenance policy.
3. What is preventive maintenance? Explain different preventive maintenance tasks.
4. Compare preventive and breakdown maintenance. Which one is better?
5. Differentiate between TPM and RCM.
6. What are the objectives and elements of a good preventive maintenance plan?
7. Discuss various types of maintenance policies.
8. What are the factors responsible for the selection of a particular maintenance policy ?
9. Discuss maintenance policy for repair Vs replacement.
10. How do you decide optimum level of preventive maintenance?

Discussion Questions

1. "Selection of maintenance policy varies from organisation to organisation". Comment.
2. "An organisation can have different maintenance policy for different systems". Discuss.
3. "Maintenance policy bears direct co-relation with the organizational goal". Explain.
4. "Maintenance policy influences work culture in an organisation". Do you agree or disagree?
5. "Contract based maintenance policy is always the best choice". Discuss.
6. "Condition based maintenance policy is an expensive proposition". Justify.

Multiple Choice Questions

1. Objectives of Preventive maintenance are not:
   (*a*) Safety. (*b*) Availability.
   (*c*) Lower cost. (*d*) Optimum utilisation.
2. Advantages of preventive maintenance are:
   (*a*) Low spares consumption. (*b*) Higher cost.
   (*c*) Lower breakdowns. (*d*) Less record keeping.
3. Maintenance policy for expensive failures should be:
   (*a*) Operate to failure. (*b*) Condition based.
   (*c*) Opportunity based. (*d*) Shutdown maintenance.
4. High MTTF (Mean Time to Failure) failures should follow:
   (*a*) Operate to failure policy (*b*) TPMC policy
   (*c*) Condition based policy (*d*) Fixed time based policy.

True/False Questions

1. Contract based maintenance is useful for large organizations.
2. Operate to failure policy is suitable for equipments with high MTBF (Mean Time Between Failures).
3. Preventive maintenance is more expensive than breakdown maintenance.
4. Opportunity based maintenance policy is suitable for inexpensive repairs.
5. For expensive failures, TPM is a good choice.
6. Planned maintenance improves plant safety.

ANSWERS

Multiple Choice Questions

1. (*c*) 2. (*c*) 3. (*b*) 4. (*a*)

True/False Questions

1. False 2. True 3. False 4. True
5. True 6. True

Caselet

Mr. Rahul Srivastava, maintenance head of a newly established pharmaceutical company is unable to decide the type of maintenance policy for

(*i*) Low cost high failure rate equipments.

(*ii*) Low cost, low failure rate equipments.

(*iii*) High cost, high failure rate equipments.

(*iv*) High cost, low failure rate equipments.

Can you select right maintenance policy for each application? Give reasons for the choice.

❑❑❑

4

Maintenance Planning and Scheduling

Learning Objectives

After reading this chapter, you will be able to:

- Understand meaning of planning and scheduling.
- Understand planning and scheduling techniques.
- Explore long-term, medium-term and short-term planning.
- Examine issues related to manpower planning, material planning, equipment replacement planning *etc.*

4.1 INTRODUCTION

Planning and scheduling are most vital aspects of maintenance management. To carry out any job, we plan requirement of material, manpower, time and estimate the total cost. Planning is an essential process of carrying-out any maintenance task. A well-planned job avoids wastage of material, man hours, and interruption or delays.

In this chapter, we will study planning and scheduling process, types of planning and various scheduling techniques.

4.2 DEFINITION OF PLANNING AND SCHEDULING

Planning is a process of estimation of various elements that are required to carry out a job. For example, to replace a spark plug in a bike, we need to procure a spark plug (material and money requirement), arrange a technician to replace it (manpower). So, unless all elements required to replace a spark plug are planned, the job cannot be carried out.

Scheduling is estimation of time and proper sequencing of planned task. For example, to carry out above mentioned task of spark plug replacement, schedule will be:

Task Sequence : (1) Procurement of new spark plug and necessary tool.
(2) Hiring of technician.
(3) Opening of existing spark plug with necessary tool by the technician.

↓

(4) Cleaning of surface and fixing of new spark plug.

↓

(5) Trial

Estimated Time = 30 minutes, Estimated { cost of spare = ₹ 200
labour charges = ₹ 50
Total cost = ₹ 250 }

4.3 OBJECTIVES OF PLANNING AND SCHEDULING

Main objectives of planning and scheduling are:

(*i*) Efficient utilisation of time, manpower and material.
(*ii*) Avoid delays and interruptions.
(*iii*) Minimise cost of repair.
(*iv*) Improved work quality.

(*i*) **Efficient Utilisation of Time, Manpower and Material:** A well-planned and scheduled job minimizes wast of time, manpower and material. Since, all the elements to carry out a job are estimated to a fair degree of accuracy, hence there is efficient utilisation of available resources.

(*ii*) **Avoid Delays and Interruptions:** With fairly accurate estimation of time required to accomplish a given task, unnecessary delays and interruptions are thus avoided.

(*iii*) **Minimize Cost of Repair:** A well planned and scheduled job requires exact amount of resources resulting in efficient utilisation of resources and hence reduced cost.

(*iv*) **Improved Work Quality:** Proper planning improves work quality by incorporating all necessary inputs.

4.4 TYPES OF MAINTENANCE PLANNING

(*i*) **Planning Based on Time Duration**

(*a*) **Short-term planning:** Short-term planning is carried out by junior management level for day-to-day activities and deals with manpower work load allocation, spares procurement, preventive maintenance *etc*. Following points are kept in mind while carrying out short-term planning:

- Requirement and availability of skilled workers for the type of job at hand.
- Requirement of anticipated spares and consumables and their availability in stores.
- Realistic estimation of time for execution of job. The time starts from stoppage of equipment under maintenance till it is handed over to production after satisfactory trial.
- Necessary safety permits *e.g.*, permit to start work from production, electrical isolation of equipments, clearance from safety department *etc.*

(*b*) **Medium-term planning:** This type of planning is done at middle management level and time duration is one to five years. It involves planning of annual/bi-annual shutdown, major overhauling, plant modification, equipment performance enhancement, estimation of manpower requirement, training of manpower, energy conservation efforts, inventory control *etc.* Following points need to be kept in mind while carrying out medium-term planning:

- For spares procurements planning, a realistic lead time should be considered.
- Manpower planning has to keep in mind current as well as future skill sets required, current attrition rate, current as well as future training need, retirements/promotion *etc.*

- Key result areas allotted to supervisors should not remain static for more than two years (key result area means alloted area of work). Areas should be allotted on rotational basis.
- After freezing mid-term planning, necessary approval from concerned authorities is obtained with regard to cost, downtime, desired result *etc*. It is preferable that budget for such jobs are planned well in advance.

(*c*) **Long-term planning:** This type of planning is carried out by senior management level for next five to ten years. These are basic strategic policy decisions regarding departmental operations. These include long-term departmental goal, manpower recruitment policy, manpower training policy, adoption of quality systems (such as ISO – 9000, ISO – 14000, Gemba-Kaizon, 5-s *etc*.), adoption of maintenance policies, system of documentation, deciding type of MIS (Management Information System), spare parts and inventory policy *etc*. The following points need to be considered for long-term planning:

- Departmental goals are to be set realistically in synchronisation with organisation goals. For example, if organisation is planning for reduction in energy consumption, this should get reflected in the departmental goal.
- All goals need to be precise and measurable.
- Maintenance policy manual should be exhaustive and leave no room for doubt. The manual should be revised every 3 to 5 years.
- The policy should take into account technological obsolescence, automation, manpower downsizing, future expansion *etc*.

(*ii*) **Planning Based on Operational Need**

(*a*) **Manpower planning:** Long-term manpower planning deals with manpower recruitment, career advancements and training. While short-term manpower planning deals with selection of crew size, allocation of job to right person and estimation of time duration required to carry out a job.

Short-term manpower planning deals with day-to-day management such as selection of crew size. Selection of crew and its size (maintenance team or gang) presents a difficult choice since manpower available is limited but jobs available are a plenty. If more crew are employed on a particular job, other jobs may suffer. Similarly, if less manpower is employed, the job at hand may get delayed or even may not get done. Hence an optimum level of crew is to be selected for the job at hand. Choice of crew size depends on the following factors:

- Criticality and quantum of job.
- Time available for the job.
- Nature of problem (whether electrical, mechanical, electronic or combination of all).
- Number of pending jobs at hand.
- Availability of required manpower.

The fundamental requirement of successful manpower allocation is:

- Ability to measure the work content.
- Estimate time duration.
- Foresee possible hindrances.
- Good understanding of interpersonal relationship.

Manpower allocation is required for:

- Routine activities such as regular shift jobs, attending breakdowns and preventive maintenance *etc.*
- Specific jobs such as shutdown jobs, major repair/overhauling/ retrofitting, emergency jobs like recommissioning of critical machines after fire or accidents.

  In case of routine activities, technicians and supervisors with average knowledge and skill level are sufficient to carry out the task. While in case of special assignments, possession of good skill and similar past experience is essential.

While allocating manpower, the following points need to be considered:

- Maintenance work is a team work and hence technicians, supervisors, engineers *etc.* need to work in a co-ordinate manner. Any problem in interpersonal relationship can ruin the team work. Hence while choosing a crew, it is necessary that all team members get along well with each other.

- Before selection of team, it is necessary to match skill and experience of level with the job at hand. For a low skill job, a high skilled person should not be deployed as it will demotivate him. The team should be a combination of low and high performers rather than all high performers or all low performers. That is, a heterogeneous group performs better.
- The planner should be able to estimate the time required for completion of job and should build some "cushion" into it for unforeseen circumstances.
- The job of individual should be clearly defined including the time limit.
- Before start of the job, the crew incharge must brief his team about the job and manner in which it is to be carried out.

(*b*) **Material planning:** It is planning of spares parts for equipments and sub-systems. Problem with the spare part planning is that, maintenance engineer does not know when and how many spares will be required. Hence the problem is—how many to stock? If the spare part is not available at the time of requirement, the downtime will increase resulting in considerable production loss. Similarly if more spares parts are stocked then there is blockage of capital. Also, cost of holding and carrying inventory increases.

To counter such uncertainty, various models and techniques are used to forecast spares requirement.

Planning of spare parts depends on the following factors:

- Lead time of spares (time elapsed between placement of order and receipt of spares).
- Number of similar equipments in operation.
- Criticality of equipment from production point of view.
- Failure pattern of equipments
- Cost and availability of suppliers.

(*c*) **Equipment replacement/overhauling planning:** Due to regular wear and tear, the life of equipment reduces. At the end of useful economic life of equipment, there are two options — either to renew the life by overhauling/refurbishing, retrofitting *etc.* or go for complete replacement of the equipment. The decision depends on the following factors:

- Remaining useful life of equipment.
- Cost of preventive and breakdown maintenance.
- Technological obsolescence.
- Cost of new equipment.
- Cost of refurbishing.

(*d*) **Maintenance performance improvement:** Periodic upgradation of maintenance system need to be done. For example, introduction of latest predictive maintenance techniques, changes in maintenance system procedures, plant system modifications to improve efficiency and operational safety, installation of efficient equipment *etc.* For this QIP (Quality Improvement Program) could be initiated so as to improve plant performance and remove plant bottlenecks.

(*e*) **Shutdown maintenance planning:** Shutdown maintenance is carried out when all equipments and associated auxiliary systems are stopped to enable maintenance team to carry out necessary repair, replacement, modification, overhauling, inter-connection *etc.* Generally, all continuous process plants such as fertilizer, petro-chemical, refineries, cement *etc.* observe annual shutdown to carry out such jobs. Since due to continuous running of plant, it is not possible to carry out maintenance jobs. Hence, all such activities including major ones such as overhauling and replacement of major equipments (Turbines, boilers, transformers, furnaces *etc.*) are carried out during shutdown.

To carry out such huge task, it requires planning for necessary spares, specialized manpower, necessary tools and tackles or even supervision by external experts. Hence, planning, scheduling and co-ordination of highest order is required to accomplish the task within a sanctioned budget, available resources and given time limit.

4.5 PLANNING AND SCHEDULING TECHNIQUES

Various network techniques are used for presenting information related to time and resources so as to assist in planning and scheduling of maintenance activities. The information represented by a network includes sequence of activities, interdependencies, inter-relationships and criticality of activities.

Objectives of network analysis are:

(*i*) Minimisation of total cost.

(*ii*) Minimisation of total time.

(*iii*) Minimisation of idle resources.

(*iv*) Minimisation of delays, interruptions and conflicts.

A good number of planning and scheduling techniques are available today. However, two such techniques *viz.*, PERT (Program Evaluation and Review Technique) and CPM (Critical Path Method) are most widely used have been discussed.

4.5.1 CPM – Critical Path Method

Critical path method was first used by "Du Pont" to schedule shutdown and restart of a chemical plant. It is a deterministic method of scheduling activities. This is a graphical method of representing inter-relationship among various activities. The CPM diagram gives time required to complete all the activities.

All the activities are indicated by a node and events are shown as straight lines between nodes. Time estimates are given for completion of activity. The longest path through the network is identified as critical path.

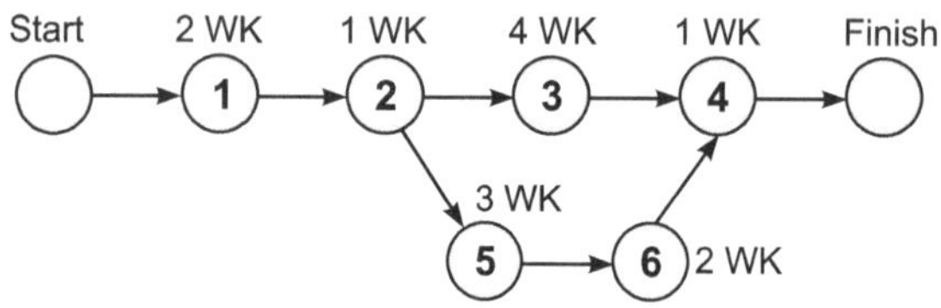

Fig. 4.1 CPM Diagram.

The diagram shows sequence of activities with completion time marked in weeks. So time required to complete task from (5) to (6) is 3 weeks.

The longest or critical path is (1)—(2)—(3)—(4).

4.5.2 PERT—Program Evaluation and Review Technique

PERT technique is used to complete project activities such as building of aircraft, submarine, cement plant *etc.* Like CPM, PERT also uses graphical technique where activities are indicated by an arrow and a node is used as a milestone.

This is a probabilistic method which gives three type of time estimates for completion of job—optimistic, pessimistic and most likely.

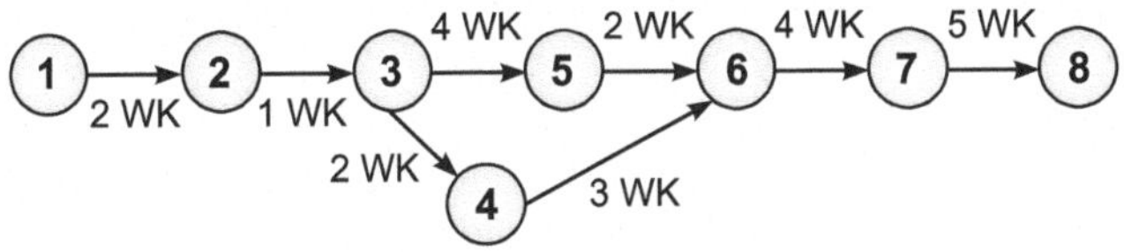

Fig. 4.2 PERT Diagram.

Activities (3)—(4)—(5) forms a parallel chain. Time required to complete a particular activity is marked in weeks.

Example:

Job: Replacement of existing coal furnace with a gas fired furnace.

Activity Code	Activity	Duration (Days)	Depends on
A	Preparation of technical specification.	12	–
B	Tender scrutiny.	26	A
C	Release of work order to vendors.	3	B
D	Supply of furnace.	60	C
E	Supply of auxiliaries and electricals.	20	C
F	Supply of pipes and pipe fittings.	10	C
G	Civil work including foundation.	15	C
H	Installation of auxiliary equipment.	6	E, F, G
I	Installation of furnace.	10	D, H
J	Testing, commissioning and trial.	4	I

CPM/PERT network for furnace replacement can be drawn as shown in Fig. 4.3.

First step is preparation of technical specifications (activity (A)) requiring 12 days. Next activity is tender processing (activity B) requires 26 days and is done after completion of activity (A). At event number 3, three activities *viz.*, E, F and G need to be carried out parallely so as to reach event numbered 8. In the meanwhile activities (C) and (D) are carried out. This goes on to reach finish at event number 11.

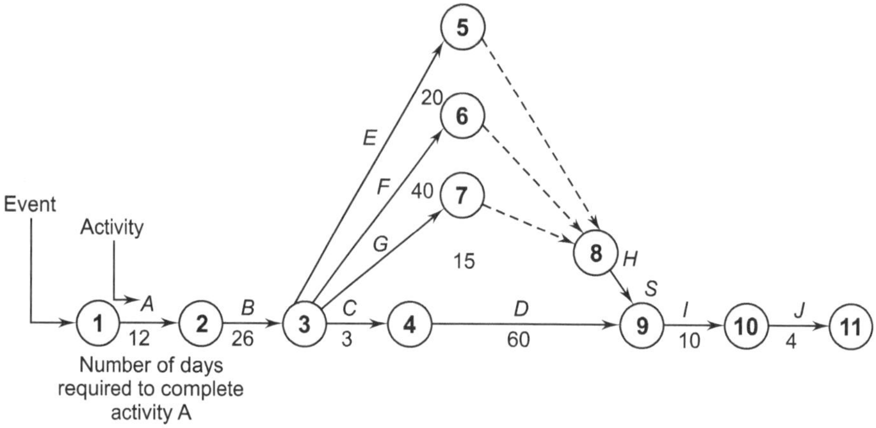

Fig. 4.3 CPM/PERT Chart for Furnace Replacement.

Critical path for this network is A-B-C-D-I-J. Events on critical path have zero extra time margin.

Summary

Planning and scheduling is an essential part of maintenance. Planning is required to estimate various elements required to carry out maintenance. A proper planning and scheduling helps efficient utilisation of available resource, reduces repair cost, improves repair quality and avoids delays. Maintenance planning can be classified based on time duration such as short-term, medium-term and long-term planning. It could be based on functional areas such as manpower planning, material planning, equipment replacement and overhauling planning, maintenance performance improvement planning *etc.*

Short-term planning is carried out for day-to-day activities. This includes daily workload allocation, preventive maintenance, small modifications *etc.* Medium-term planning is carried out by middle management level and is for one to five year duration. This includes annual shutdown, major overhauling, equipment replacement, manpower planning and training, equipment performance enhancement, inventory control. While long-term planning is carried out by senior management level and is for five to ten

Contd...

years duration. This includes, organisational goal, major policy decisions, adoption of quality systems (ISO-9000, OHSAS-18000) *etc.*

Various network techniques are available for planning and scheduling jobs. Out of all available techniques, most widely used are PERT (Program evaluation and Review Technique) and CPM (Critical path method). Application of these techniques help to minimize cost, time overrun and delays. It also provides interdependence of various tasks. CPM is a deterministic method while PERT is probabilistic method.

In short, planning and scheduling techniques help to utilize available resources efficiently.

Review Questions

1. Consolidate on scheduling, planning, monitoring and updating of maintenance activities for a manufacturing organisation.
2. What are the objectives of maintenance planning and scheduling?
3. Enlist various types of maintenance planning.
4. How manpower planning is done in maintenance department?
5. What are fundamental factors on which successful manpower allocation depends?
6. Why good material planning is essential for successful preventive maintenance plan?
7. What are various planning and scheduling techniques used in maintenance management?
8. Explain critical path method with an example.
9. What are the differences between CPM and PERT techniques?
10. How maintenance crew size is selected?

Discussion Questions

1. "Planning is a probabilistic method of estimating maintenance resources". Comment
2. "Decision about maintenance crew size is always complex". Discuss.
3. "Material planning plays vital role on success of preventive maintenance plan". Justify.
4. "Quality of maintenance is dependent on manpower planning". Do you agree?
5. "Replacement and overhauling decisions are most complex decisions". Discuss.

Multiple Choice Questions

1. CPM is a

(*a*) Deterministic method. (*b*) Probabilistic method.
(*c*) Pessimistic method. (*d*) Optimistic method.

2. PERT is a

(*a*) Deterministic method. (*b*) Probabilistic method.
(*c*) Pessimistic method. (*d*) Optimistic method.

3. A good manpower planning requires

(*a*) Ability to measure work content.
(*b*) Estimate time duration.
(*c*) Understanding of human inter-personal relationship.
(*d*) All of these.

4. Decision for equipment replacement depends on

(*a*) Remaining useful life of equipment.
(*b*) Cost of new equipment.
(*c*) Cost of preventive of breakdown maintenance of equipment.
(*d*) All of these.

5. Spares planning depends on

(*a*) Lead time required. (*b*) Cost of spare.
(*c*) Availability. (*d*) None of these.

True/False Questions

1. Monitoring is not part of maintenance planning.

2. Planning can never be deterministic.

3. CPM can be applied to shutdown maintenance.

4. Overhauling is always cheaper option than replacement.

5. Plant bottlenecks can be removed through performance improvement.

ANSWERS

Multiple Choice Questions

1. (*a*) **2.** (*b*) **3.** (*d*) **4.** (*d*) **5.** (*a*)

True/False Questions

1. False **2.** True **3.** True **4.** False **5.** True

CASELET

Mr. Jai Singh is a maintenance planning supervisor. He is supervising boiler errection work. Due to transporter strike the Forced Draught (FD) fan and Induced Draft (ID) fan and motor could not reach site on time. The errection is getting delayed. What are options available to him for:

(*i*) Completing the job on time.

(*ii*) Crashing the total time available for errection of FD and ID fans.

❑❑❑

5

Maintenance System Operation, Documentation and Regulatory Compliance

Learning Objectives

After reading this chapter, you will be able to:

- Understand maintenance operation system.
- Understand systems and procedures related to materials job work order and new capital equipment purchase and disposal *etc*.
- Examine maintenance documentation system.
- Comprehend regulatory compliance.

5.1 MAINTENANCE OPERATION SYSTEM

Running a department or an organisation or home requires certain systems, procedures, rules *etc*. Presence of such systems, procedures and rules remove doubts, conflicts, ambiguity and makes decision making easier and faster. For maintenance organisation, following systems and procedures are followed (these are indicative and not exhaustive):

(*i*) Systems and procedures related to material purchase, inspection, rejection, issue and return.

(*ii*) Systems and procedures related to job work order issue (fabrication and manpower hire) and cancellation.

(*iii*) Systems and procedures related to new capital equipment purchase and disposal.

(*iv*) Systems and procedures related to safety and emergency conditions.

(*v*) Systems and procedures related to complaint handling and their redressal (*i.e.*, attending breakdown calls) and so on.

These systems and procedures are described in detail:

5.1.1 Systems and Procedures Related to Materials

For purchase of new items such as bearings , lubricating oil, equipment spare parts *etc.*, a document named purchase indent is raised. Format of a typical indent is given in Table 5.1.

Indenter (person who raised the indent) fills up a document (Format as shown Table 5.1) and gets authorization from department head and head of the organisation depending upon the cost of the item. Signing authority is defined for monetary limit *e.g.*, for items costing upto Rs 25000/- department head is authorized to sign while for more than this amount, General Manager (operations) or President's signature is required *etc.*

Table 5.1 Format for Purchase Indent

ABC Ltd
PURCHASE INDENT

Deptt .... Cost Code................... Date ....

Required for...................

S. No.	Item Code	Item Description	Qty reqd.	Cost

Approx. Total Cost =

Indentor Deptt. Head GM(o)

After authorisation, one copy of indent is given to purchase department for procurement. Second copy is given to stores and third is retained by the indenter.

After arrival of the material in stores, it is inspected by the indenter and if found satisfactory with respect to specification, the item is issued from store using material receipt. Generally material receipt can be authorized by engineers and above rank. First copy of MR (Material Receipt) goes to EDP (Electronic Data Processing Deptt), second copy to accounts/finance, third copy to stores and fourth copy is retained by the indentor.

Similarly, if any material already issued from the stored is to be returned then MRN (Material Return Note) is issued to the store along with material.

If an item received from a supplier is found to be unsatisfactory, the same is rejected upon inspection by the indentor and reason for rejection is mentioned in the GIN (Goods Inward Note). GIN is prepared for every incoming material by the stores department.

5.1.2 Systems and Procedures Related to Job Work Order

When certain items are to be fabricated on contract basis then job work order are issued to the contractor or fabricator. Job work order contains type of job, quantity, drawing of item to be fabricated, raw material (either supplied by contractor or by maintenance department). The job work order is authorised and released by maintenance head. Format of job work order is given in table 5.2.

Table 5.2 Format for Job Work Order

ABC Ltd.
WORK ORDER

Department............................ Order No.............. Date...................

Name of Contractor.. Delivery date................

Raw Material Supplied by ....

S. No.	Item Description	Rate	Qty.	Remarks

Total Cost =

(Engineer) Deptt. Head (Maintenance Head)

Job work order is also raised for unskilled labour requirement. Such manpower may be required on daily basis or during major maintenance job. The work order is raised for labour and is released by maintenance head.

5.1.3 Systems and Procedures related to New Capital Equipment Purchase and Disposal

When a new capital equipment is to be purchased (excluding for green field projects), the following steps are taken:

(*i*) A proposal form is filled up with details such as reason for purchase, cost of equipment, payback period *etc.* and submitted to the top management for approval.

(*ii*) Once approval is granted, quotations are invited from suppliers (minimum three or more quotations are required) based on specifications.

(*iii*) After receipt of quotations these are scrutinised technically as well as commercially (cost, delivery, payment terms, guarantee/ warrantee clauses *etc.*) doubt/clarification is sought from suppliers (if required).

(*iv*) A comparative statement is prepared based on all quotations received.

(*v*) Technical and commercial negotiations are carried out.

(*vi*) Based on final negotiation, a revised comparative statement is prepared where in the preferred supplier is indicated.

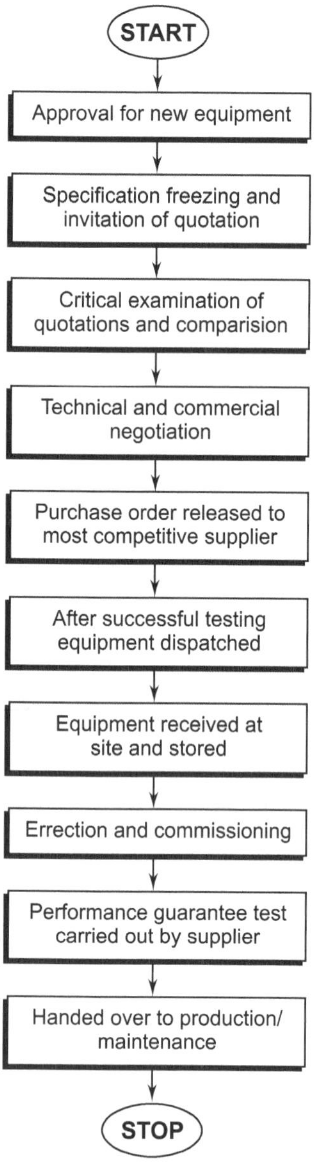

Fig. 5.1 Flowchart for Purchase of New Capital Equipment.

(*vii*) A purchase order (P.O.) is released by the procurement department with negotiated terms and conditions.

(*viii*) After receiving intimation from supplier equipment test is carried out at supplier's premises and witnessed by buyer's engineer.

(*ix*) Equipment is dispatched after successful witnessing of test.

(*x*) Upon receipt of equipment at site, it is unloaded and stored for errection and commissioning or taken to the site directly.

(*xi*) After commissioning and performance guarantee test, equipment is handed over to production/maintenance.

This is explained in Fig. 5.1 through a flow chart.

5.1.4 Systems and Procedures Related to Disposal of Equipment

Need for disposal of a capital equipment arises due to the following reasons:

(*i*) Completion of useful economic life of equipment.

(*ii*) Change of technology.

(*iii*) Requirement of increased production.

(*iv*) Damage to existing equipment due to fire, explosion, accident or catastrophic failure.

(*v*) Equipment lying idle due to abandoned project.

Under all above conditions, the equipment is disposed off depending upon its physical and operational condition and remaining life. The method chosen for disposal are following:

(*i*) *Sold as scrap*: If the equipment is beyond economic repair or has completed useful economic life, it is sold as scrap.

(*ii*) *Competitive sales bids*: Competitive sales bids are invited or auctioned off, if the equipment is in working condition and has useful economic life left.

(*iii*) *Standby*: Used as standby or stored for future projects.

5.1.5 Systems Related to Safety and Emergency Conditions

Before carrying out maintenance activities, certain permissions are needed under specific conditions. Various permissions (called safety permits) required are the following (list is indicative):

(*i*) for welding and gas/arc cutting → Hot job permit

(*ii*) for breaking of line or opening of a closed vessel → Line breaking/vessel opening permit

(*iii*) for starting civil construction work → Construction permit (*iv*) for working in sub-station → Substation work permit

(*v*) For electrical maintenance job → Electrical isolation permit

(*vi*) For working inside a closed vessel or confined space → Confined space entry permit

These safety permits are issued by safety/production department before the start of maintenance work. The basic purpose of safety permit system is prevent occurrence of accidents in work place.

Procedures related to emergency situations are detailed in emergency plans and maintenance manual. This is explained in detail in industrial safety management, chapter 10.

5.1.6 Systems and Procedures Related to Complaint Handling

Whenever a maintenance technician is required to carry out a job, a memo or maintenance request is raised by concerned plant shift engineer with date and time and given to maintenance shift engineer. In case of emergency, such as breakdown or fire or accidents, request is conveyed telephonically followed by the memo. After completion of job, the memo is signed off by concerned plant shift engineer with date, time and remarks, if any.

All the memos received during a shift are entered in the daily log book of maintenance department. Usually old memos are stored for three months and thereafter disposed off.

5.2 MAINTENANCE DOCUMENTATION

Maintenance documentation is vital to all maintenance activities. Documentation acts as a basis for maintenance information system which indicates current status of plant and machinery.

Maintenance records are numerous, only few of them are discussed below:

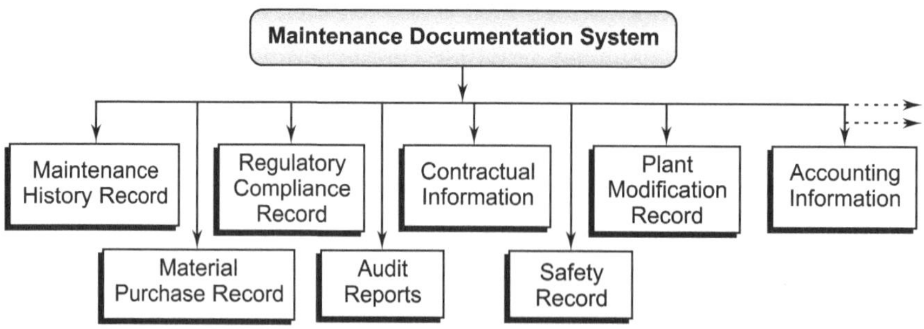

Fig. 5.2 Classification of Maintenance Documentation System.

(*i*) **Maintenance history record:** It contains the following documents:

(*a*) Plant and machinery detail (specifications, drawings, parts lists, manuals *etc.*).

(*b*) Equipment history cards.

(*c*) Equipment breakdown record.

(*d*) Shutdown maintenance record.

(*e*) Preventive maintenance record *etc.*

(*ii*) **Material purchase record:** It contains the following:

(*a*) Spares purchase record.

(*b*) Consumables purchase record.

(*c*) Capital equipment purchase record.

(*d*) Scrap auction record.

(*e*) Equipment disposal record.

(*iii*) **Regulatory compliance record:** It contains the following:

(*a*) Electrical safety	— License for generator sets.
	— License for L.T. installation.
	— License H.T. installation.
	— Earth test reports.
(*b*) Environmental	— Water consent.
	— Air consent.
	— Hazardous solid waste disposal consent.
(*c*) Fired pressure vessels	— License for boilers.
(*d*) Petroleum products	— License for storage of high speed diesel, light diesel oil, furnace oil, liquified petroleum gas *etc.*
(*e*) Pressure vessels and lifting Tools and tackles, Goods and Passenger lifts.	— Certificates from designated authorities for lifts (passenger and goods), lifting tool and tackles (EOT crane, winches, chain pulleys *etc.*).

(*iv*) **Audit reports:** It contains all types of audit reports such as maintenance audit, safety audit, equipment performance audit *etc.*

(*v*) **Contractual information:** It contains details of repair work orders, job orders, work order for unskilled manpower, annual maintenance contracts *etc.*

(*vi*) **Safety record:** It contains all safety related documents such as accident report, safety meeting report *etc.*

(*vii*) **Plant modification record:** It contains details of all plant modification carried out till date.

This list is indicative and not exhaustive. Other records are — performance monitoring and equipment replacement record, energy consumption record, daily and monthly reports, equipments catalogue, departmental correspondence, manpower training record, leave and overtime record, test reports such as lubricating oil test report, treated effluent test report, transformer oil test report *etc.*

5.3 REGULATORY COMPLIANCE

Regulatory compliance refers to adherence to various laws enacted by central as well as state governments for manufacturing establishments. It is mandatory for owner/occupier of industrial establishment to comply with the rules and regulations. We shall confine the discussions of various acts applicable to maintenance and utilities section. However, there are other acts which are applicable to personnel department, stores, finance and accounts *etc.*

Following are major acts applicable to maintenance and utilities department.

(*i*) Acts related to environment:

- (*a*) Water (Prevention and Control of Pollution) Act, 1977.
- (*b*) Air (Prevention and Control of Pollution) Act, 1981.
- (*c*) Environment Protection Act, 1986.
- (*d*) Hazardous Waste (Management and Handling) Rules, 1989 and 1991.

(*ii*) Acts related to general safety:

- (*a*) The Factories Act, 1948.
- (*b*) Manufacture, Storage and Import of Hazardous Chemicals Rules, 1989.
- (*c*) Explosive Act, 1884.

(*iii*) Acts related to electrical safety

- (*a*) The Indian Electricity Rules, 1956.

(*iv*) Acts related to pressure vessel:

(*a*) Static and Mobile Pressure Vessels (unfired) Rules, 1981.

(*b*) Indian Boiler Act, 1923.

(*c*) Indian Boiler Regulations, 1950.

(*v*) Acts related to use and storage of fuel:

(*a*) Petroleum Act, 1934.

(*b*) Petroleum Rules, 2002.

The objectives of above mentioned Acts, salient provisions and obligations are discussed in brief:

I. Acts Related to Environment

(*a*) **The Water (Prevention and control of pollution) Act, 1974**
***Objective*:** The purpose of this act is to prevent and control water pollution.

Scope of work related to maintenance department:

(*i*) Filing of returns in specified forms to State Pollution Control Board annually.

(*ii*) Testing of effluent water sample from a lab authorised by Pollution Control Board for various parameters laid down by State Pollution Control Board and Central Pollution Control Board and submission of same along with annual return.

(*b*) **The Air (Prevention and control of pollution) Act, 1981**

***Objective*:** The purpose of this act is to prevent and control air pollution.

Scope of work related to maintenance department

(*i*) Filing of returns in specified forms annually to State Pollution Control Board.

(*ii*) Testing of sample of polluted air discharged into atmosphere e.g. from stacks of boilers, diesel generating sets furnaces *etc.*, from a lab authorised by State Pollution Control Board. This test report is submitted with annual return.

(*c*) **Environment Protection Act, 1986**

***Objective*:** It is a general legislation for environmental protection. It provides for the protection and improvement of environment and the prevention of hazards to human beings, plants and property.

Scope of work related to maintenance department:

The act defines procedure for taking samples and standards for emission of environmental pollutants. In certain cases, information need to be furnished to authorities.

(*d*) **Hazardous Waste (Management and Handling) Rules, 1989**

***Objective*:** To impose restrictions on handling of hazardous substances and to establish a control mechanism for the management of hazardous waste.

Scope of work related to maintenance department:

Occupier of the factory has to obtain authorization for handling of hazardous waste. Disposal site need to be identified for proper disposal of hazardous waste.

The Act requires maintaining record of all hazardous waste, submission of returns, reporting of accidents, application for importing hazardous waste as raw material *etc.*

II. Acts Related to General Safety

(*a*) **Factories Act, 1948**

***Objective*:** It is a labour welfare legislation with prime objective of protecting workmen against industrial and occupational hazards.

Scope related to maintenance development:

(*i*) Testing of hoists, lifts, chains, ropes, lifting tackles *etc.* by competent authority (declared by state factory inspectorate) annually. Submission of such test reports to factory Inspectorate along with the annual returns.

(*ii*) Testing of non-fired pressure vessels by competent authority annually. Submission of such test reports to Factory inspectorate along with the annual returns.

(*iii*) Taking general safety measures such as fencing of machines, maintaining proper illumination levels, provision of fire safety, formation of safety committee *etc.*

(*b*) **Manufacture, Storage and Import of Hazardous Chemicals Rules, 1989**

***Objective*:** To identify major accident hazards and take adequate steps to prevent major accidents.

Scope related to maintenance department:

(*i*) To carry out safety audit and submit report to concerned authorities.

(*ii*) To prepare and keep up-to-date on-site emergency plan.

(*iii*) To conduct mock drill once in a six months of the on-site emergency plan.

All above actions are to be carried out with the support of safety department.

(*c*) **Explosive Act, 1884**

This Act regulates manufacture, possession, use, sale, transport, import and export of authorized explosives included in the list published by Central Govt.

License is required for storage of LPG, Naptha *etc.* and is governed by this Act. The license is granted by Chief Controller of Explosives.

It requires periodic testing of storage vessels (bullets) and submission of report.

III. Acts Related to Electrical Safety

(*a*) **Indian Electricity Rules, 1956**

Objective: To regulate generation, transmission supply and use of electrical energy from safety point of view.

Scope related to maintenance department:

(*i*) Only licensed contractors holding certificate of competency should be engaged for installation/maintenance work.

(*ii*) All new installations are inspected and approved by electrical inspector of state government before commissioning.

(*iii*) Subsequent inspection and testing is required at an interval not exceeding five years.

IV. Acts Related to Pressure Vessels

(*a*) **Static and Mobile Pressure Vessels (unfired) Rules, 1981**

***Objective*:** To regulate construction, fitment, storage, loading/transport and inspection of unfired vessels in service of compressed gases. Under these rules, it is obligatory to obtain approval and license from chief controller of explosives prior to construction/alteration and commissioning /operation of any storage vessel and allied facilities.

Scope related to maintenance department:

(*i*) Hydraulic testing of all pressure vessels containing corrosive or toxic gases once in three years by a competent agency duly approved by the chief controller of explosives.

(*ii*) To obtain license for existing pressure vessels from chief controller of explosives.

(*iii*) To obtain approval prior to construction/alteration, commission/ operation of any pressure vessel from chief controller of explosives.

(*b*) **Indian Boilers Act, 1923**

(*c*) **Indian Boilers Regulations Act, 1950**

Objective: To ensure proper standards for design, construction, materials and registration of boilers for over all safety and pollution control.

Scope related to maintenance department:

(*i*) All new boilers are inspected and registered and license issued before commercial operation can be commenced.

(*ii*) License for the boiler need to be renewed after every 12 months from state factory inspectorate.

V. Acts Related to Use and Storage of Fuel

(*a*) **Petroleum Act, 1934, and**

(*b*) **Petroleum Rules, 2002.**

Objective: To regulate import, transport, storage, production, refining and blending of petroleum products.

Scope related to maintenance/store department:

(*i*) License is required for storage of petroleum products *i.e.*, Naptha, light diesel oil, high speed diesel oil, *etc.*, above a specified quantity. Permission is granted by Chief Controller of Explosives.

Various Acts and Rules described are applicable to manufacturing units in the country. The list is indicative and not exhaustive.

Summary

Every maintenance department has rules, procedures and policies for various activities such as material purchase, job work orders, capital equipment purchase, complaint handling *etc.* A clear and unambiguous rule and procedure helps to remove doubts or conflicts thereby resulting in faster and smoother decision making.

Documentation is most vital as it forms basis for maintenance Information system. Various records are equipment history, Job work orders, audit reports, regulatory compliance record, material purchase record *etc.* Now-a-days, all documents are computerised so that their retrieval and storage is easier.

Regulatory compliance is another critical area. Maintenance department has to comply with various state and central laws. The applicable laws are related to environment, safety, electricity usage, storage of fuel, storage and handling of hazardous chemicals, boilers *etc.* Periodic return has to be filed and inspection is also carried by the competent authority. If found complying with the law, a license or permit is issued for specified duration. This license or permit need to be renewed periodically.

There are other laws applicable to personnel department, stores, finance and accounts *etc.* However, maintenance sections has to maintain record of all applicable laws. Non-compliance or violation of laws may lead to penalty, closure of plant or even imprisonment.

Hence, documentation and regulatory compliance forms a very vital part of entire maintenance management system.

Review Questions

1. Explain the need for maintenance documentation.
2. What is the need for written down maintenance systems and procedures?
3. What do you mean by work order? How is it raised?
4. Explain "Work to permit system". What is the need for such systems?
5. What is the procedure for the purchase of a new capital equipment?
6. How does permit system reduce work place accident ?
7. What is the meaning of regulatory compliance ?
8. Give examples of regulatory compliance required for electrical, pressure vessels and environmental pollution.
9. Why returns need to be filed for regulatory compliance ?

10. Who issues work permits for:

(*i*) hot job (welding/cutting) in a refinery.

(*ii*) electrical maintenance in a fertilizer plant.

Discussion Questions

1. "Maintenance Documentation is a necessary evil". Justify.

2. "A written down policy helps in quicker decisions". Comment.

3. "Work to permit system is essential for a maintenance deptt". Do you agree?

4. "A complex maintenance policy manual creates more confusion". Comment.

Multiple Choice Questions

1. Maintenance documentation is necessary for

(*a*) Record keeping. (*b*) Analysis and monitoring.

(*c*) Information. (*d*) For PM.

2. Work-to-permit is not necessary for

(*a*) Attending preventive maintenance.

(*b*) Attending breakdown maintenance.

(*c*) For welding job in a confined area.

(*d*) For electrical isolation.

3. Maintenance history cards are essential for

(*a*) Specifications. (*b*) Past breakdown record.

(*c*) Major modification record. (*d*) All of these.

4. Performance monitoring record contains –

(*a*) Test reports. (*b*) Energy conservation project reports.

(*c*) Equipment replacement details. (*d*) All of these.

True/ False Questions

1. Maintenance documentation is required only for regulatory purpose.

2. Work-to-permit system is not required for attending breakdown.

3. Work-to-permit system reduces accidents.

4. Maintenance procedures hinders work progress.

ANSWERS

Multiple Choice Questions

1. (*b*) **2.** (*b*) **3.** (*d*) **4.** (*d*)

True/False Questions

1. False **2.** True **3.** True **4.** False

Caselet

Mr. Naresh Kansal is a junior engineer responsible for regulatory compliance of maintenance department. He failed to file previous year's return for environmental pollution. The test reports are also not available for waste water, solid waste and air pollution (due to negligence). An inspection is due from State Pollution Control Department. The pollution control department inspector is corrupt.

Discuss options available to Mr. Kansal. What is the best path?

❑❑❑

6

Project and Tendering Management

Learning Objectives

After reading this chapter, you will be able to:

- Understand project management and its elements.
- Explore site mobilisation plan and project control procedure.
- Understand tender and its types.
- Understand tendering process and source selection.

6.1 INTRODUCTION

A project is a goal-oriented and time-limited activity requiring use of various resources and skills. A project could be building a dam over a river, construction of machinery foundation and shed, procurement, installation and commissioning of a steam turbine-generator, overhauling and retrofitting of a boiler, building of a ship or an aircraft, *etc.* The nature of projects mentioned above may be different but they have a common feature. This common feature is the accumulation and utilization of resources such as men, material, machines and money for the achievement of a goal. It differs from maintenance activities due to its uniqueness *i.e.*, a project is a non-repetitive and one-shot job requiring various skills and resources. It is a time and cost bound activity spread over months or even years.

Management of project activities require different skills compared to maintenance. The focus of this chapter will be to understand project process in totality. Tendering is an integral part of any project activity and is described in detail.

6.2 ELEMENTS OF PROJECT

A project has the following characteristics:

(*i*) Project objective.

(*ii*) A resource constraint.

(*iii*) Interdisciplinary approach.

(*iv*) A time constraint.

(*i*) **Project objective:** Any project clearly defines its final goal. It could be construction of a store room, efficiency improvement of a boiler (from 85% to 89%), development of an imported spare substitute, installation and operation of a machinery *etc.* The objective is defined in clear terms in respect of final goal. The goal could be say, raising efficiency of an existing boiler from 85% to 89% *i.e.*, in measurable terms.

(*ii*) **A resource constraint:** A project always operates under various resource constraints (finance, time, manpower, material and equipment). All the resources are defined in clear terms for the completion of a project.

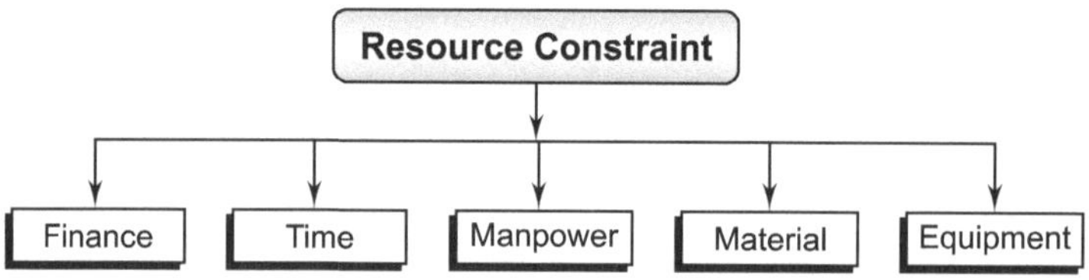

Fig. 6.1 Resource Constraints in a Project.

(*iii*) **Interdisciplinary approach:** A project is not a discipline specific activity rather require participation of various disciplines. For example, in the installation and commissioning of a boiler project, a civil engineer is required for foundation and super structure, a mechanical engineer is required for errection, testing and commissioning, an electrical engineer is required for various electrical machines such as motor, crushers, conveyors, lighting, power distribution *etc.* and instrumentation and control engineer is required for various boiler control (boiler pressure, level, water and steam flow *etc.*). Hence, it is an amalgamation of various engineering as well as non-engineering disciplines (finance, contract, human resource, materials *etc.*).

(*iv*) **A time constraint:** All project activities are time bound *i.e.*, they have a fixed start date and finish date.

6.2.1 Steps in Project Management

Various steps in project management are depicted through the flow chart shown in Fig. 6.2.

(*i*) **Project definition and scope:** Project definition and scope includes clear identification of scope of activities to be undertaken vis-à-vis achievements of final goal. For examples, for setting up a grassroot refinery from scratch, definition of project includes – 15 million tones per annum crude distillation capacity, capable of processing wide variety of crude, producing specific quantity of distillates at specified parameters. The scope of project may include-land acquisition, construction of civil structure, procurement, installation and commissioning of specified equipments and associated auxiliaries including pipelines, cables etc. and finally commercial operation.

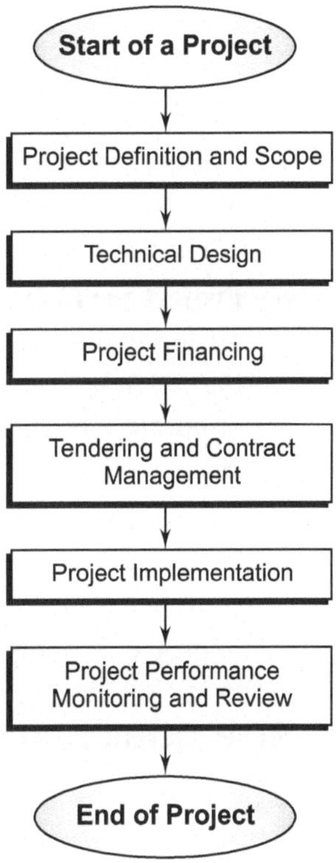

Fig. 6.2 Flowchart of a Project.

(*ii*) **Technical design:** Technical design includes selection of technology, design of process, equipment specification and selection, technical support (manpower, installation *etc.*), material selection *etc.*

Generally, technical design is broken down into two distinct activities *viz* basic engineering and detailed engineering. Basic engineering takes a macro view while detailed engineering takes a micro view of the project.

(*iii*) **Project financing:** After detailed technical design, next step is financing of the project. Project funds are obtained from either internal or external sources. Internal sources of finance are cash reserves of company, new share capital *etc.*, while external sources of finance are bank loan, equity participation by financial institution, public offers, collaboration with technology partner with equity participation, private finance *etc.*

(*iv*) **Tendering and contract management:** A substantial portion of project is executed by outside parties *viz* consultant, contractors, third party inspectors, technical experts *etc.* So for procurement of all materials, bids are invited through tenders and after scrutiny and negotiation, contract is awarded. Similarly for availing services, contracts are awarded to the party.

(*v*) **Project implementation:** Project implementation forms biggest chunk of the job. It requires co-ordination amongst all service providers. The project has to meet time deadline and budget constraint.

(*vi*) **Project performance monitoring and review:** Once the project is off-the-ground, regular monitoring of progress of project is carried out. Purpose is to ensure that job progress as per schedule and any obstacles or hurdles are checked and removed through review.

(*vii*) **End of project:** After completion of all project activities, the performance guarantee test (plant is run at designed capacity for a certain duration) is carried out. Thereafter, plant is handed over for commercial operation. After settlement of all outstanding dues, closure of project is completed.

6.3 SITE MOBILISATION PLAN

Before the start of project activities, site mobilisation plan is furnished by the contractor which contains the following details:

(*i*) **Site mobilization schedule:** A schedule showing datewise deployment of initial manpower, materials, machines *etc.*, at site.

(ii) **Site organization chart:** An organisational chart of all manpower working at site.

(*iii*) **Manpower mobilization schedule:** Schedule of manpower deployment in tune with progress of project.

(*iv*) **Equipment mobilization schedule:** Schedule of all critical machinery deployment.

(*v*) **Material delivery schedule:** Schedule of all materials such as piping, structural steels, cables, building raw material, *etc.*

(*vi*) **Program of work:** Chart showing achievement of major milestones.

6.4 PROJECT CONTROL PROCEDURE

A contractor has to furnish the following details for the purpose of control:

(*i*) Project organogram:

(*a*) Organisation chart at corporate and site office with contact details.

(*b*) Bio-data of key personnel.

(*ii*) Project Communication system:

(*a*) Project Communication/co-ordination procedure.

(*b*) Documentation system

(*iii*) Project quality assurance system:

(*a*) Project quality plan.

(*b*) Job procedure for civil, mechanical and electrical and instrumentation work.

(*c*) Third party inspection agencies.

(*iv*) Project execution strategy:

(*a*) Project execution methodology.

(*b*) List of critical items and schedule for procurement.

(*c*) Construction methodology.

(*v*) Project control procedure:

(*a*) Next 90 days work schedule.

(*b*) Work breakdown structure.

(*c*) Key project milestones.

(*d*) Procurement procedure and methodology.

6.5 TENDER

Tender is defined as an offer in writing which includes price among others, to invite bids to execute works, supply goods or associated service. It contains all relevant details of the job to be executed such as specifications, quantity, delivery schedule, payment terms, taxes and duties, warranty/guarantee clauses etc. Based on broad outlines in tender document, contract is drafted accordingly.

6.6 TYPES OF TENDER

There are various types of tender, some of them are following:

(*i*) Open tender.

(*ii*) Limited tender.

(*iii*) Single tender.

(*iv*) Global tender.

(*i*) **Open tender:** This type of tender is chosen when wide range of sources/contractors are available. Tenders are advertised through print and electronic media (news papers, magazines, trade journals, television, company websites, business portals *etc.*) open tender is open to everyone, irrespective of whether supplier is registered with the organization or not. Normally, 30 days are given to send the quotation.

(*ii*) **Limited tender:** This type of tender is chosen when sources of supply are few and definitely known or if it is not in public interest to call for open tenders due to security reasons (*e.g.*, defence procurement) or when the requirement is extremely urgent. So only selected parties are approached to offer bids (bid is a document in response to a tender).

(*iii*) **Single tender:** This type is used for single source items (proprietary in nature) and only single response is available despite previous advertisements or when source is specified by the user.

(*iv*) **Global tender:** This type of tender is applicable to items of foreign origin. Global tender is advertised in international media.

6.7 TERMS ASSOCIATED WITH TENDERING

(*i*) **Bid:** It is an offer document deposited with tenderer containing all details as per tender document. It is the willingness by the supplier to execute the job with terms and conditions as per tender document.

(*ii*) **Earnest money deposit:** All tenderers offering their quotations in response to open tender enquiry are required to deposit earnest money. It is normally 2% of the estimated value of the tender in the form of bank guarantee. This amount is returned to unsuccessful bidders within 15 or 30 days of purchase decision. In case of successful bidder, it is adjusted against security deposits. Purpose of earnest money deposit is to prevent non-serious bidders from making infructuous offer. If the tenderer withdraws or amends or derogates, the amount is fore-feited.

(*iii*) **Security deposit:** Security amount not exceeding 5% of the amount is taken from successful bidder before the release of order. Purpose of security deposit is to ensure successful execution of contract.

(*iv*) **Force-majeure clause:** It means inability of supplier to supply goods due to situation beyond his control. Such as riot, earthquake, flood, fire, strike, lockout *etc.*

(*v*) **Warranty clause:** It is the assurance from supplies to rectify defects in the material supplied within specified period. Period of warranty could be six months, one year *etc.*

(*vi*) **Arbitration clause:** Arbitration is private settlement of dispute by having a neutral third party imposing a decision. It is alternate dispute redressal method without involving court. It is governed by Arbitration and Conciliation Act, 1996.

(*vii*) **Liquidated damage clause:** Liquidated damage or LD clause refers to a condition when buyer imposes a penalty of 2 to 5% of total cost of contract on supplier for not meeting contract conditions.

(*viii*) **Payment terms:** It specifies duration in which percentage of total amount is to be paid. For example, 10% of total amount as advance, remaining 90% after completion of job.

6.8 TENDERING PROCESS

Tendering process consists of the following steps:

(*i*) **Estimation of requirement:** It involves writing specifications of material required, estimation of quantity of material, terms and conditions such as payment terms, guarantee/warrantee, taxes/duties, *etc.*

(*ii*) **Invitation of bids through tender:** The tender document is advertised

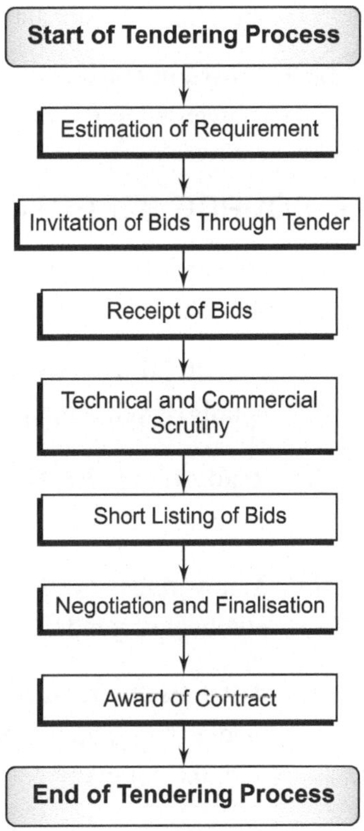

Fig. 6.3 Flowchart of Tendering Process.

in print media (news papers, magazines, trade journals, or pasted on notice board of the company) and electronic media (company website, business portals, television *etc.*). The suppliers or bidders send their bids in response to the tender within stipulated period.

(*iii*) **Receipt of bids:** All bids received from suppliers are opened on a fixed date and time.

(*iv*) **Technical and commercial scrutiny:** All bids received are thoroughly scrutinized *w.r.t.* to technical parameters and commercial terms and conditions by an appointed team.

(*v*) **Shortlisting of bids:** Based on scrutiny, the most favourable bid is selected. The basis for selection could be lowest quoted price or any other parameter.

(*vi*) **Negotiation and finalisation:** Negotiations are carried out with the short listed bidder and the bid is finalized.

(*vii*) **Award of contract:** After finalization of bid with the bidder, contract is awarded to the bidder.

6.9 SOURCE SELECTION

Selection of vendors (suppliers) with potential to successfully execute supplies against order placed are done from the following sources:

(*i*) **Industrial directories:** These are specialized directories containing details of vendors under different heads.

(*ii*) **Trade journals:** These are published by various trade associations containing details of vendors.

(*iii*) **Yellow pages:** Yellow pages directories contain addresses and contact numbers of vendors.

(*iv*) **Advertisement in print media:** Advertisements are issued in print media (newspapers, magazines, trade journals *etc*) for empanelment with the company.

(*v*) **Advertisement in electronic media:** Advertisements are issued in electronic media (company website, business portals *etc.*) for empanelment.

6.10 PERFORMANCE EVALUATION OF REGISTERED VENDORS

The performance of registered vendor is monitored regularly and evaluated based on the following major parameters:

(*i*) Delivery schedule.

(*ii*) Consistency in quality.

(*iii*) Competitive cost.

Apart from above, the following aspects are also considered:

(*i*) Number of order placed and executed.

(*ii*) Sales turnover and income-tax paid.

(*iii*) Number of times part supplies made.

(*iv*) Number of times orders cancelled or supplies rejected.

(*v*) Number of times items replaced after rejection.

Summary

Project is a one shot, time limited, goal-oriented activity requiring use of various resources. Project work starts after acceptance of project feasibility report. First, the scope of project is defined followed by technical design, project financing, tendering and contract management, project implementation, project performance monitoring, project completion and financial closure of project.

Technical design includes the basic engineering and detailed engineering. This activity includes selection of technology, design of process, equipment specification drafting and selection, technical support *etc.*

Funds for project are obtained from various sources such as company cash reserve, new share capital, bank loan, public offer, equity participation by financial institutions *etc.*

Majority of project work is executed by outside parties. So tenders are floated followed by negotiation and award of contract.

Next step is implementation of project within time limit and budget. The entire project work is closely monitored and reviewed so as to remove any obstacle or dispute.

In any project work, tendering forms an important and vital part. Generally, four types of tendering procedure is followed *viz* open tendering, limited tendering, single tendering and global tendering.

Contd...

The bidder invites tender for a variety of jobs from eligible suppliers. The prospective supplier has to deposit earnest money. Tenders so received are scrutinized technically and financially. Thereafter bid with lowest value is short listed. The party concerned is called for negotiation and finalization. Thereafter, contract is awarded.

In short, project is a sequential activity starting from project feasibility report and ending with financial closure of the project.

Review Questions

1. Draw the flowchart of project process.
2. Why is tendering an important process of project?
3. What is a tender? Discuss various types of tender.
4. Define terms— (*a*) earnest money (*b*) security deposit.
5. Draw the flowchart of tendering process.
6. Discuss various characteristics of a project.
7. Why interdisciplinary approach is necessary in a project?
8. Explain project control procedure.
9. How performance evaluation of registered vendors is carried out?
10. What is site mobilization plan?

Discussion Questions

1. "Tendering is a tedious and time consuming process". Comments.
2. "It is difficult to verify credentials of a tenderer". Do you agree?
3. "Project is 50% good planning and 50% execution". Discuss.
4. "It is always preferable to breakdown the complete project into set of packages such as civil works, piping works, cabling *etc.*, for timely completion of job". Do you agree?
5. Elaborate "cost, quality and timely completion are three fundamental columns of a successful projects management".

Multiple Choice Questions

1. Open tendering is applicable when
   (*a*) Sources of supply are limited. (*b*) Source of supply are plenty.
   (*c*) For proprietary items. (*d*) None of these.

2. Purpose of security deposit is to
   (*a*) Ensure successful execution.
   (*b*) Present non-serious bidders for participating.
   (*c*) None of these.
   (*d*) All of these.
3. Vendors are evaluated based on
   (*a*) Delivery schedule. (*b*) Competitive cost.
   (*c*) Consistent quality. (*d*) All of these.
4. Project financing is carried out through:
   (*a*) Bank loan. (*b*) Cash reserve of company.
   (*c*) Consistent quality. (*d*) All of these.
5. Earnest money is deposited in the form of
   (*a*) Cash. (*b*) Cheque.
   (*c*) Draft. (*d*) Bank guarantee.

True/False Questions

1. Earnest money cannot be refunded if the tenderer does not turn-up.
2. Security deposit is forefeited if the tenderer fails to supply goods.
3. It is not possible to ensure quality from a single source supply.
4. Source selection and verification poses most difficult challenge.

ANSWERS

Multiple Choice Questions

1. (*b*) **2.** (*a*) **3.** (*d*) **4.** (*d*) **5.** (*d*)

True/False Questions

1. True **2.** True **3.** True **4.** True

❑❑❑

7

Contract Management

Learning Objectives

After reading this chapter, you will be able to:

- Understand about contract management process.
- Explore the types and basic elements of contract.
- Understand monitoring, enforcement, and arbitration in contract management.

7.1 INTRODUCTION TO CONTRACT MANAGEMENT

An organization needs to purchase various materials required for production as well as for non-production and project purposes. Apart from material requirement, services of skilled experts are also required. The services could be from a simple communication system repair (telephone, fax, internet *etc*) to more complicated equipment overhauling and periodic service. All these activities involve good amount of money transfer from buyer to seller. This amount could be from few thousand rupees to crores of rupees. Hence, to safeguard the interest of both buyer and seller, an agreement need to be signed which clearly spells out terms and conditions. The purpose is to make both parties adhere to agreed terms and conditions so that both parties benefit from the deal and there is no room for dispute.

However, improper management of contract results not only in financial loss but also may lead to delays, disputes, poor quality *etc*. Hence, to avoid such pitfalls, it is necessary to understand the whole process of contract management.

A typical example of contracts in maintenance area could be annual maintenance contract (AMC) for computers, communications systems, drinking water system *etc*. and more complicated ones such as major equipment (turbine, boiler *etc*.) installation and commissioning, failure investigation of critical equipments, maintenance/safety audits, critical spares development *etc*. Such contracts being more technical in nature requires help of technical as well as finance and account personnel.

7.2 DEFINITION OF CONTRACT

A contract is a process that enables contracting parties to meet their obligations in order to deliver objectives required from contract. The main aim of contract management is to obtain agreed services/goods and achieve a value for money. Indian Contract Act, 1872, Indian Sale of Goods Act, 1930 and Indian Arbitration Act, 1940 form the basis for all contracts.

Contract is defined as an agreement between, two or more consenting parties to enable transfer of goods and services for monetary consideration or otherwise. The whole process of contract management involves tendering, negotiations, ordering supply of material and services, payment and monitoring. Failure to observe clauses of contract leads to financial losses, non-receipt of material, receipt of sub-standard material, delay in completion of work *etc*. Hence, organizations lay down rules and guidelines for contract administration.

7.3 TYPES OF CONTRACT

There are various types of contracts applicable to specific need. These are explained below.

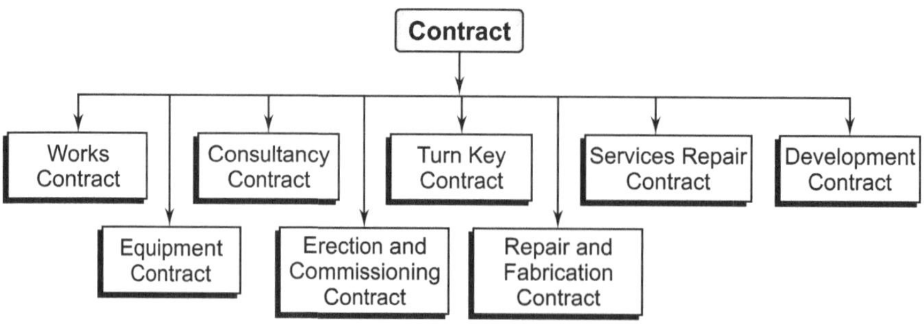

(*i*) **Works contract:** This type of contract is applicable to civil construction such as buildings, roads, machinery foundation *etc.*

Salient features of such contracts are following:

(*a*) *Estimate:* Costing is done on the basis of estimated work. But in practice, actual work exceeds the estimated work. Hence, provisions need to be built in contract to tackle such situations.

(*b*) *Piece-rates:* Piece rates are applicable to smaller items. So contract should have provision for piece rates. Generally, contractors quote low rates for smaller jobs and high rates for large volume of job.

(*c*) *Obligation of buyer and seller:* Works contract requires certain facilities to be provided by the buyer such as water and electricity for construction, drinking water, labour quarter, canteen facility, cranes, compressed air, drawings *etc.* The contract should clearly mention scope of facilities so as to avoid delays and disputes.

(*d*) *Labour requirement:* This type of contract requires large number of labourers. It is cheaper to employ local labourers. However, if outside contractor is employed then it becomes difficult to recruit local labourer due to opposition from local labour contractors. Hence, this aspect needs to be built into contract.

(*ii*) **Equipment contract:** All capital equipments such as Turbines, Boilers, Furnaces, Transformers, Reactor vessels, *etc.* are erected and commissioned through this type of contract.

Salient features of this type of contracts are following:

(*a*) *Specifications*: Specifications mentioned in the contract must be clear, unambiguous and in detail. This will prevent contractors to avoid contractual obligations in the guise of vague specifications.

(*b*) *Performance guarantee and testing*: The contract must clearly spell out performance guarantee parameters and conditions under which these are to be obtained. This is due to the fact that performance of equipment are affected by ambient temperature, relative humidity, elevation above mean sea level, presence of dust *etc.* For example, power output of gas turbine is severely affected by ambient temperature. Higher the ambient temperature above specification, lower is the power output. Similarly cooling and hence power output of transformers is affected by elevation. Higher the elevation above specification, lesser the cooling and hence lesser is the output.

A performance warrantee bond is obtained from supplier which can be revoked in case of unsatisfactory performance or a penalty clause can be built into contract.

Generally, suppliers provide 12 months warranty from date of commissioning or 18 months from date of dispatch (whichever is earlier). If delay in commissioning are expected then warranty must be modified accordingly.

(c) *Maintenance support*: All equipment manufactures have to provide spares and service support during the life time of equipment. So care must be taken to built in such clause into the contract. This is due to fact that spare parts are proprietary in nature and are not freely available in market or are expensive to develop. Hence, the contract must ensure uninterrupted supply of spares and service during the life time of equipment.

(iii) **Consultancy contract:** Consultancy contracts are required when large projects such as fertilizer, refinery, cement plant, petro-chemical complex are to be set up. For such jobs, a large number of consultants are hired *e.g.*, drafting of specifications, development of process design, inspection of capital equipment at manufacturer's premises, procurement of material, errection and commissioning supervision *etc*.

Salient features of such contracts are following:

(*a*) *Role and responsibilities of consultant*: Role and responsibility of consultants must be clearly defined in the contract in quantifiable terms.

(*b*) *Background of consultant*: Background of consultant must be known in terms of past track record, capabilities , reputation, fees charged *etc* before recruiting.

(*c*) *Fees charged*: A clear criteria for the fees being charged must be decided either lump sum or a fixed percentage of overall project cost or a fixed amount after completion of certain milestone.

(*d*) *Terms and conditions of Payment*: Terms and condition of payment must be linked to completion of certain section of plant or reaching certain milestone or simply time based. However, payment must be in proportion to the job completed.

(iv) **Errection and commissioning contracts:** The type of contract is used for commissioning of plants and utilities. Usually separate contractor is hired for this purpose but sometimes original equipment

manufacture hires a contractor and carries out the job under their supervision.

Examples of such contracts are errection and commissioning of boilers, turbines, furnaces, pressure vessels, piping, cabling, lighting, electrical distribution *etc.*

Salient features of this type of contract are:

(*a*) *Background of contractor*: The contractor must have good technical ability, excellent past track record and sound financial capability.

(*b*) *Performance warranty and guarantee*: The contract should clearly mention performance parameter at site conditions. This could be due to difference in design specifications and actual site conditions.

(*c*) *Synchronisation with other plant activities*: All jobs carried out in the site bear a relationship with other activities. For example, unless equipment foundations is ready, equipment cannot be installed, similarly, unless equipment is installed, piping and cabling work cannot be started. Hence, all activities are so planned that nobody sits idle.

(*v*) **Turn-key contracts:** As the name indicates, the buyer has to turn the key 'on' to get the system running. In this type of contract, all jobs are performed by consultants and contractors. It includes civil work, electrical, instrumentation, piping, equipment errection *etc.*, so it is an all inclusive job contract.

Salient features of this contract are:

(*a*) *Ability of the contractor*: Since this type of contract involves large volume of job hence the contractor must be financially sound. Also the contractor must be capable to take complete responsibility of design, procurement, errection, commissioning and load trial.

For example, BHEL undertakes turnkey projects in the field of power plants. BHEL has ability and experience in design, supply, errection, commissioning and load trial of complete power plant. Moreover, BHEL is a manufacturer of plant equipment. Hence, such contractors are suitable for handling large volume of job.

(*b*) *Past experience of contractor*: The contractor must have sufficient experience of handling such jobs in the past successfully in a time bound manner.

(c) *Performance warranty and guarantee*: Due to turn-key contract, the contractor may have a tendency to hide shortcomings which can be eliminated by clearly drafting performance warranty and guarantee clauses.

(d) *Liquidated damage clause*: To prevent delay and poor quality of work, LD clause must be incorporated into the contract.

(*vi*) **Repair and fabrication contract:** These are job order contracts in which design/sketch is supplied by the buyer. The order could be with or without material. It could be repair of existing equipment. Examples of such contract are piping and support fabrication, cabling and cable tray fabrication, sheet metal duct fabrication (for air conditioning), fabrication of chimney and vessels, fabrication of jigs and fixtures *etc.*

Salient features of this contract are:

(a) *Background of fabricator*: The fabricator must have good track record in terms of timely delivery, quality of job, reliability and good financial background.

(b) *Technical ability*: The fabricator must be skilled enough to fabricate jobs exactly as per given design or sketch. He should possess necessary tools and tackles and skilled manpower.

Repair contract must clearly spell out scope of dismantling transportation, testing *etc.* to avoid dispute at later stage.

(*vii*) **Development contract:** This type of contracts are awarded for the development of new items for the first time. For example, development of special purpose machine (SPM) for packing. Being developmental in nature, final success is not guaranteed.

Salient features of this contract are:

(a) *Actual cost estimation*: Being undertaken for the first time, it is difficult to arrive at exact cost. So, to take care of uncertainty, vendors have tendency to inflate cost. This can be taken care of by good negotiation.

(b) *Changes in specification*: The specification changes from initial to final stage due to trial and error nature of job. Hence, contract must flexible to accommodate this feature.

(c) *Delivery time*: It is difficult to set exact delivery date due to developmental nature of job. The contract must mention a realistic date based on agreement with vendor.

(*d*) *Scope for worst possible outcome*: Due to developmental nature of job, provision need to be made for worst possible outcome *i.e.* failure of product.

(*viii*) **Service and repair contract:** All annual maintenance contract (AMC) for computers, water coolers, telephones, photo copying machines, transport *etc*. fall into this category.

In this type of contract, exact nature of job cannot be estimated. Hence, a minimum charges plus cost per visit are charged. In case of transport, it is predecided charge per km and number of vehicles used. The salient features of this contract are:

(*a*) *Vendor reliability*: It is necessary to judge reliability from past experience and current customers. Being a routine service, reliability is of utmost importance.

(*b*) *Service quality*: Quality of service is another criterion. The contract must have clause for dissolution in case of poor quality of service and reliability.

7.4 BASIC ELEMENTS OF CONTRACT

A contract is a legal document between buyer and seller with respect to certain terms and conditions. These terms and conditions form basic elements of any contract. Unless drafted clearly, these could lead to avoidable disputes and delays.

The basic elements of contract are following:

(*a*) Price.

(*b*) Payment terms.

(*c*) Delivery schedule.

(*d*) Taxes and duties.

(*e*) Scope of work.

(*f*) Changes in scope of work.

(*g*) Performance evaluation.

(*h*) Measure of quality of product.

(*i*) Penal provisions.

(*j*) Dispute redressal.

(*k*) Dissolution of contract.

(*a*) **Price:** Price is decided by negotiation. It must include any hidden cost such as packing and forwarding, freight insurance, taxes and duties *etc.* The contract must include price-escalation *e.g.*, due to reduction/increase in raw material cost, changes in structure of taxes and duties, changes in currency exchange rate (for imported items) *etc*. Hence, quoted price must be all inclusive.

(*b*) **Payment terms:** The contract must clearly mention payment terms *e.g.*, payment could be related to various stages of progress such as 10% of total amount as advance, 80% after dispatch and remaining 10% within three months of receipt of goods. Payment terms must be kept in mind considering future cash flows of the organization. One must restrict the advance payment to bare minimum.

(*c*) **Delivery schedule:** The delivery schedule must be clearly mentioned along with penal provisions so that the supplier sticks to the schedule. This is often the cause of delay and dispute.

(*d*) **Taxes and duties:** Taxes and duties must be clearly mentioned in contract as these vary from state to state. The clause must include provision for changes in taxes and duties.

(*e*) **Scope of work:** All tender and contract documents must be specific in the scope of work. This often leads to delays and disputes.

(*f*) **Changes in scope of work:** The contract must include provisions for changes in scope of work. This is generally the case, as scope sometimes gets enhanced. Only larger changes must be paid for.

(*g*) **Performance evaluation:** Performance parameters and their method of evaluation must be clearly specified. These have final bearing on output and price.

(*h*) **Measure of quality of product:** The quality to be purchased must be measurable in definite terms *e.g.*, 10 micron thick hot dip galvanized iron sheet or pickling, phosphating and primer coat with three coats of paint (with colour shade and grade number).

(*i*) **Penal provisions:** These include Liquidated Damage (LD) and performance guarantee clauses. It allows buyer to impose penalty upto the level mentioned in contract for delay, poor material quality, poor workmanship, part material supply *etc.*

(*j*) **Dispute redressal:** The contract must have clause of redressal of disputes either through court or through arbitration.

(*k*) **Dissolutions of contract:** The contract must have clause for dissolution of contract.

7.5 MONITORING AND ENFORCEMENT OF CONTRACT

Any contract howsoever good it may be, needs proper monitoring and enforcement. Drafting of good contract is 50% of the job done, balance is successful execution of contract.

There are three basic parameters of good contract management:

(*i*) Quality of work.

(*ii*) Completion within scheduled time.

(*iii*) Completion within budgeted cost.

So monitoring requires a close tab on these three parameters.

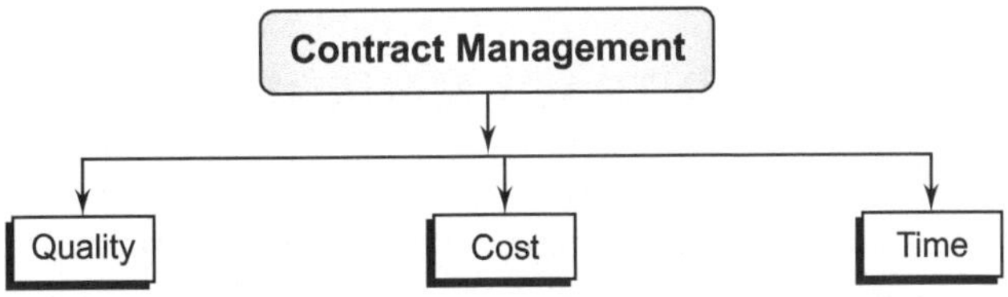

Fig. 7.1 Parameters of Good Contract Management

Following points need to be considered for contract enforcement.

(*i*) For large project, a project monitoring team must be constituted who are well conversant with various clauses of contracts and must be responsible for successful closure of project.

(*ii*) The team must be flexible enough to solve minor issues and avoid disputes at all costs.

(*iii*) The team must also be able to understand and appreciate genuine difficulties faced by the contractor and try to provide support wherever possible.

7.6 ARBITRATION

Arbitration is private settlement of disputes by having a neutral third party imposing the dicision. It is an alternate method of dispute redressel and avoids going to court. The arbitration proceedings are conducted as per Arbitration and Conciliation Act, 1996 and Indian Arbitration Act, 1940. Comparison of arbitration and court proceeding are given ahead.

Table 7.1 Comparison of Arbitration and Court Proceedings

Arbitration	Court Proceedings
1. It goes by agreement between the parties.	1. It goes as per law.
2. There is freedom to choose arbitrator, venue and time frame.	2. No such freedom.
3. Consumes less time.	3. Time consuming process.
4. Privacy is maintained.	4. No privacy.
5. It is a cheaper method of dispute settlement.	5. It is an expensive method of dispute settlement.
6. The award is final and binding.	6. Court decision is appealable.

CASE STUDIES IN CONTRACT MANAGEMENT

Real life case studies help to understand and appreciate the importance of subject matter in a better way. The case studies show where the mistakes were made and what were their consequences. A little alertness and thoughtfulness at the time of contract drafting and subsequently its implementation, goes a long way in achieving goals of contract and avoiding disputes.

Case Study

1. This case study pertains to a water treatment project which involved supply, errection and commissioning of pressure sand filters for process water purification.

   Quotations were received from reputed suppliers. The contract was awarded to the lowest bidder.

   When the project was complete, it was observed that instead of diaphragm type valves as mentioned in quotation and purchase order, party had supplied and fitted butterfly valves (cheaper and not suitable for frequent operation). When the party was asked to replace butterfly valves with diaphragm types, it took refugee under the logic that the design has been changed. Since full payment was already made against receipt of material, nothing much could be done.

   Here, it is necessary that a fixed percentage of total order amount must be retained till successful load trial.

2. This case study refers to construction of five foundations for steam turbines. The design was supplied to the contractor by the buyer. The contract was with material.

Contd...

The project got delayed due to delay in providing water, electricity and other amenities at site by the buyer. The contractor then claimed extra payment for the delay resulting in bitter relationship between contractor and buyer.

This situation could have been easily avoided if the buyer had provided amenities as agreed in the contract.

So it was a case of poor monitoring and implementation of contract.

3. This case study refers to a developmental contract. The contract was awarded to a party for development and fabrication of aluminium strip welding machine (a special purpose machine). Accordingly, 15% of the contractual amount was paid as advance. Later on, neither the machine nor the money could be recovered as the vendor had disappeared without trace.

So it is a case of improper background verification of the vendor.

4. This case study pertains to supply, errection and commissioning contract. A contract was awarded to a reputed party for supply, errection and commissioning of a standby lubricating oil cooler for a 3MW gas turbine. This party had earlier supplied air handling unit (AHU) coils. The AHU coils were functioning satisfactorily for the past three years.

After installation, commissioning and load trial of lubricating oil cooler, it was observed that the oil cooler is unable to maintain desired oil temperature in peak summer. On being pointed out to the supplier, the supplier took refuge in the logic that clear specifications were not mentioned in the contract (of oil flow rate and temperature). So the very purpose of standby cooler was lost and it was run in parallel with existing cooler during peak summer months.

So it was a case of incomplete specifications.

Summary

Contract is an agreement between two or more consenting parties to enable transfer of good and services for monetary considerations or otherwise. There are number of types of contract in engineering field. For example, works contract, equipment contract, consultancy contract, errection and commissioning contract *etc.* Basic understanding of terms and conditions of contract are essential for successful enforcement. Each type of contract has its own peculiarities and need to be dealt with accordingly. A poorly drafted

Contd...

contract is a source of dispute. Hence, basic clauses need to be understood properly and woven into contract so as to minimize vagueness and hence chances of dispute.

All the contracts must clearly define certain clauses such as price, payment terms, product quality, delivery schedule, scope of work, penal provisions, termination of contract, dispute redressal *etc.*

Next comes, monitoring and enforcement of contract. There are three critical parameters that need to be considered cost, time and quality. A contract monitoring team (for large contracts) monitors the project progress periodically. It also sorts out difficulties or disputes so that project milestones are achieved within targeted time and budget.

After completion of all project activities to the satisfaction of all parties concerned, all dues are settled and project closure is achieved.

Review Questions

1. Define "contract" in the ambit of Indian Contracts Act, 1872.
2. What are various types of contracts in engineering applications ?
3. What are most crucial factors for successful implementation of any contract?
4. What are various dispute redressal mechanism available ?
5. Why comprehensive specifications are absolutely necessary for procurement and commissioning of capital equipments ?
6. Explain the following terms as applicable to contract management:
   (*i*) Performance guarantee.
   (*ii*) Liquidated Damage.
   (*iii*) Taxes and duties.
   (*iv*) Scope of work.
   (*v*) Arbitration.
7. How would you measure the quality of service rendered?
8. Discuss fundamental parameters for good enforcement of a contract.
9. How a "Development contract" is different from a "construction contract"?
10. What are the factors to be considered for selection of a good contractor ?

Discussion Questions

1. Discuss, "Arbitration of disputes in a contract is the fastest way of redressal of grievances".
2. Elaborate, "Disputes arise due to non-cooperating nature of parties concerned".
3. "Liquidated Damage clause is least effective", do you agree?
4. "A proper implementation and monitoring of contract is absolutely necessary for the successful closure of a contract". Comment.
5. Discuss, "A clear demarcation of scope of activities is a must to avoid future disputes".

Multiple Choice Questions

1. Works contract deals with
   (*a*) Development of new parts. (*b*) Consultancy.
   (*c*) Service and repair. (*d*) None of these.
2. Essential parameters for contract monitoring are –
   (*a*) Time. (*b*) Quality of work.
   (*c*) Cost. (*d*) All of these.
3. Force majeure conditions are:
   (*a*) Earthquake. (*b*) Flood.
   (*c*) Strike/lockout. (*d*) All of these.
4. Turn-key contracts refer to
   (*a*) Service contract. (*b*) Development contract.
   (*c*) Consultancy contract. (*d*) None of these.

True/False Questions

1. Court cases are only solution for settlement of disputes.
2. Any contract can be terminated with the mutual consent of both parties.
3. Success of a contract depends on clarity and comprehensiveness.
4. Terms and conditions of payments must be tied to achievement of certain milestone in the project.

ANSWERS

Multiple Choice Questions

1. (*d*) **2.** (*d*) **3.** (*d*) **4.** (*d*)

True/False Questions

1. False **2.** True **3.** True **4.** True

CASELET

Mr. R.S. Gupta is an engineer in an engineering firm. He had awarded a development contract to a local party for an imported air filter. Import substitution of filters was not possible due to non-availability of filter fabric.

The contractor developed the proto-type filter after one year (time permitted in contract was six months). Upon trial, the filter failed within a week. As per contract, there is no mention of liquidated damage clause. The amount paid till date was close to Rs. One Lakh. The contractor is demanding more money to develop new proto-type.

Mr. Gupta is in a dilemma. Discuss:

(*i*) What are the options open to Mr. Gupta?

(*ii*) Should he re-negotiate the contract with LD clauses?

(*iii*) Should he select a new contractor?

(*iv*) Should he abandon the project altogether?

Your answers should be based on valid reasons.

❑❑❑

PART III

Operational Aspects of Maintenance Management

Strategic Outline

This part deals with various operational aspects of maintenance management. It begins with spare parts and inventory management. Purpose, function and importance of inventory is explained followed by inventory classification, inventory models, various inventory control techniques, obsolescence and preservation of spares.

Next, material handling system discusses purpose, classification, selection and performance, evaluation of MHS. At the end, bulk material handling is discussed in detail.

Chapter on Industrial safety management discusses various accident models, accident costs and prevention. Also causes, classification of fire, fire detection system have been discussed including electrical safety.

Industrial disaster management discusses, types of disaster, risk and hazard management and disaster preparedness.

Asset replacement decisions begin with asset life cycle, asset depreciation, failures and replacement policies including life extension strategies.

Shutdown maintenance covers shutdown planning, scheduling, monitoring including plant start up.

Computerised maintenance management system discusses basic of CMMS, various modules, software selection and implementation.

Contd...

Maintenance audit deals with types, purpose and its methodology. Financials of maintenance management deals with classification of cost, maintenance budget, maintenance information system and performance indices.

Advances in maintenance management discusses latest maintenance concepts such as total productive maintenance, Reliability Centred Maintenance, Benchmarking, Value Analysis, Reverse Engineering and Anticipatory Failure Determination.

Chapters in this section

8

Spare Parts and Inventory Management

Learning Objectives

After reading this chapter, you will be able to:

- Understand inventory models.
- Understand different inventory control techniques.
- Understand obsolescence and ways to avoid it.
- Examine ways of preserving spares.

8.1 INTRODUCTION

Inventory is defined as idle resource having economic value. It includes raw materials, semi-finished goods and maintenance spares and consumables.

Purpose of keeping inventory is to provide a cushion against material non-availability, long lead time, faster consumption rate, minimize cost of procurement *etc.* So inventory acts as cushion against stock-out.

Here, focus of attention will be on maintenance spare parts in this chapter. Maintenance spare parts possess an unique characteristic unlike production raw materials. That is, the consumption pattern of maintenance spares is highly unpredictable. Hence, it is not known when a particular spare will be required. This puts in a fix as to stock or not to stock. But if the stock-out cost is very high then inventory need to be maintained. Again, if the spare is not used for a long time then it represents blockage of capital apart from other inventory carrying costs such as space required, obsolescence, pilferage, deterioration *etc.*

Hence, management of maintenance spare parts presents a very difficult, challenging and complex process. However, various mathematical models have been used to predict the spares requirement only to a certain degree of success.

8.2 PURPOSE OF INVENTORY

Following are the purpose of keeping inventory:

(*i*) To meet anticipated demand.

(*ii*) To protect against stockout.

(*iii*) To provide hedge against price increase.

(*iv*) To take advantage of quantity discount.

(*v*) To maintain continuity of operation

(*vi*) To take advantage of order cycles to minimize purchasing and inventory cost.

8.3 FUNCTIONS OF INVENTORY

Following are the functions of inventory:

(*i*) *To serve as cushion*: Inventory acts as cushion against demand fluctuations.

(*ii*) *To provide production economy*: Bulk purchase helps to get quantity discounts.

(*iii*) *To maintain efficient production flow*: Inventory helps to maintain smooth production flow without causing any interruption.

8.4 IMPORTANCE OF INVENTORY

Following are the importance of inventory:

(*i*) It provides protection against uncertainties of demand and supply.

(*ii*) It helps in better utilization of men, machines and materials.

(*iii*) It provides operational economy by allowing to avail quantity discounts.

8.5 TYPES OF INVENTORY

Following are major types of inventory:

(*i*) Raw material.

(*ii*) Work in progress or semi-finished goods.

(*iii*) Finished goods.

(*iv*) Production consumables.

(*v*) Maintenance spares.

(*vi*) Maintenance consumables.

(*vii*) General consumables.

8.6 DISADVANTAGES OF INVENTORY

Following are disadvantages of inventory:

(*i*) **Carrying cost:** It is the cost of holding the inventory for a length of time. It includes cost of capital blocked, interest paid on capital, obsolescence, pilferage, damage *etc.*

(*ii*) **Ordering cost:** It is the cost of ordering and requiring inventory which includes cost of stationary and postage, cost of phone/fax, wages of clerical staff, cost of supervision *etc.*

(*iii*) **Higher cost of purchase:** It is possible that material is purchased at a time when the prevailing market rates are high which leads to higher cost of purchase.

8.7 BENEFITS OF INVENTORY CONTROL

Following are the benefits:

(*i*) **Less storage space required:** Lesser the inventory, lesser is the space required to store it.

(*ii*) **Reduced labour is required to control it:** Lesser inventory means lesser numbers of persons required to maintain it.

(*iii*) **Easy retrieval of spares:** Due to less number of items stocked, it is easier to retrieve an item when demanded. Hence, response time is reduced.

(*iv*) **Reduction in maintenance repair cost due to reduction in minimum/maximum level:** Regular maintenance consumables such as nut-bolts, flexible PVC pipes, *etc.*, are maintained on a reorder level, once the stock reaches a critical point. If less inventory is maintained then lesser level of stock need to be maintained. This helps in lowering maintenance repair cost.

(*v*) **Saving through better utilisation of money:** Money saved by maintaining lesser inventory can be put to better use.

8.8 HOW MUCH TO ORDER?

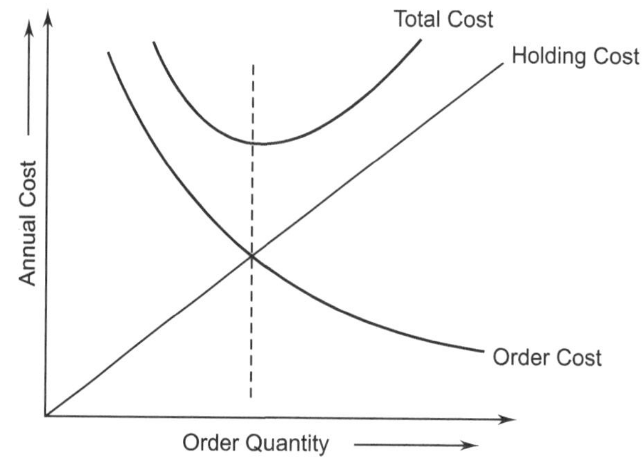

Fig. 8.1 Optimum Order Quantity

As shown in the above graph, with increase in order quantity, cost of holding inventory increases. The order cost decreases initially due to volume discounts but thereafter it increases.

The total cost is minimum when holding cost and order cost curves intersect each other.

8.9 CLASSIFICATION OF INVENTORY

Following are the classifications:

(*i*) **Raw materials:** Raw materials are used for producing finished goods. For example, cotton and yarn are raw material for textile mill. Similarly rubber is the raw material for rubber parts manufacturing factory or crude is the raw material for refineries *etc.*

(*ii*) **Machinery and equipment and parts:** All machinery equipments and parts which are used for plant operation are stored in the store. For example, electric motors, fans, pump spares, tooling (punches, dies *etc.*).

(*iii*) **Consumable items:** These are used for production and maintenance purposes. For example, oils, coal, lubricants, paints, stationary items, nuts and bolts, filters *etc.*

(*iv*) **Chemicals:** These chemicals are required by production and maintenance departments. For example, process oils for rubber industry, hydrochloric acid and caustic soda for water treatment plant, chemicals for boilers, *etc.*

(*v*) **Inflammable items:** These items are highly susceptible to fires such as petrol, kerosene, benzene, xylene, paints *etc.*, due to their hazardous nature, they are stored far away from the main building.

(*vi*) **Fuel stock:** These are used directly for production and are consumable. For example, fuels used in furnace, oven, boilers *etc.* These are coal, high speed diesel, LSHS Oil (Low Sulphur Heavy Stock), Furnace oil, LDO (Light Diesel oil), LPG (Liquefied Petroleum Gas) *etc.*

(*vii*) **Packaging material:** These include all kinds of packing materials used for packing of final product.

For example, paper, cardboard, straw, sawdust, kraft boards, boxes, gates, drums, bottles, bags *etc.* It will also include material required for protective coating to final product *e.g.*, wax, grease, oils, poly ethylene sheet, *etc.*

(*viii*) **General items:** These include those items not included in any classification such as brooms, uniform for staff, safety shoes, helmets *etc.*

(*ix*) **Scrap material:** These are materials left as waste from production process. It may also include chemicals whose expiry date is over. These are disposed off as scrap.

(*x*) **Finished and semi-finished goods:** Finished goods are items that are ready for sale while semi-finished goods are items that are in the production-loop (work-in-progress items). Semi-finished products are taken from store by production department for further processing.

(*xi*) **Dead-stock items:** These are items which are not used for a long time and there are no chances of future use.

8.10 INVENTORY MODELS

Various inventory models are used to maintain stocks of specified items so that they are available whenever a demand is received from the customer. In order to ensure continuous availability, the stocks are to be replenished from time to time. Also, to minimise inventory cost, optimum level of inventory need to be maintained so that there is no stock-out situation.

There are two types of inventory models—deterministic and probabilistic:

(*i*) Deterministic Inventory Model
   (*a*) Economic order quantity.
   (*b*) Production quantity or economic batch quantity.
   (*c*) Annual estimate system.

(*ii*) Probabilistic Inventory Model
   (*a*) Fixed level system.

8.10.1 Deterministic Inventory Model

(*i*) **Economic Order Quantity:** It is the quantity which when ordered at a time results in least total annual cost to the organization. If larger quantity is ordered at a time, large capital will be blocked and inventory carrying cost will go up though ordering cost will be low. But if order quantity is small, the frequency of ordering will go up though inventory carrying cost will go down. Therefore, when a balance between ordering cost and inventory cost is struck, the quantity calculated is called economic order quantity (EOQ).

Assumptions in EOQ model are following:
   (*a*) Annual demand of item is known.
   (*b*) Demand is constant throughout the year.
   (*c*) Lead time does not vary.
   (*d*) Order is required in a single delivery.
   (*e*) There are no quantity discounts.

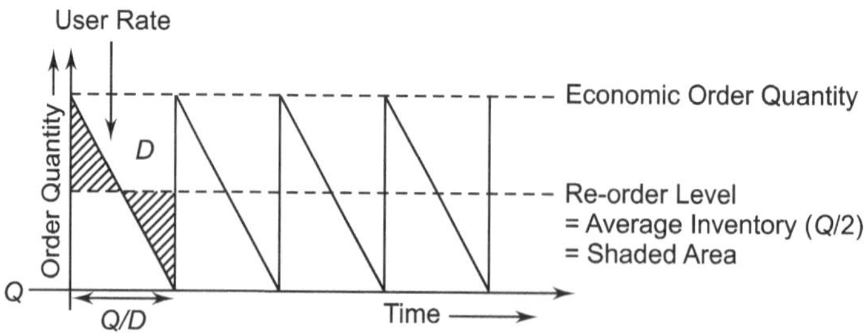

Fig. 8.2 Inventory Level Profile for EOQ Model

Let Q = Order quantity

D = Annual demand or usage rate

H = Holding or inventory carrying cost

P = Procurement cost

T = Total annual cost

Total, Inventory cost (T) = Inventory Procurement cost + Inventory Carrying cost

Now,

Inventory procurement cost = Procurement cost × No. of order/period

$$= P \times \frac{D}{Q}$$

$$\left(\because \text{ Time interval between two deliveries} = \frac{Q}{D}\right)$$

$$\left(\therefore \text{ Frequency of delivery} = \frac{1}{\text{Time Period}} = \frac{D}{Q}\right)$$

And Inventory carrying cost = carrying cost/unit × Average inventory

$$= H \times \frac{Q}{2}$$

$$\left(\because \frac{Q}{2} = \text{Average Inventory}\right)$$

$$\therefore \text{ Total cost } T = \frac{PD}{Q} + \frac{HQ}{2}$$

For minimum cost $\frac{dT}{dQ} = 0$

$$\therefore \quad \frac{dT}{dQ} = \frac{-PD}{Q^2} + \frac{H}{2} = 0$$

or

$$\frac{H}{2} = \frac{PD}{Q^2}$$

$$\Rightarrow \quad Q = \sqrt{\frac{2PD}{H}}$$

$$Q = \sqrt{\frac{2 \times \text{Procurement cost} \times \text{demand}}{\text{Carrying cost}}}$$

(*ii*) **Economic Batch Quantity or Economic Production Quantity (EPQ):** This model is valid for batch production unlike EOQ model which is valid for continuous production. It is based on the assumption that capacity to produce an item exceeds the demand rate (or usage rate) and orders are received incrementally during the production.

Assumptions of EPQ models are:

(*a*) Annual demand is known.

(*b*) Demand is constant.

(*c*) Production rate is constant.

(*d*) Lead times do not vary.

(*e*) There are no quantity discounts for bulk purchase.

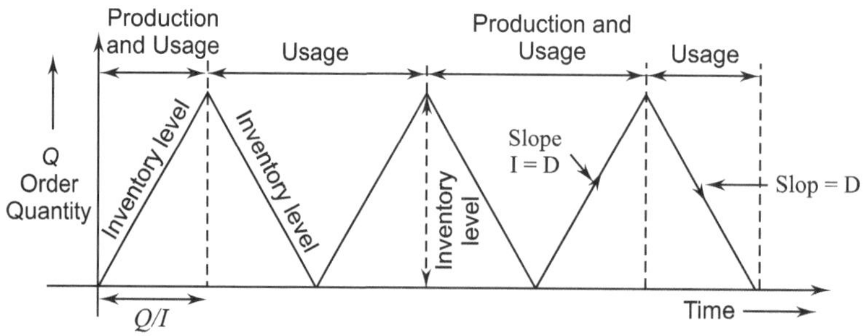

Fig. 8.3 Inventory level profile for EPQ model

Let Q = Order Quantity

M = Maximum Stock Level

D = Annual demand rate

I = Inventory level

H = Holding or carrying cost

P = Procurement cost

Slope of inventory build-up = $I - D$

$$\text{Also, slope of inventory build-up} = \frac{M}{Q/I}$$

$$= \frac{MI}{Q}$$

$$\therefore \quad I - D = \frac{MI}{Q} \quad \text{or} \quad M = \frac{(I-D)}{I}$$

$$\text{Average inventory level} = \frac{M}{2}$$

$$= \frac{Q(I-D)}{2I}$$

Now, Total cost = Holding cost + procurement cost

$$T = \text{H} \times \frac{Q(I-D)}{2I} + \text{P} \times \frac{D}{Q}$$

For minimum total cost $\frac{dT}{dQ} = \text{O}$

$$\frac{dT}{dQ} = \frac{H(I-D)}{2I} - \frac{PD}{Q^2} = \text{O}$$

or

$$Q = \sqrt{\frac{2PD}{H\left(1-\frac{D}{I}\right)}}$$

or

$$\text{EPQ} = \sqrt{\frac{2 \times \text{Procurement cost} \times \text{demand}}{\text{Carrying cost}\left(1-\frac{\text{Demand}}{\text{Inventory Level}}\right)}}$$

(iii) **Annual Estimate System:** In this method, a fixed time (usually one year) is used for replenishment of items. For the purpose of convenience and for uniformly spreading the purchase work, different groups of items are replenished in different specified months. This method helps to combine the demands of different plant locations and enable to make one centralized purchase. This gives quantity discount for bulk purchase.

Inventory carrying cost can be minimized if the order delivery is planned in number of phases. This system is followed for general consumable items such as nuts, bolts, stationary, oil and lubricants *etc.*

Now, we will discuss some specific cases in deterministic inventory model by introducing concept of buffer stock (cushion), shortage, penalty *etc.* This will provide more realistic picture of inventory management.

Case I

Assumptions

(*i*) Supply of items is instantaneous (*i.e.*, no lead time required).

(*ii*) Consumption is continuous.

(*iii*) A certain level of reserve or buffer stock need to be maintained.

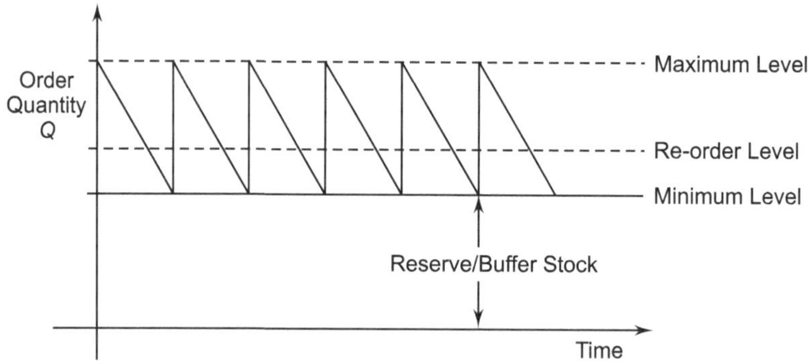

Fig. 8.4 Inventory level profile with buffer stock requirement

Let Q = Order Quantity

R = Reserve stock

D = Annual demand rate

C = Unit cost

H = Holding or carrying cost

P = Procurement cost

T = Total cost

$$\text{Inventory holding or carrying cost} = \text{H} \times \left(R + \frac{Q}{2}\right) \times C$$

$$\text{Set up cost} = \frac{P}{Q} \times D$$

$\therefore$ Total cost T = Material Cost + Holding cost + ordering cost

$$= C \times D + H \times \left(R + \frac{Q}{2}\right) \times C + \frac{P}{Q} \times D$$

$$= C\,.\,D + \frac{PD}{Q} + H.R.C + H \cdot \frac{Q}{2} \cdot C$$

Let $\frac{R}{Q} = f$ (*i.e.*, R be fraction of Q)

$$T = C.D + \frac{P}{Q} \cdot D + H.Q.f.C + H \cdot \frac{Q}{2} \cdot C$$

For minimum total cost $\frac{dT}{dQ} = 0$

$$\frac{-P}{Q^2} D + H.f.c + \frac{H \cdot C}{2} = 0$$

or $$Q = \sqrt{\frac{2PD}{C \cdot H(1+2f)}}$$

Case –II

Assumptions

(*i*) Supply is continuous.

(*ii*) Consumption is continuous.

(*iii*) No buffer stock.

(*iv*) Penalty for shortage.

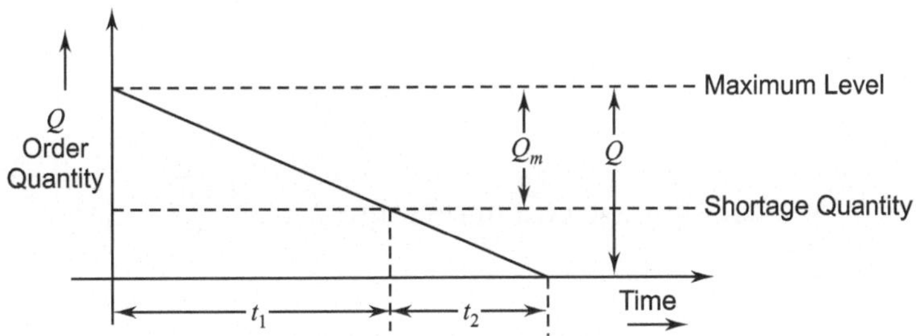

Fig. 8.5 Inventory level profile with no buffer stock and penalty for shortage

Let Q = Order Quantity

Q_m = Maximum Inventory level

D = Annual demand rate

C_1 = Penalty/Unit

H = Holding or carrying out

t_1 = Time during which demand met from stock

P = Procurement cost

t_2 = when shortage occurs

C = unit cost

Total cycle time $t = t_1 + t_2$

$$\text{No. of orders placed/year} = \frac{D}{Q}$$

$$\text{Now} \quad \frac{t_1}{t} = \frac{Q_m}{Q}$$

$$t_1 = \frac{Q_m}{Q} t \quad \text{and} \quad t_2 = \frac{Q - Q_m}{Q} \cdot t$$

$$\text{Average inventory during time } t_1 = \frac{Q_m}{2}$$

$$\text{Average inventory during time } t_1 = \frac{Q_m}{2} \times C \times \frac{t_1}{t}$$

Average inventory carrying cost for n orders placed annually

$$= H \cdot \frac{Q_m}{2} \cdot C \cdot t_1 \cdot n$$

$$\text{Average no. of units short during time } t_2 = \frac{Q - Q_m}{2}$$

$$\text{Average shortage cost during time } t_2 = \frac{Q - Q_m}{2} \cdot C_1 \cdot t_2$$

$$\text{Annual shortage cost} = \frac{Q - Q_m}{2} . C_1 . \ t_2 . \ n$$

$\therefore$ Total annual cost = Material cost + Inventory carrying cost + set-up cost + shortage cost

$$T = C.H. \frac{Q_m}{2} .t_1 n + C \cdot D + \frac{P}{Q} \cdot D + \frac{Q - Q_m}{2} \cdot C_1 \cdot t_2 \cdot n$$

Substituting values of t_1 and t_2

$$T = C \cdot D + \frac{P}{Q} \cdot D + C \cdot H \cdot \frac{Q_m}{2} \cdot \frac{Q_m}{2} \cdot t \cdot n + \frac{Q - Q_m}{2} \cdot C_1 \frac{(Q - Q_m)}{Q} \cdot t \cdot n$$

Let $t.n = 1$ year

$$T = C \cdot D + \frac{P}{Q} \cdot D + C \cdot H \cdot \frac{Q_m^2}{2Q} + \frac{(Q - Q_m)^2}{2Q} \cdot C_1$$

For total minimum cost $\frac{dT}{dQ} = 0$

Case (A) $CHQ_m - QC_1 + Q_mC_1 = 0$

$$Q_m = \frac{QC_1}{CH + C_1}$$

Since $\frac{C_1}{CH + C_1} < 1$ $\therefore$ $Q_m < Q$

Case (B) $Q = \sqrt{\frac{2PD}{CH}} \sqrt{\frac{C_1 + CH}{C_1}}$

since $\sqrt{\frac{C_1 + CH}{C_1}} > 1$ $\therefore$ $Q > \sqrt{\frac{2PD}{CH}}$

8.10.2 Probabilistic Inventory Model

(*i*) **Fixed Level Model:** This is a statistical method of fixing upper limit of stock holding. This method is used for replenishment of maintenance spares (especially insurance spares). In the previous models *viz.*, EOQ and EPQ, the demand is regular and known. But in case of insurance spares, the demand is not regular and the quantity required cannot be estimated. This is due to the typical failure pattern of spares.

To estimate such requirement, the probability distribution of failures of spares is used to fix the upper limit of stock holding.

Different types of spares follow different probability distribution functions (*e.g.*, Normal, Bi-nomial, Poission, Weibull, Exponential, Releigh *etc.*). So based on the probability distribution of failures of spares, future consumption rate is predicted.

For example, for a slow moving insurance spare, poission distribution is followed:

$$p(n, \lambda t) = \frac{(\lambda t)^n e^{-\lambda t}}{n!}$$

Where $p(n, \lambda t)$ = Probability of n failures during a period of t hours when they occur at a constant rate of λ failure/hour

λ = Failure rate.

n = Number of failures in time t.

λt = Expected number of spares.

8.11 TYPES OF MAINTENANCE SPARES

Maintenance spares are of following types:

(*i*) **Insurance Spares:** Insurance spares are very slow moving or non-moving spares having very low expected demand. The cost of insurance spares is a considerable percentage of the equipment itself. For example, in diesel generating set, crankshaft is an example of insurance spares.

Cost of stock-out of insurance spares is extremely high and have a long lead time. The spare may or may not be required during entire life of the equipment. Other examples of insurance spares are blades of turbine, pump shaft of boiler feed pump, critical non-standard motor *etc.*

Characteristics of insurance spares are following:

(*a*) These are very expensive.

(*b*) Have long lead time.

(*c*) These may or may not be required during whole life time of equipment.

(*d*) Generally purchased along with capital equipment, as they may not be available after 10 or 15 years.

(*e*) These are non-moving items.

(*ii*) **Ordinary Spares:** These are spares of machines which are slow moving and medium value items. These spares are available with medium lead time (typically 2 to 3 months). The rate of consumption is low and is known.

(*iii*) **Consumables:** These are fast moving, low value items such as filters, bearings, V-belts *etc*. The rate of consumption of these items is fairly constant. There are available either ex-stock or with little lead time.

8.12 CONTROL OF SPARES

Various techniques are available for control of spare parts.

These are following:

(*i*) A-B-C analysis.

(*ii*) V-E-D analysis.

(*iii*) F-S-N analysis.

(*iv*) S-D-E analysis.

(*v*) H-M-L analysis.

(*vi*) G-O-L-F analysis.

(*vii*) S-O-S analysis.

(*viii*) X-Y-Z analysis.

(*i*) **A-B-C Analysis:** This techniques is based on Pareto principle. Wilfredo Pareto, an Italian economist, concluded after a study that 80% of the wealth of the country is held by 20% of the people. This principle is used in spare-parts management. As majority of the items are not significant and only few items are expensive.

In 1948, then chief materials manager of General Electric found that top 20% of items contributed to almost 80% of total cost. Next 30% of items contributed to 5% of the total cost. Remaining 50% of the items contributed to 5% of the total cost.

Accordingly, all stock items are categorised into A-class, B-class and C-class items.

(*a*) ***A-Class items:*** These items together form 80% of annual consumption value of all stock items, these may form only 20% of the total numbers of stock items. All insurance spares fall under this category.

(*b*) ***B-Class items:*** These are items whose consumption value form 15–20% of all items. These are ordinary spares used in preventive and breakdown maintenance.

(*c*) ***C-Class items:*** All balance stock items though large in number forms 5-10% of the total value. These are consumables spares such as nuts, bolts, V-belts, filters *etc.*

For good inventory management, it is necessary to tightly control A and B class items leaving '*C*' class items to operating level,

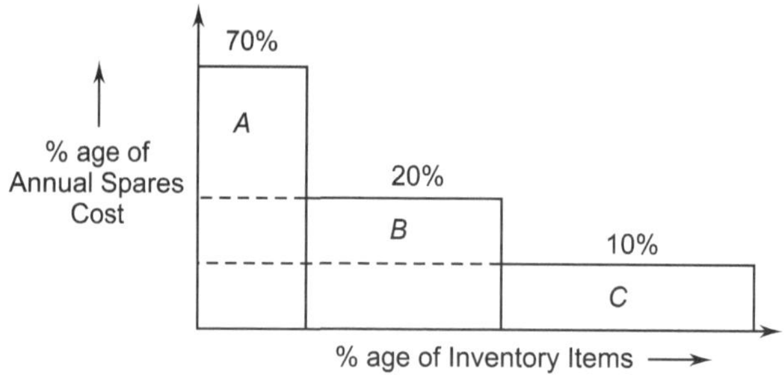

Fig. 8.6 *A-B-C* Analysis.

(*ii*) **V-E-D Analysis:** In this technique, the items are classified as V–Vital, E–Essential and *D* – Desirable.

"Vital" are those items whose stock-out cost is very high. "Essential" are those items which have potential to stop operation in near future and rest are "Desirable".

This categorisation takes into account, the degree of stock-out cost due to non-availability of items. This classification enables management to focus attention on just few items where maintenance of high service level is necessary.

(*iii*) **F-S-N Analysis:** In this technique, items are classified as *F*-fast moving, *S*-show moving and *N* – non-moving.

This categorisation is based on rate of consumption of items. Fast moving items are consumables which are low value items. Slow moving items are ordinary spares having moderate value. Non-moving items are insurance spares having high value.

F-S-N analysis helps to identify slow moving items and dead stock. Purpose of this technique is to control obsolescence.

(*iv*) **S-D-E Analysis:** In this technique, items are classified as '*S*' - Scarce, '*D*'-Difficult and '*E*'-Easy to get. Items are categorised based on difficulty in procurement.

Purpose of this technique is to calculate lead time required for procurement and hence decide buying strategy.

(*v*) **H-M-L Analysis:** Items categorised are - '*H*' - High value, '*M*' - medium value and 'L' - low value, categorisation is based on unit price of items.

(*vi*) **G-O-L-F Analysis:** Items categorised in this technique are '*G*' - Govt, '*O*'- Ordinary, '*L*'-Local and '*F*' - Foreign categorisation is based on sources of material.

(*vii*) **S-O-S Analysis:** '*S*' stands for seasonal and '*OS*' - off seasonal. This classification is based on nature of supplies and used for deciding spares holding strategies.

(*viii*) **X-Y-Z Analysis:** This analysis is based on value of items in storage and is used for inventory control purpose.

8.13 INVENTORY CONTROL

Objectives of inventory control is to achieve satisfactory level of customer service while keeping inventory cost within reasonable limits.

Following techniques are used for inventory control:

(*i*) Control on introduction of new items.

(*ii*) Proper care in forecast of annual demand.

(*iii*) Use of appropriate replenishment policies.

(*iv*) A-B-C/V-E-D/F-S-N analysis.

(*v*) Standardisation.

(*vi*) Variety reduction.

(*vii*) Value analysis.

(*viii*) Disposal of overstocks.

(*ix*) Surplus review.

(*x*) Scrap disposal.

(*i*) **Control on introduction of new items:** New items should be introduced only when there is going to be regular demand or is required to be stocked as insurance spares.

(*ii*) **Proper care in forecast of annual demand:** If the quantity forecast is on higher side than the actual consumption, then, this results in excess inventory. On the contrary, if it is on lower side, then there would be stock-out leading to stoppage of work. Statistical techniques allow forecast with a fair level of accurary.

(*iii*) **Use of appropriate replenishment policy:** The objective of replenishment policy should be to ensure availability of material in accordance with demand pattern without creating excess inventory.

(*iv*) **A-B-C/V-E-D/F-S-N analysis:** As discussed earlier, these techniques can be used to check excess inventory.

(*v*) **Standardisation:** By standardising machinery in terms of make, model *etc.*, variety of spares can be reduced. This helps is controlling excess inventory.

(*vi*) **Variety reduction:** Variety reduction can be done is number of ways *e.g.*, instead of using large number of oils and greases of different makes and grade, variety can be reduced by using equivalent grades of one or two makes, replacement with existing indigenous substitute (import substitution) *etc.*

(*vii*) **Value analysis:** It refers to analysis of an item w.r.t. its functionality and utility. It has been found that an average item has 20% extra cost built into it due to inclusion of many features which hardly contribute to any useful function. Value analysis aims to achieve elimination of this unnecessary cost and thereby lower the cost of material without sacrificing quality or its function.

(*viii*) **Disposal of overstocks:** In order to control inventory carrying cost, it becomes necessary to control stock levels and devise policies to deal with this excess stock. This apart from releasing funds, will also result in minimisation of loss of material due to obsolescence, pilferage, damage *etc.*

(*ix*) **Surplus review:** Any item not demanded by the customer for certain length of time is categorised as surplus. If the surplus material is carried for long, the inventory carrying cost is likely to exceed the cost of material. A constant watch is needed on those items which are not demanded for period of say 12 months or more so as to identify items which are likely to become surplus. This helps in finding alternate use for the material and also in preventing further procurement.

(*x*) **Scrap disposal:** The material which are no longer required for the purpose for which they were procured are termed as 'scrap'. This includes all obsolate materials and all types of waste generated during the course of production and maintenance. Other examples of scrap are turning and boring generated in workshops (metal turnings), wornout and demaged components, condemned plant and machinery, used oil, empty containers, drums *etc.*

The scrap is disposed-off by three methods:

(*i*) By auction.

(*ii*) Sale by inviting tender.

(*iii*) By direct sale.

(*i*) **By auction:** It is the most extensively used method of scrap disposal. Auction is conducted either by the organisation or by auctioneers.

Before conducting an auction, advertisement are issued in newspapers with the details of date, time and venue of auction. Auction conditions are explained to all bidders. Sale of lot is made to the highest bidder. The supervising officer decides beforehand a reserve price which is based on quality of scrap, previous auction rate and recent trend. Successful bidder has to pay earnest money on the spot and balance payment within 10 to 20 days. The earnest money is liable to be forefeited if the buyere fails to pay the balance amount and take delivery of sold material.

(*ii*) **Sale by inviting tender:** This procedure is time consuming and is resorted to if suitable bids are not forthcoming in auctions. On finalisation of tender, a formal agreement is entered to between the buyer and the seller.

(*iii*) **By direct sale:** Certain items such as empty oil barrels, empty acid jars, scrap timber etc. are sold by direct sale.

8.14 CAUSES OF OBSOLESCENCE OF SPARES

Various causes of obsolescence are:

(*i*) **Change in product design:** Change in design of a product renders existing spares as obsolete. For example, a change in design of a filter will necessitate change in filter housing.

(*ii*) **Technological innovation:** Technology is changing very fast causing frequent changes and modification of product. This leads to spares obsolescence.

(*iii*) **Faulty Purchase:** A wrong pruchase of equipment results in generation of obsolete material.

(*iv*) **Improper storage:** Improper storage may lead to deterioration and damage. For example, certain chemicals need to be stored at certain temperature and humidity, and if conditions are not maintained, the chemicals get spoiled. Similarly, glass items if not properly stored will lead to damage.

8.15 PRESERVATION OF SPARES

In material stores, a wide variety of items are stored. All items requires different storage conditions *w.r.t.* temperature, humidity, dust, vibration, static charge *etc.* Any change in storage condition may render these items unsuitable for use. Hence proper care must be given to preservation of spares which will help in reducing scrap generation.

Following guidelines will help in proper storage of materials:

(*i*) All bearings should be greased and packed in grease proof paper.

(*ii*) Rubber items must be dusted with french chalk.

(*iii*) Sintered bush bearings should be soaked in warm oil for 24 hours, once in a year.

(*iv*) Electrodes should be kept in their original pack in cool and dry place.

(*v*) All switchgear parts must be enclosed in plastic bags.

(*vi*) Glass sheets must be kept vertical and protected from sunlight.

(*vii*) All water treatment plant resins must be kept in moist condition in plastic bags.

(*viii*) All electronic cards must be kept in cool, dry and dust-free environment.

(*ix*) All oil drums must be kept in horizontal position.

(*x*) All sheet metal items must be kept with a layer of oil so as to protect them from corrosion.

(*xi*) All explosive chemicals must be kept in a seperate store away from main store.

Summary

Spares-parts form a significant amount of capital that remains locked-up. However, one can ignore it as the stock-out cost could be too high. So, the aim of inventory management is to devise ways and means so as to minimise holding cost but prevent chances of stock-out.This is not as simple as it looks due to complex nature of spare parts consumption in maintenance. For a production shop having constant rate of consumption, it is easier to forecast requirement. However, maintenance spare pose a complex problems due to various reasons such as cost, proprietory nature, unexpected requirement, chances of obsolescene and degradation *etc.*

Inventory models both deterministic and probabilistic are discussed. Deterministic models ar economic order quantity, economic batch quantity and annual estimate system. Fixed level system is discussed in probabilistic model. For estimation of future spares need, failure distribution pattern of component need to be known. This is difficult and hence certain probability distribution functions such as weibull, normal, Reileigh *etc.* are used to approximate failure distribution pattern.

Contd...

After procurement of spares, next is to control inventory so as to minimise holding cost. Various techniques such as ABC, VED, FSN, SDE *etc.* are used.

Important areas of spares parts management are — variety reduction, value analysis, surplus review, disposal of overstocks and obsolescene management.

Next is proper preservation of spare parts. This is due to the fact that different type of spares need different environment and care *e.g.* Certain chemicals need to be kept under controlled temperature and humidity conditions.

Hence, spare parts management of maintenance presents a unique problem and a careful planning, procurement and review can improve organisational profitability.

Review Questions

1. Define inventory. What is the purpose of keeping inventory?
2. Explain importance and functions of inventory.
3. Discuss various types of inventory in a manufacturing organisation.
4. Explain classificaion of inventory.
5. Why inventory models are required? Explain any one method of deterministic inventory model.
6. Explain terms:
   (*i*) Buffer stock.
   (*ii*) Lead time.
   (*iii*) Inventory holding cost.
   (*iv*) Stock-out cost.
7. Discuss role of inventory in maintenance management.
8. What are various methods used for inventory control of spares?
9. Why is inventory control necessary for profitability of an organisations?
10. What are various reasons for obsolescence of spare ? How are these controlled?

Discussion Questions

1. "Inventory carrying cost is the single largest component affecting profitability of an industrial enterprise". Justify.
2. "Stock-out cost of critical insurance spares justifies large inventory carrying cost". Comment.

3. "Inventory management in the maintenance department is a complex job". Explain.
4. "It is easier to control production inventory than maintenance inventory". Discuss.
5. "A careful planning and periodic review of inventory can avoid obsolescence cost of spares to a large extent". Do you agree?
6. "Cannibalisation of discarded machinery can help to reduce spares cost of old running machinery". Comment.

Multiple Choice Questions

1. Function of inventory is to:
   (*a*) Provide buffer stock.
   (*b*) Provide economics of scale.
   (*c*) Maintain smooth production flow.
   (*d*) All of these.
2. Dead stock items are those which are:
   (*a*) Damaged.
   (*b*) Not being consumed for a long time.
   (*c*) A category items.
   (*d*) None of these.
3. Re-order system is applicable to:
   (*a*) Batch system.
   (*b*) Continuous production system.
   (*c*) Both.
   (*d*) None of these.
4. V-E-D analysis is based on:
   (*a*) Cost of spares. (*b*) Service level.
   (*c*) Lead time. (*d*) Source of material.
5. Standardisation means
   (*a*) Reduction of spares quantity. (*b*) Improvement in design.
   (*c*) Variety reduction. (*d*) None of these.

True/False Questions

1. Optimum ordering quantity is based on minimum of total cost (holding and ordering cost).
2. Economic batch quantity is used for continuous process plants.

3. A-B-C analysis is most versatile method of inventory management.
4. Obsolete spares may result due to faulty purchase or insufficient specifications
5. Value analysis is not useful for variety reduction of spares.

ANSWERS

Multiple Choice Questions

1. (*d*)	**2.** (*b*)	**3.** (*b*)	**4.** (*b*)	**5.** (*c*)

True/False Questions

1. True	**2.** False	**3.** True	**4.** True	**5.** False

❑❑❑

9

Material Handling Systems

Learning Objectives

After reading this chapter, you will be able to:

- Understand basic process of material handling.
- Understand design and evaluation of material handling systems.
- Understand bulk handling process.

9.1 INTRODUCTION

Material handling is the art and science of moving, storing, protecting and controlling material. The moment, raw material enters a plant till it is dispatched for sale, material passes through various types of material handling systems. Although material handling systems do not add any value to the material being handled, but cost of handling is 20–30% of cost of product.

Various material handling systems are used in factories, godowns, airports, construction sites, super markets *etc.*

Material handling systems are becoming part of automated production systems, improving system speed, efficiency and reducing accidents.

9.2 PURPOSE OF MATERIALS HANDLING SYSTEM

These are following:

(*i*) To maintain or improve quality.

(*ii*) To promote plant safety.

(*iii*) To improve productivity.

(*iv*) To control inventory.

(*v*) To help smooth movement of material through various work stations.

(*vi*) To reduce fatigue to workers.

(*vii*) To provide better working conditions.

(*viii*) To reduce cost of transportation.

(*ix*) To increase plant capacity.

9.3 CLASSIFICATION OF MATERIAL HANDLING EQUIPMENT

Following are the classification of Material Handling Equipment:

(*i*) Lifting Equipment—These are following:

(*a*) Pulley.

(*b*) Chain type manual hoist.

(*c*) Electrical hoist.

(*d*) Preumatic hoist.

(*e*) Jib crane.

(*f*) Gantry crane.

(*g*) Electrical overhead travelling crane.

(*h*) Elevator.

(*i*) Jack.

(*j*) Winche.

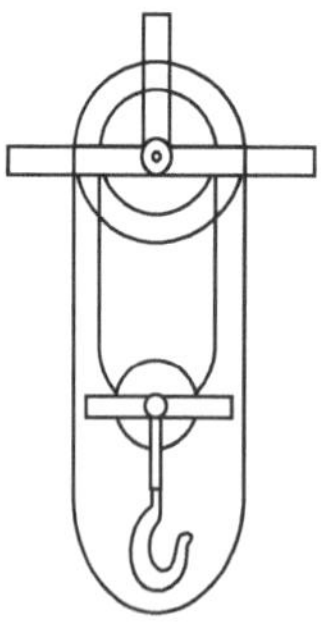

Fig. 9.1 Pulley

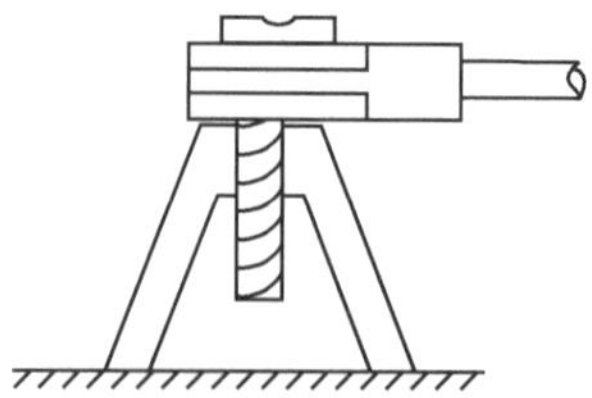

Fig. 9.2 Screw Jack

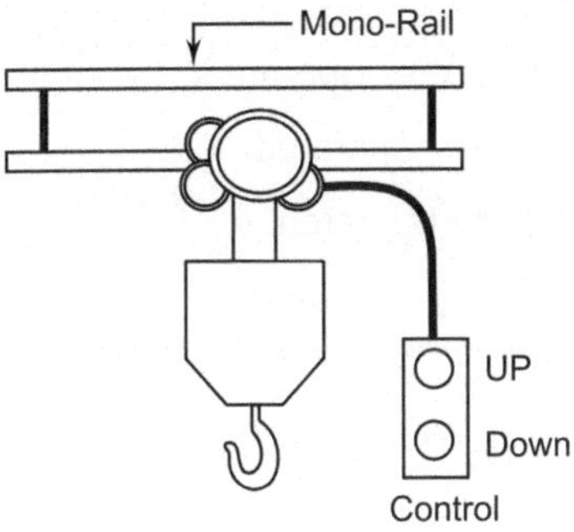

Fig. 9.3 Electric Hoist

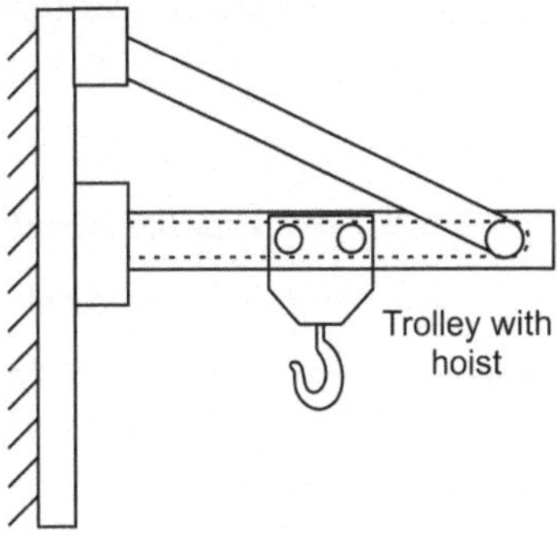

Fig. 9.4 Jib Crane

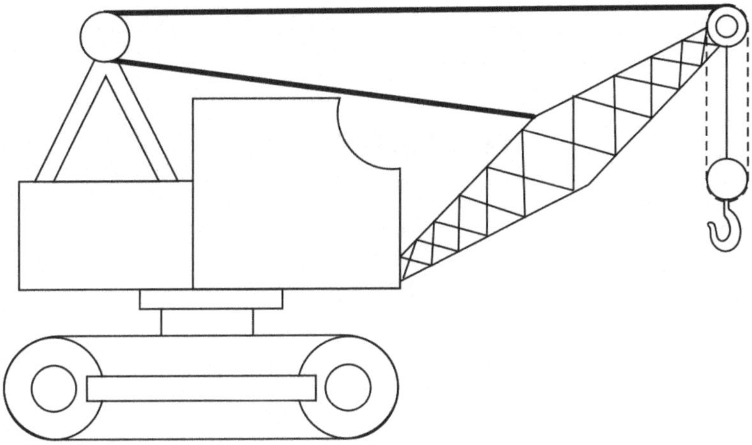

Fig. 9.5 Mobile Crane.

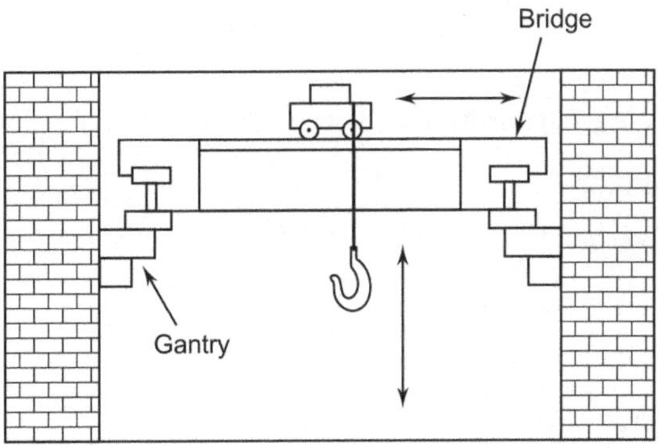

Fig. 9.6 Electrical Overhead Travelling Crane.

(*ii*) Transportation Equipment

(*a*) *Forklift Truck*: Diesel engine or battery operated.

(*b*) *Platform Truck*: Hydraulic or battery operated.

(*c*) *Tow-Truck*: Diesel engine or battery operated.

(*d*) *Manual Trolley*: Hydraulic or battery operated.

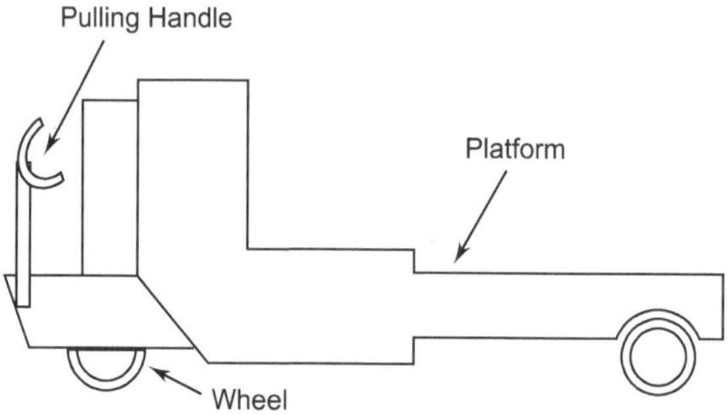

Fig. 9.7 Hydraulic Trolley.

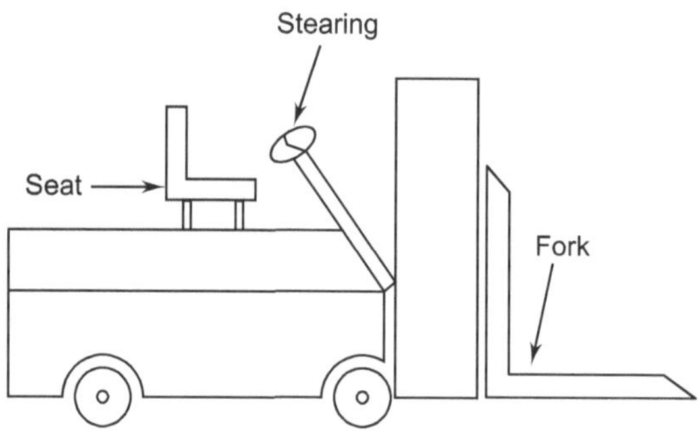

Fig. 9.8 Forklift Truck.

(*iii*) Conveying Equipments: These are following:

(*a*) Hopper.

(*b*) Overhead Conveyor.

(*c*) Belt Conveyor.

(*d*) Rollar Conveyor.

(*e*) Screw Conveyor.

(*f*) Flat Conveyor.

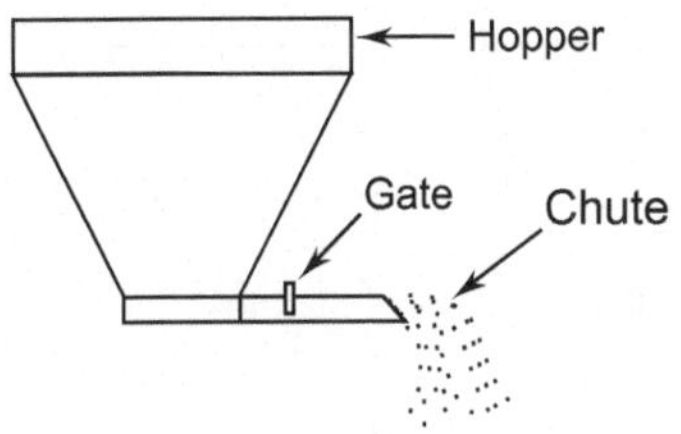

Fig. 9.9 Hopper

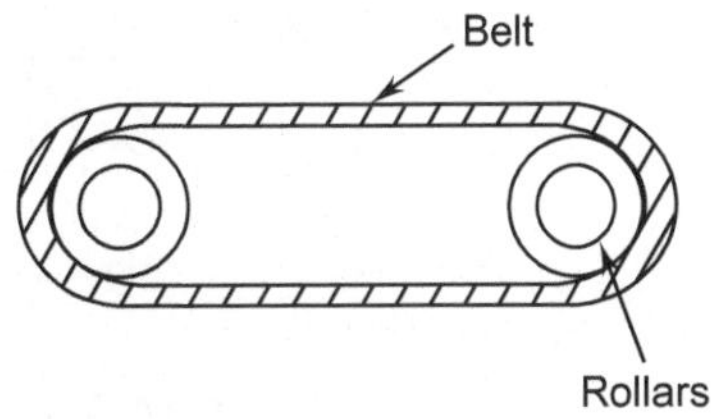

Fig. 9.10 Belt Conveyor

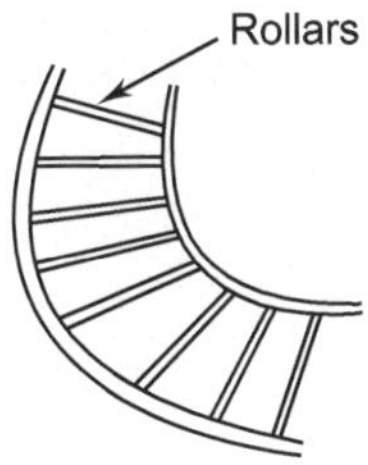

Fig. 9.11 Rollar Conveyor

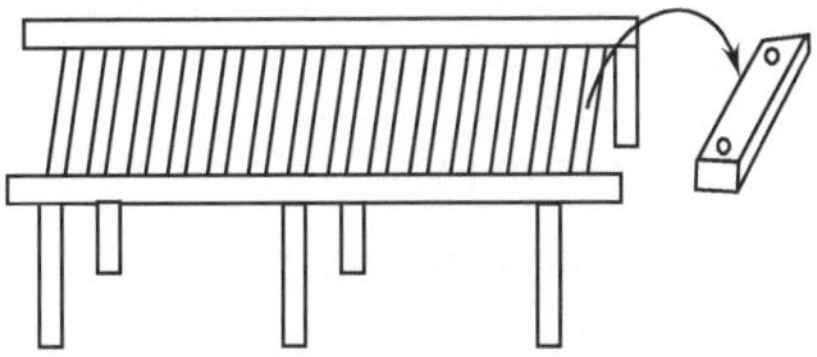

Fig. 9.12 Flat Conveyor

(*iv*) **Storage Equipment:** These material handling equipments store or stack-up material in a heap, in a vertical rack or bins *etc*. These equipments are— stacker (used for making heap in coal handling yard), storage carousal, racks, bins, drums, barrels *etc.*

(*v*) **Positioning Equipment:**These are used to position material in a particular place or in a particular direction such as turntable, rotary index table, industrial robot, *etc.*

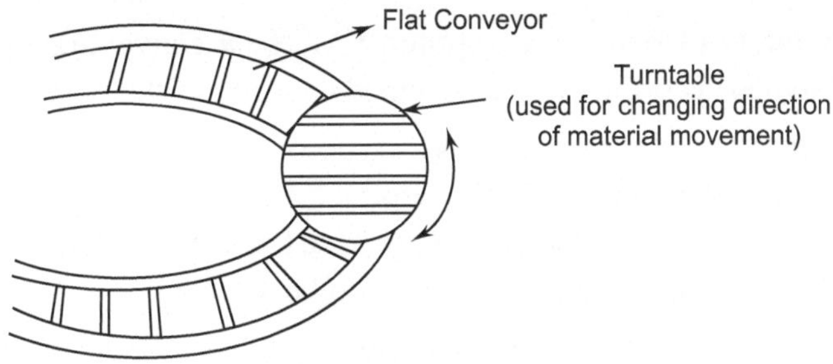

Fig. 9.13 Turntable

(*vi*) **Unitising Equipment:** These are unit load formation equipments such as pallets, bins, cartons, crates *etc.* They convert small items into one big unit which allows easy handling and storage.

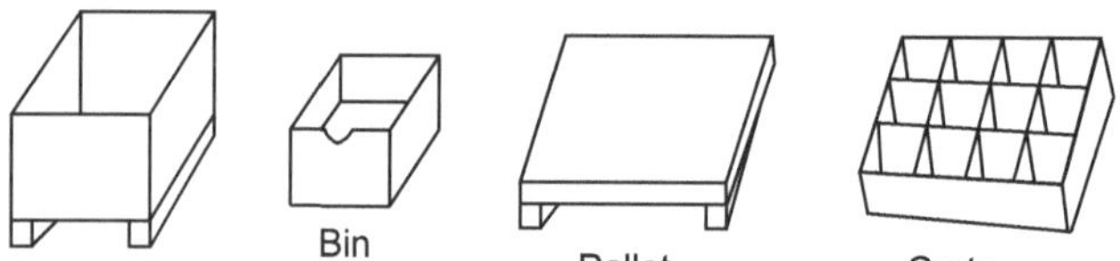

Fig. 9.14 Types of Unitising Equipment.

9.4 SELECTION OF MATERIAL HANDLING EQUIPMENT

Following factors may be considered for selection of material handling equipments:

(*i*) Type of material to be moved.
(*ii*) Type of plant layout.
(*iii*) Capacity of process equipment.
(*iv*) Type of production process.
(*v*) Cost of material handling equipment.
(*vi*) Safety feature of equipment.
(*vii*) Type of material flow – continuous/intermittent

(*i*) **Type of material to be moved:** It depends upon shape, size, weight and nature of material (solid/liquid/gas).

(*ii*) **Type of plant layout:** Plant layout post limitation of space, height *etc.* So material conveying equipment must meet these conditions. For example, fork lift truck requires more space for movement than a belt conveyor.

(*iii*) **Capacity of process equipment:** Material conveying equipment is designed according to the designed capacity of the plant.

(*iv*) **Type of production process:** Type of material conveying method selected depends upon the production process. For example, for conveying iron ore, belt conveyor have to be chosen and for conveying refrigerators, slat conveyors have to be selected.

(*v*) **Cost of material handling equipment:** It is the single most important factor that dicides the choice of material conveying equipment. A simple and cheaper method of material conveying system is always preferable.

(*vi*) **Safety features of equipment:** While selecting a material handling equipment, safety condition of plant and process need to be considered.

(*vii*) **Type of material flow (continuous/intermittent):** Depending upon type of flow, material handling equipment is chosen. For example, conveyors are meant for continuous flow of material but fork-lift trucks, tow trucks *etc.*, carry material in batches.

9.5 CONCEPT OF CONTAINERIZATION

It is easier to move material by grouping them together rather than moving individually. For example, instead of moving household items such as cups, plates, forks individually, these can be placed in a cardboard box and then transported.

Same concept is used in containerization whereby items are grouped together and moved as single unit. These units can be moved by fork-lift truck and can be stored one above other in the storage area.

Advantages of Containerization

These are following:

(*i*) It is possible to handle more material in a given time.

(*ii*) Cost of handling is reduced.

(*iii*) Better utilisation of storage space.

(*iv*) Faster and safer material movement.

(*v*) Lesser chances of material damage during handling

The unit load is placed on a flat board called pallet. This can be then easily moved by fork-lift trucks. These pellets are made of plywood or polymer.

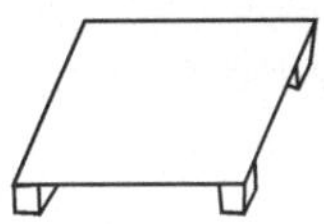

Fig. 9.15 Flat Wooden Pellet

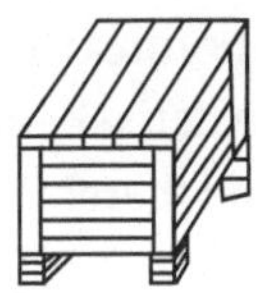

Fig. 9.16 Box Type Pellet

9.6 PRINCIPLE OF DESIGN OF MATERIAL HANDLING SYSTEM

Cost of material handling system consists of the following:

(*i*) Initial cost of equipment.

(*ii*) Operating cost.

(*iii*) Maintenance cost.

(*iv*) Direct and indirect labour cost.

(*v*) Miscellaneous cost such as training, insurance, *etc*.

While designing a material handling system, not only the cost should be optimum but also a lot of other factors need to be considered. These are following:

(*i*) **Simplification of system design:** This is foremost step which involves in checking the following questions:

(*a*) Is the movement necessary?

(*b*) Can this be eliminated?

(*c*) Can it be combined with some other process?

This helps to reduce cost, make layout uncluttered and reduces process time.

(*ii*) **Use of gravity principle:** Wherever possible, gravity should be used for movement of material *e.g.*, material can freely flow from higher location to a lower location through a chute without use of energy.

(*iii*) **Efficient utilisation of available space:** The space available for material movement must be utilized to the optimum with sufficient space for manpower movement and maintenance.

(*iv*) **Use of unit load principle:** Wherever possible material should be combined to form a single unit load rather than individual items.

(*v*) **Automation:** Automation in material handling system helps to increase productivity, reduce labour cost, improve safety and improve product quality. For example, sequential operation such as periodic start/stop *etc*. can be controlled by using PLC (Programmable Logic Controller).

(*vi*) **Ergonomics:** Ergonomics is the art and science of human comfort. The material handling systems must be designed keeping human comfort in mind thereby minimizing worker fatigue and accidents.

9.7 Criteria for Performance Evaluation of Material Handling System

A group of factors help to evaluate performance of a given system. There are following:

(*i*) **Response time:** It is the required in moving a unit item from source to destination.

(*ii*) **Quantity delivered:** It is the amount of material handled per unit time.

(*iii*) **Lifecycle cost:** It is the total cost of ownership from purchase of equipment to its final disposal.

(*iv*) **Space utilisation:** It is the total plant area occupied by material handling equipments.

(*v*) **System expandability:** It is the ability of the system to modify existing material handling system in order to meet future need.

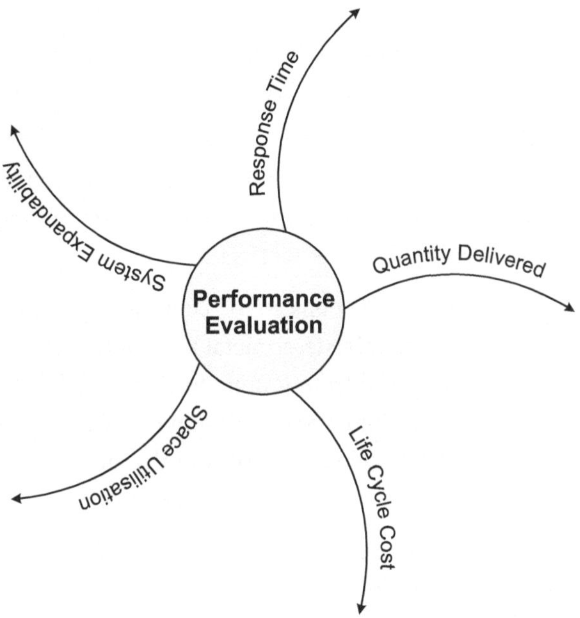

Fig. 9.17 Performance Evaluation of MHS

9.8 BULK MATERIAL HANDLING

This method is used for conveying large quantity of material to short/long distances or heights. For example, Soda Ash (Sodium carbonate), Talc, Pulverised coal, Carbon black, flour, detergents *etc.*, are moved by bulk material handling system.

9.9 CLASSIFICATION OF BULK MATERIAL HANDLING SYSTEMS

Bulk material handling systems are classified into two categories:

(*i*) Mechanical conveying.

(*ii*) Pneumatic conveying.

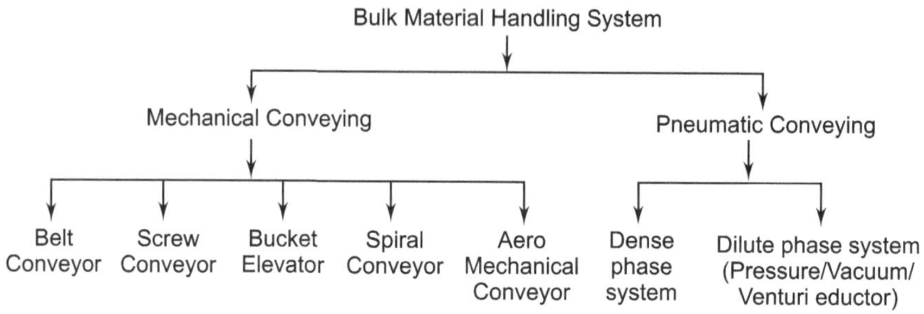

(*i*) **Mechanical Conveying:** It is used for conveying solid, powdered and granular material over short distance and upto 8m elevation from ground level. The various mechanical conveying systems are following:

(*a*) *Belt/screw/Bucket elevator*: Belt and Bucket conveyor is used for conveying solid, non-lumpy material while screw conveyor is used for conveying dry bulk material in powdered form.

(*b*) *Spiral conveyor*: It consists of centreless flexible angle rotating within a flexible plastic or metal tube. It is suitable for lifting dry powdered material from ground to 25 m height. Speed of rotation of spiral range from 300 to 500 rpm depending upon the nature of product. Main advantages of spiral conveyors are—minimum moving parts, dust and noise free operation and reliable operation.

(*c*) *Aero-Mechanical conveyor*: These are used where large volume of material is to be conveyed over large elevation or height. These are used to transport material upto 20 m elevation and upto 40 tons/hr capacity.

The system operates with air-streams in closed tubes. The material is fed into updraft tube where it gets entrained in the air-stream generated and the material discharged at the discharge head on the top while the air-stream returns through downdraft tube and keeps re-circulating.

The air stream is established by means of a series of evenly spaced discs fitted on a wire rope which moves at high velocities within up and downdraft tube as endless assembly.

This can be used to handle materials such as carbon black, Flour, Sugar, Lime, Talc, Detergents *etc.*

(*ii*) **Pneumatic Conveying:** This method of conveying is suitable for transfer of powder type material over long distances (more than 50 m) and at a higher rate (more than 5 tons/hr).The main advantages of these types of conveying are:

(*a*) Clean outside environment due to closed system.

(*b*) Absence of moving parts.

(*c*) Contamination free product.

Types of Pneumatic conveying system are following:

(*i*) Positive pressure system.

(*ii*) Negative pressure or vaccum system.

(*iii*) Combination system.

(*iv*) Closed loop/open loop system.

(*v*) Dilute phase system.

(*vi*) Dense phase system.

(*i*) **Positive Pressure System:** In this system, material is transported with the help of positive pressure of conveying medium such as air or gas. It is suitable for high capacity and multiple discharge points.

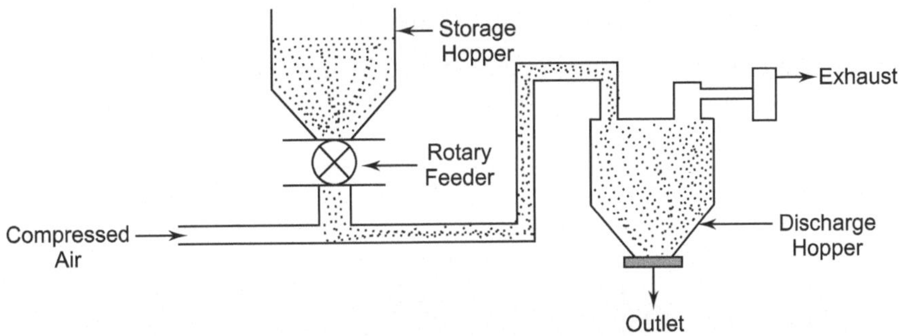

Fig. 9.18 Positive Pressure System

(*ii*) **Negative Pressure or Vacuum System:** In this system, material is continuously sucked into pipeline and transported under negative pressure. It is suitable for low capacity (5–10 Tons/hr) and short distances (upto 100 m). The material is conveyed with multiple pick-up points and is used for unloading of bulk material from Rail, Road vehicles or ships using a hose connection and quick-couplers.

(*iii*) **Combination System:** This system performs two operations simultaneously. One part of the system operates in vaccum and other part under pressure. It is used for smaller capacity (upto 5 tons/hr) and smaller distance (upto 50 m).

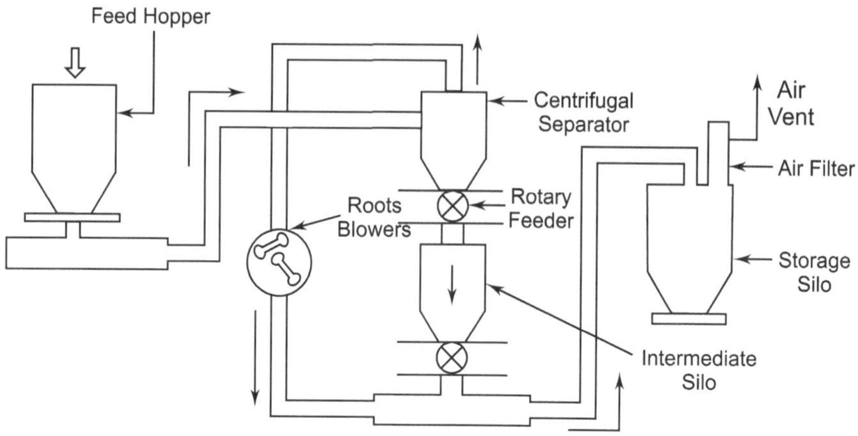

Fig. 9.19 Combination System.

(*iv*) **Closed Loop and Open Loop System:** A closed loop system is one in which conveying medium after being exhausted from material separator is returned to the inlet of fan blower compressor. Such systems are adopted for materials which degrades or reacts or forms explosive mixture with atmospheric oxygen. In such cases, Nitrogen or other inert gases are used for conveying.

An open loop system is one in which conveying medium after being exhausted from material separator is vented out to atmosphere.

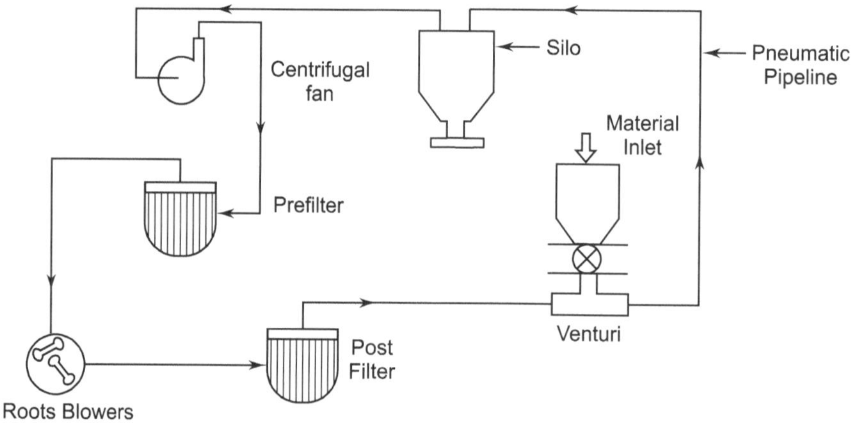

Fig. 9.20 Closed Loop System.

(*v*) **Dilute Phase System:** Dilute phase conveying refers to stream flow conveying with low material to air-ratio. The material is conveyed at high velocity and is uniformly suspended in the stream. It may be positive or negative pressure system.

(*vi*) **Dense Phase System:** Dense phase system is used to reduce unit power consumption by allowing a higher material to air ratio (30 to 150) and to reduce pipeline wear and tear.

In this system of conveying, velocity is insufficient to keep material uniformly suspended in the conveying medium. The conveying medium enters pressure vessel fed with material to be conveyed. The conveying medium first fluidizes the material and then carries it though pipes to the destination.

Table 9.1 Comparison of Positive and Negative Pressure System

Parameter	Positive Pressure System	Negative Pressure System	
		Dilute phase	Dense phase
Operating Pressure	0.406 to/1.03 bar	1 to 8 bar	1 to 8.5 bar
Pick-up Velocity	7.5–40 m/sec	7.5–40 m/sec	0.25–10 m/sec
Material to Air ratio	1 to 5	3 to 10	10 to 25
Capacity	0 to 10 Tons/hr	0 to 100 Tons/hr	0 to 100 Tons/hr
Conveying distance	0 to 100 m	0 to 200 m or more	0 to 200 m or more

9.10 SELECTION OF EQUIPMENT AND SYSTEMS FOR BULK MATERIAL HANDLING

Selection of equipment and systems for bulk handling depends on the following factors:

(*i*) Low cost automation.
(*ii*) Efficiency of handling.
(*iii*) Reliability of equipment.
(*iv*) Installation and operating cost.
(*v*) Clean operating environment.
(*vi*) Suitability to handle required material.
(*vii*) Best available technology and equipment.

(*i*) **Low cost automation:** All bulk handling systems are automated systems. The cost of automation should be low as higher cost leads to higher maintenance and upgradation cost.

(*ii*) **Efficiency of handling:** The proposed system should be efficient enough to handle desired material at required speed.

(*iii*) **Reliability of equipment:** The system components must be reliable so as to reduce unexpected downtime and cost of maintenance.

(*iv*) **Installation and operating cost:** The material handling system should have lower installation and operating cost. The specific energy consumption should be as low as possible.

(*v*) **Clean operating environment:** The material handling system should provide clean/pollution free environment.

(*vi*) **Suitability to handle required material:** The proposed material handling system must be designed to handle required material *e.g.* coal, lime, talc, detergent *etc.* Because different materials require different type of material handling system.

9.11 PARTS OF BULK HANDLING SYSTEM

Following are the parts or sub-systems of bulk material handling system:

(*i*) Material receipt and storage.

(*ii*) Material issue system.

(*iii*) Day storage and material processing system.

(*iv*) Product storage and dispatch.

(*i*) **Material Receipt and Storage:** The material receipt is in various packed forms such as bags of 25 to 75 kg or big bags of 500 to 1500 kg or inbulk by road or rail system. In case of liquid material, it is through rail and road tankers of 10 m^3 to 60 m^3 capacity.

The bags are unloaded either manually through fork lift trucks through inclined belt conveyors.

When product is received in bulk rail/road tankers, the storage system is different from above. The unloading plant requires storage silos of capacities ranging from 50 to 500 cubic meters depending upon flowability and characteristics of material and buffer storage requirement.

(*ii*) **Material Issue System:** The method of issuing stored material for production differs. When storing product in small bags, issue can be through intermediate storage silo from where material is transported pneumatically to plant usage area. This would require an automatic bag slitting machine or self contained dust-proof bag dump station to empty bags and transfer contents to the storage silo.

When material is received in bulk tankers, the system is greatly simplified and requires piping connections to transfer material from tanker to warehouse silo (*e.g.*, Naptha, Diesel, LPG *etc.*).

(*iii*) **Day Storage:** Because of plant demands in the production process, it generally becomes necessary to use raw material in specific quantities and at specific times, depending on production process. Many industries such as paint, food, building material, glass and ceramics, soaps *etc.* require mixture of various powders in pre-determined quantities. The system comprises of weighing, batching, mixing and blending.

(*iv*) **Product Storage and Despatch:** The final product is stored and despatched in bags, containers, carbuoys, drums *etc.*

9.12 VARIOUS OTHER BULK MATERIAL HANDLING EQUIPMENTS

Apart from bulk material handling systems discussed earlier, there are large variety of systems available to suit wide range of industries. However, some commonly used bulk handling equipments are given below:

(*i*) Vibrating Screens.

(*ii*) Crushers—Cone, Jaw and Impact.

(*iii*) Hammer and Bawl mills.

(*iv*) Wagon Tipplers.

(*v*) Stacker—Reclaimers.

(*i*) **Vibrating Screens:** Vibrating screens are used for seggregation *e.g.* pulverised coal is passed through screens to remote higher size particles. These screens make vibratory or reciprocating motion to sieve out higher size particles.

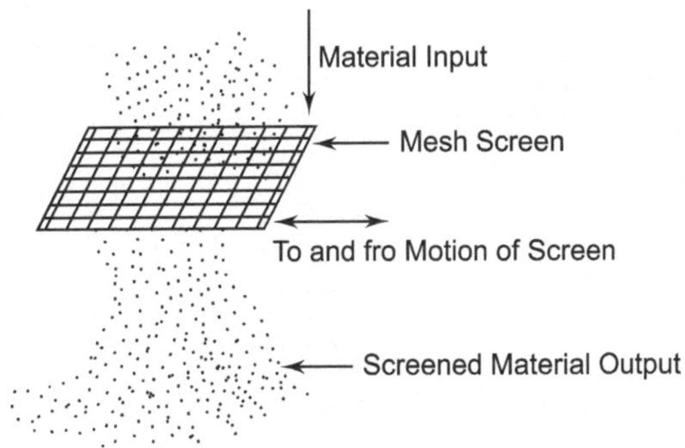

Fig. 9. 21 Vibrating Screen.

(*ii*) **Crushers:** Crushers are used to crush large lumps of material into smaller size *e.g.*, coal is crushed into smaller sizes to suit requirement in the process. Various types of crushers are cone crusher, jaw crusher or impact crusher.

(*iii*) **Hammer and Bowl Mills:** These are used to convert smaller aggregates into powdered form. The aggregates are powdered by either hammer action or by through balls of managnese steel.

(*iv*) **Wagon Tipplers:** Wagon tipplers are need to unload rail wagons into the material handling yard. This system unloads the rail wagons by tilting it and transfers its content into the storage space.

(*v*) **Stacker—Reclaimers:** These are used to store material in stacks in open yards. It consists of a giant wheel with buckets attached to it and can be swivelled in different directions.

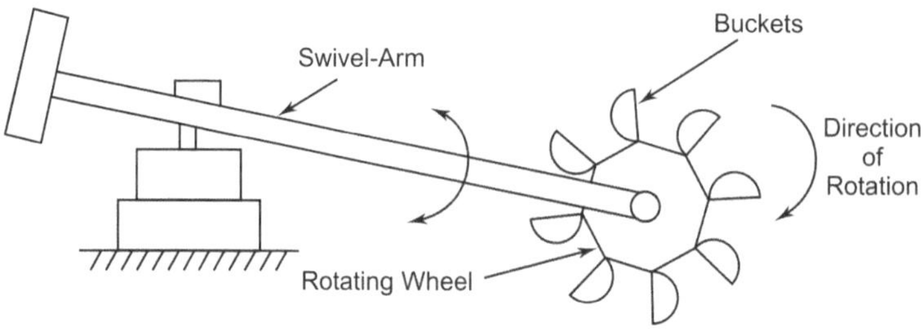

Fig. 9.22 Stacker—Reclaimer.

9.13 EXAMPLE OF LIQUID AND SOLID MATERIAL UNLOADING SYSTEM

A typical storage and transfer system of fuel oil is given below:

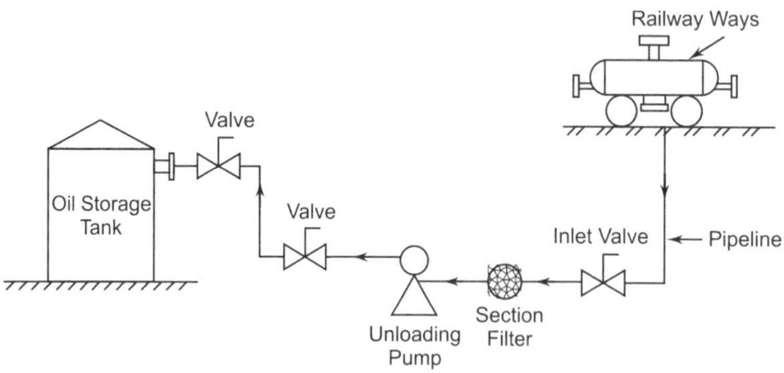

Fig. 9.23 Fuel oil Unloading System.

Note: Fuel oil lines and equipments are steam/electric traced and insulated. These are not shown here.

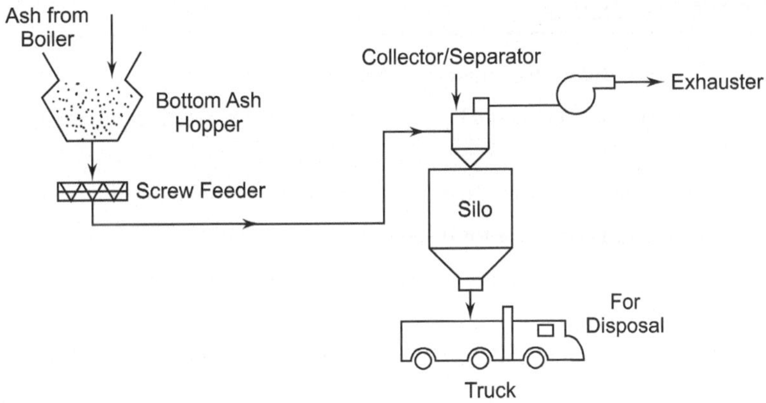

Fig. 9.24 Boiler Ash Disposal System

Summary

Material handling is the art and science of moving, storing, protecting and controlling material. Material handling starts from entry of raw material and ends with despatch of material. Material handling systems are used in factories, godowns, airports, construction sites, super-markets *etc.*

Material handling equipments are primarily of three types viz lifting type, conveying type and transportation type. Selection of material handling equipment is based on several factors such as type of material to be moved, plant layout, type of production process, capacity of process *etc.* To speed-up the material movement, materials are moved in group or contanerisation concept is used *e.g.*, pallets, crates *etc.* This helps to move more material at a given time.

These systems are designed based on operating cost, maintenance cost, labour cost, initial cost of equipment *etc.* However, any design adopted must be simple, use gravity principle, use available space efficiently and use unit load wherever possible.

Bulk material handling systems are used for conveying large quantity of material to large distances or heights. Such systems use either mechanical conveying or pneumatic conveying. These systems are fully automated. These systems are extensively used to convey, carbon black, sugar, talc, detergents, ash in thermal power plants, *etc.*

Other bulk handling equipments used are crushers, stackers—reclaimers, hammer and bowl mill, wagon tipplers *etc.* Material handling systems help to speed up material movement and are designed as per the process requirement.

Review Questions

1. Discuss classification of material handling systems.
2. How will you select material handling equipments for a refrigerator manufacturing unit?
3. What is containerisation? How is it helpful in material handling operations?
4. What are various criteria for design of material handling systems?
5. Discuss various parameters necessary for performance evaluation of material handling system.
6. What is bulk material handling? Discuss mechanical and pneumatic conveying.
7. Discuss various parameters for selection of equipments for bulk material handling.
8. Describe in detail various parts of bulk material handling system.
9. Differentiate dense phase and light phase system of pneumatic conveying.
10. How ergonomic in design of material handling systems helps in cost reduction and accident prevention?

Discussion Questions

1. "Material Handling system constitutes a significant part of plant and machinery in a continuous process plant". Justify
2. Elaborate—"A wrong selection of material handling system increases product cost".
3. Explain "Containerisation reduces wastage, damage and improves operational efficiency".
4. "Automation of plant and machinery reduces cost of material handling". Comment
5. "Material Handling system if properly designed helps to reduce cost by preventing wastage and damage of material moved, minimising accident risk and improving operational efficiency". Elaborate.

Multiple Choice Questions

1. Purpose of material handling is to:

   (*a*) Improve productivity. (*b*) Reduce worker fatigue.

   (*c*) Promote plant safety. (*d*) All of these.

2. Selection of material handling equipment does not depend on:

   (*a*) Cost. (*b*) Type of production process.

   (*c*) Type of plant layout. (*d*) None of these.

3. Ergonomically designed material handling system helps to:
   (*a*) Reduce maintenance cost. (*b*) Reduce worker fatigue.
   (*c*) Reduce energy cost. (*d*) None of these.
4. Positive pressure pneumatic conveying system is suitable for:
   (*a*) Low capacity and multiple discharge points.
   (*b*) Low capacity and single discharge point.
   (*c*) High capacity and multiple discharge points.
   (*d*) High capacity and single discharge point.
5. Closed loop system uses the following gas for conveying:
   (*a*) Oxygen. (b) Air.
   (*c*) Nitrogen. (d) Carbon dioxide.

True-False Questions

1. Choice of material handling systems is dependent on type of production process.
2. Use of gravity principle in movement of material causes wastage of material handled.
3. Automation of plant and bucket elevators are used for conveying solid and non-lumpy material.
4. Belt, screw and bucket elevators are used for conveying solid and non-lumpy material .
5. Dilute phase conveying is used with negative pressure system.

ANSWERS

Multiple Choice Questions

1. (*b*) **2.** (*d*) **3.** (*b*) **4.** (*c*) **5.** (*c*)

True/False Questions

1. True **2.** False **3.** True **4.** True **5.** False

❑❑❑

10

Industrial Safety Management

Learning Objectives

After reading this chapter, you will be able to:

- Develop basic concept of safety.
- Analyse causes of accidents and ways to prevent them.
- Understand accidents caused by fire and electricity and their prevention methods.
- Understand fire detection and fire fighting system.

10.1 INTRODUCTION

The modern safety movement was started in 1912 in USA. This led to the formation of National Safety Council in USA.

Millions of accidents take place in the industry every year leading to injury, disablement or even death. Statistics show that 98% of accidents can be prevented by observing safe practices.

According to factories Act 1948, an industrial accident is defined as, "an occurrence in an industrial establishment causing bodily injury to a person which makes him unfit to resume his duties in the next 48 hours".

Workmen's compensation Act 1923 defines industrial injury as, "a personal injury to an employee which has been caused by an accident or an occupational diseases and which arises out of or in the course of employment and which would entitle such employee to compensation under the Workmen's Compensation Act, 1923".

10.2 NEED FOR SAFETY

Following are the need for safety in work place:

(*i*) To protect human beings from hazards.

(*ii*) To prevent avoidable pain and sufferings.

(*iii*) To prevent damage to plant and machinery.

(*iv*) To increase production rate.

(*v*) To prevent employee unrest or industrial strife.

10.3 ACCIDENT MODELS

Accident models help to understand cause-effect relationship. These are explained below:

(*i*) **Behaviour Models (Motivation-reward-satisfaction model):** Operational safety of employee is dependent upon his level of motivation and ability to perform. Ability is a function of training and motivation and depends upon several factors. But employee is primarily affected by his boss and top management attitude to safety.

(*ii*) **Human Factor Model—The Forrel Theory:** This theory of accident causation was put forward by Dr. Russel Forrel, professor of human factors at the university of Arizona, USA. The theory states that accidents are results of a casual chain in which human error is one of the factors. This human error in turn is caused by one of the three situations—(a) mismatch between load bearing capacity of human being and the actual subjected load (b) incorrect response of the human to the situation he is subjected to (c) incorrect activity performed by the human beings either due to lack of knowledge or taking deliberate risk.

(*iii*) **Incident Causation Model (The Peterson accident):** This theory is an adoption and extention of Forrel theory. Only difference is that it allows two possible causes for accidents *i.e.* human error and/or system failure. So accidents take place due to one of the factors or both.

There are three major factors causing accidents viz. (a) overload (b) traps and (c) decisions. Overload is the capacity-mismatch as defined in Forrel's theory. Decision refers to situations when employee commits human error due to conscious or unconscious decisions.

The model suggests that employees tend to perform unsafe acts due to their perception of low probability of accidents taking place or due to low potential cost to them of the accident.

(*iv*) **System Model:** Any system comprises of pre-elements such as man, machinery and environment. In order to achieve a task, there results an interaction amongst man, machinery and environment.

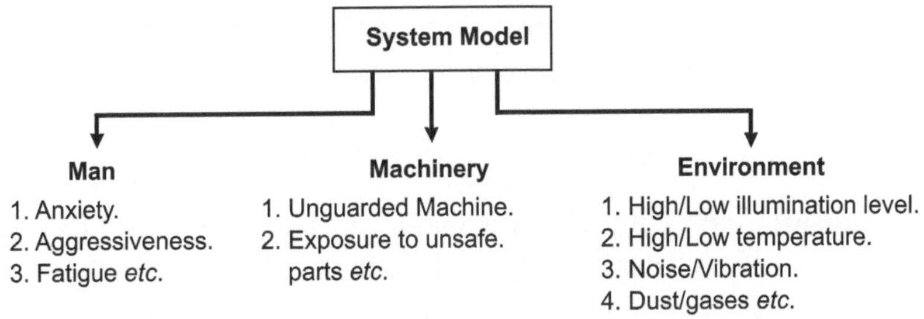

Fig. 10.1 System Model.

The presence of stress leads to accident if the information available about pre-elements *i.e.* man, machinery and environment are inadequate. Based on the information available human beings take decisions and risks to perform the task. However, if the decision is based on inadequate information and the risk is high, accident may take place.

10.4 CAUSES OF ACCIDENTS

Accidents result due to combination of several factors which may vary from situation to situation. Causes of all accidents can be categorised into three major heads (a) unsafe conditions (b) unsafe acts and (c) chance occurrences.

(*a*) **Unsafe Conditions:** These constitute major causes of accidents. There are following:

(*i*) Defective equipment.

(*ii*) Unguarded or improperly guarded equipments.

(*iii*) Shop floor congestion.

(*iv*) Over loading.

(*v*) Unsafe material storage.

(*vi*) Inadequate or absence of safety devices on machines.

(*vii*) Poor environmental factors such as high noise level, low illumination level, presence of dust/gases, very low/high ambient temperature, high humidity, poor ventilation etc.

(*viii*) Poor house keeping.

(*ix*) Hazardous nature of job.

(*x*) Tight work schedule leaving litttle time for rest pauses.

(*xi*) Psychological environment such as emotional imbalance, pre-occupation, fatigue, tiredness, overload, job monotony, lack of self-confidence, frustration related to job/family life, adjustment problems with fellow employees *etc*.

(*b*) **Unsafe Acts:** Unsafe acts undertaken by employees results due to lack of knowledge/skill, bad attitude and physiological defects, etc. Such acts are following:

(*i*) Operating equipment without protective gears.

(*ii*) Failing to warn fellow employees to possible dangers.

(*iii*) Very high/very low speed of operation.

(*iv*) Bypassing safety devices mounted on the equipments.

(*v*) Improper lifting of materials.

(*vi*) Working with machinery having unguarded moving parts.

(*vii*) Distracting fellow employees by talking, shouting, teasing or abusing.

(*viii*) Carelessness or indifference in handling accident prone jobs.

(*c*) **Chance Occurrences:** Accidents take place due to situational reason despite taking adequate safety measures. During various studies it was found that certain employees are more prone to accidents compared to their fellow employees. Also chances of occurrence of accident are more during night shift than any other shifts. Such types of accidents can be controlled by proper awareness and training.

10.5 ACCIDENT COST

An accident costs enormously in terms of direct and indirect cost. The direct cost of accidents are following:

(*i*) Loss of life (which cannot be estimated in monetary terms).

(*ii*) Loss of wages.

(*iii*) Cost due to loss of goods and services.

(*iv*) Cost of medical aid.

(*v*) Cost of compensation to affected employee.

(*vi*) Cost of poor quality due to induction of substitute employee.

(*vii*) Cost of repair of machinery.

Indirect cost of accidents are following:

(*i*) Time lost by other employees who stop work because of accident.

(*ii*) Cost of psychological effect on workers due to fear and low morale.

(*iii*) Loss of profit on the production which otherwise would have been produced.

(*iv*) Overhead costs *e.g.* expenses incurred on lighting, heating, rent *etc.*

(*v*) Cost of assisting injured employee, accident investigation, arranging replacement, training new employees and attending hearings conducted by Govt or other officials.

(*vi*) Cost of rework.

(*vii*) Cost of wages paid to worker during confinement.

10.6 ACCIDENT PREVENTION

Accident Prevention comprises three fundamental steps *viz*:

(*a*) Eliminating unsafe conditions.

(*b*) Eliminating unsafe actions.

(*c*) Discovering causes of accidents.

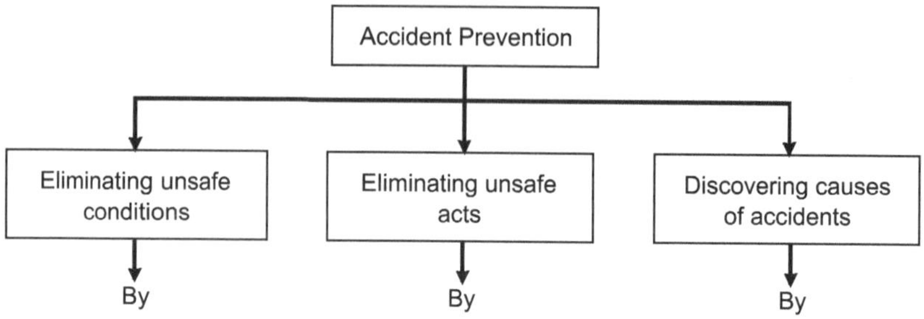

Eliminating unsafe conditions — By

1. Proper guarding/ masking of all machines/equipments/ entry and exit routes.
2. Safe design in machinery construction and operation.
3. Better environmental control *viz.* house-keeping, illumination, ventilation *etc.*
4. Observance of safe procedure, methods *etc.*
5. Eliminating unsafe operation.
6. Provision of safety devices such as eye shower, first-aid box *etc.*

Eliminating unsafe acts — By

1. Safety education and training.
2. Enforcement of discipline.
3. Supervision.
4. Use of personal protective safety gears.
5. Use of work permit system.
6. Insistence on Standard operating procedures for all plant/machineries.

Discovering causes of accidents — By

1. Accident investigation.
2. Job safety analysis.
3. Safety audit and survey.
4. Proper plant and machinery preventive maintenance.
5. Creation of safety awareness about past accidents.
6. Job rotation.

Fig. 10.2 Accident Prevention Methods

10.7 INFLUENCE OF ENVIRONMENTAL FACTORS ON ACCIDENTS

Various environmental factors such as noise, lighting, temperature, humidity, dust, gases *etc.* in the work place affect the probability of occurrence of accidents.

(*a*) **Effect of Noise and Vibrations:** Presence of noise and vibrations in work place induce fatigue in the worker. This reduces efficiency and leads to poor attention on the job. This increases chances of accident.

(*b*) **Effect of Lighting:** Insufficient level of illumination causes eye strain resulting in worker fatigue. A higher level of illumination causes glare and discomfort. Hence, improper level of illumination has direct bearing on occurrence of accidents.

(*c*) **Effect of Humidity:** High level of humidity promotes higher perspiration leading to discomfort and weakness.

(*d*) **Effect of Temperature:** Presence of higher or lower temperature in the working area leads to discomfort and drastic reduction of efficiency.

(*e*) **Effect of Dust and Gases:** Presence of dust and gases reduces efficiency and increase fatigue.

10.8 MEASUREMENT OF ACCIDENTS

Statistical measures are devised to measure frequency and severity of accidents. The following indices are used:

(*i*) **Frequency Rate:** It indicates numbers of injury occurred in relation to man hours worked. It is defined as number of disabling injuries per one million man-hours worked.

$$\therefore \text{ Frequency Rate} = \frac{\text{No. of disabling injuries}}{\text{No. of Man-hours worked}} \times 10,00,000$$

(*ii*) **Severity Rate:** It measures the severity of injury sustained. It is defined as total number of days lost due to accidents per one-million man-hours worked.

$$\therefore \text{ Severity Rate} = \frac{\text{Days lost per year due to accidents}}{\text{No. of Man-hours worked}} \times 10,00,000$$

(*iii*) **Frequency-Severity Rate:** It is defined as given below:

Frequency Severity Rate = Frequency rate × Severity rate

(*iv*) **Cost-Severity Rate:** $\frac{\text{Total cost of accidents}}{\text{Total man} - \text{hours worked}} \times 10^6$

(*v*) **Incident Rate:** It is the ratio of injuries to the number of employees during a certain period. It is expressed as number of accidents per 1000 persons.

(*vi*) **Fatal Accident Frequency Rate:** It is the number of deaths due to industrial injuries expected in a group of 1000 people per 10^8 man-hours.

10.9 ACCIDENT INVESTIGATION

Occurrence of accident is investigated by a safety officer of the organisation. Purpose of investigation is to find out causes of accident and thereby devise systems/procedures for prevention of recurrence. Also, investigation helps to estimate extent of demage/loss to assets and helps to claim insurance benefits.

Safety officer is intimated about the occurrence of accident. He collects necessary information about the accident by visiting the site, collecting specimens, talking to persons present at that place, and scanning logbook *etc.*

After collection of data from all possible sources, possible causes of occurrence of accidents are analysed.

Thereafter, a most possible/definite cause is zeroed-on based on scientific investigation. The report is then submitted to top management.

Later on, the findings are discussed with heads of departments and safety committee members and preventive actions are taken. Awareness is spread amongst the workforce through the safety committee.

10.10 SAFETY LEGISLATION

A number of legislations govern the safety of plant and personnel in India. These are enumerated below:

(*i*) The Factories Act, 1948.

(*ii*) Indian Explosives Act, 1884.

(*iii*) Mines Act, 1952.

(*iv*) Indian Boilers Act, 1923.

(*v*) Indian Electricity Rulers, 1956.

(*vi*) Petroleum Act, 1934.

(*vii*) Workmen's Compensation Act, 1923.

Details of the Acts are beyond the scope of this book. For details, bare acts could be referred to.

10.11 FIRE AND MACHINERY INSURANCE

Fire insurance is method of compensating for loss due to fire. Insurance company charges a premium to insure an intended physical asset such as Buildings, Machinery, Stock Materials, Fixtures *etc.* Fire insurance acts as a safety net.

Similarly, all capital equipments such as Boilers, Steam Turbines, Engines, Electric Generators, Process Machinery *etc* are insured against unexpected breakdowns.

10.12 FIRE SAFETY

It is said that "Fire is a good servant but very bad master". Crores of properties are lost due to industrial fires every year. Initialisation of any fire requires three agents — (a) oxygen (b) source of ignition and (c) combustible material. Removal of any one agent puts-out the fire. Hence, all fire extinction strategies are based on this principle.

10.13 CAUSES OF FIRES

Fires are caused by variety of reasons, such as:

(*i*) Ignition of combustible material due to nearby open flame.
(*ii*) Electrical short circuit.
(*iii*) Electrical circuit overloading
(*iv*) Overheating of electrical appliances.
(*v*) Excess temperature rise of material.
(*vi*) Spontaneous ignition due to dust explosion in mines.
(*vii*) Ambient temperature higher than auto-ignition temperature of material (e.g. Hydrogen catches fire automatically at 560°c and above)
(*viii*) Ignition of volatile air-vapour mixture due to excess heat.

And so on.

10.14 CLASSIFICATION OF FIRE

Fires are classified as per following:

(*a*) **Class-A Fires:** Fire involving organic materials such as wood, paper, cloth, rubber, plastic *etc.*

(*b*) **Class-B Fires:** Fires involving flammable liquids or vapours such as petrol, naptha, LPG, paints, varnish *etc.*

(*c*) **Class-C Fires:** Fires involving gaseous substances such as natural gas.

(*d*) **Class-D Fires:** Fires involving combustible metals such as sodium, phosphorous, magnesium, zinc *etc.*

10.15 FIRE-FIGHTING SYSTEMS

Various fire fighting extinguishers use three techniques- *viz.* (1) Starvation — removal of combustible material (2) Smothering—reduction of oxygen supply and (3) Cooling—reduction of temperature to below that is necessary for fire to continue.

(*i*) **Fire Extinguishers:** Fire extinguishers are of following types:

(*a*) *Water Extinguishers:* These extinguishers are filled with water and non-corrosive anti freeze agents. The pressure required to expel the water is provided by the gas stored in a cartridge (gas cartridge type) when it is pierced. In other type of water extinguishers *viz* stored pressure type, a safety pin is removed and a value is depressed and a jet of water is directed at the fire. The water is pressurised by introducing air into the container. These extinguishers can be used on class-A fires only.

(*b*) *Foam Extinguishers:* It consists of container having foam compound sealed inside a plastic bag which is ruptured on release of carbon-dioxide from a cartridge. Carbon-dioxide expels water and foam. This is called mechanical type foam extinguisher. Another variant chemical foam type is no longer used. These can be used on class A, class -B and class -D fires.

(*c*) *Carbon-dioxide Extinguishers:* It consists of steel cylinder filled with liquid carbon-dioxide approximately 2/3 rd of its volume. When discharge horn is released, carbon dioxide gas comes out. It can be used for class-B and C fires.

(*d*) *Dry Chemical Powder Extinguishers:* These are filled with free flowing, non-toxic, non-conducting powders such as Potassium Oxlate, sodium bi-carbonate *etc*. The expellent gas is stored in the extinguisher body. This can be used on C-class fires.

(*ii*) **Fire Water Pumps:** Fire water pumps take water supply either from a reservoir or overground tank. Various types of pumps used are:

(*i*) Motor driven hydrant water pump.

(*ii*) Motor driven spray water pump.

(*iii*) Diesel engine driven hydrant water pump.

(*iv*) Diesel engine driven spray water pump.

(*v*) Jockey pumps.

(*iii*) **Hydrant Water System:** Water is kept in a pressurised state in fire pipe line system. At the time of need, valve nearest to the point of fire can be opened and with the help of hose reel, fire can be extinguished.

(*iv*) **Spray Water System:** This is an automatic system having a detector pipe work system charged with either compressed air or pressurised water. The detector can be quartz bulb, heat detector, flame sensors *etc*. It provides high and medium velocity water spraying system.

(*v*) **Sprinkler System:** In this system, Quartz bulb heat detectors are mounted directly on the spray nozzle. When bulb breaks due to the effect of heat, spray starts and extinguishes localised fire.

(*vi*) **Foam System:** Fuel oil system are provided with fixed foam system. Foam concentrate is kept in tanks. On detection of fire, foam pump starts and a mixture of water and foam goes to fuel oil tanks where it is converted into foam by foam maker. This foam is sprayed over the surface of oil and fire is extinguished.

(*vii*) **Inert Gas Fire Extinguishing System:** This system consists of mixture of inert gases (Carbon-dioxide, Nitrogen, Argon) in a specified ratio in a battery of cylinders.

First cylinder of each battery has a pilot value which is actuated on detection of fire by smoke detectors and inert gas is released in the affected area. This reduces concentration of oxygen thereby extinguishing fire. This system is used in process control rooms of refinery, petrochemical plants, fertilizer plants *etc.*

(*viii*) **Carbon Dioxide Fire Extingushing System:** In this system, CO_2 is stored in gaseous and liquid form. When fire is detected, system automatically releases CO_2. This system is used in gas turbine enclosure.

10.16 FIRE DETECTION SYSTEM

Various types of fire detection systems are explained below:

(*a*) **Smoke Detection System:** This system detects both visible and invisible products of combustion before flame appears. The electronic detection system, detects this as smoke or heat alarm signal and transmits it to fire alarm panel. There are of two types — optical smoke detectors and ionisation smoke detector. Former detects visible products of combustion and latter detects invisible products of combustion.

(*b*) **Quartz-bulb Heat Detectors:** It consists of a glass bulb containing a liquid, with rise in temperature, the liquid expands thereby shattering glass bulb, which removes seal around orifice of sprayers. Due of this, pressure inside pipe network is reduced which is picked up as fire alarm. It is used as fire detection system in transformers, oil tanks, cable galleries, coal conveyors *etc.*

(*c*) **Linear Heat Servicing Cable:** It consists of pair of specially insulated wires. At a certain temperature, insulation losses its property and wires get shorted thereby generating a fire signal. This system is used in cable galleries, fuel tanks *etc.*

(*d*) **Electrical Spot Heat Detectors:** These are thermo-couples which generate electrical signal upon sensing high temperature. These are used in burners, battery rooms *etc.*

(*e*) **Infra-red Heat Detectors:** These detectors pick up infra-red from smouldering or burning sources thereby generating a fire signal. These are used for coal conveyors, air preheaters *etc.*

(*f*) **Leak Detectors:** These detectors sense leakage of a particular gas such as natural gas *etc* and generate fire signal. These are used in gas turbine enclosures.

(*g*) **Flame Detectors:** These detectors pick up ultraviolet rays to generate fire signal.

10.17 ELECTRICAL SAFETY

Electricity constiutes one of the major sources of fire. Also direct electrocution may result in death of a person. Hence, knowledge of safe aspects of handling of electricity is necessary to avoid accidents.

10.18 CAUSES OF ELECTRICAL FIRES AND THEIR PREVENTION

Possible causes of electrical fires are following:

(*a*) Incorrect value of fuse.

(*b*) Improper earthing.

(*c*) Faulty joints or loose connection.

(*d*) Defective wiring.

(*e*) Poor quality of wiring, switches, sockets *etc*.

(*f*) Poor electrical maintenance.

(*g*) Explosive ambient atmosphere.

Following actions are to be taken in case of electrical fires:

(*a*) Switching off of supply to affected area.

(*b*) Extinction of fire using either carbon-dioxide or dry chemical powder extinguishers or sand.

(*c*) Removal of smouldering material from the spot.

10.19 HAZARDOUS ZONE CLASSIFICATION

A hazardous area is defined as an area where explosive atmosphere may be expected and as such require precautions for use of electrical apparatus. Installation of such electrical apparatus is not allowed in surrounding explosive atmosphere.

Hazardous area classifications are based on risk of fire and explosion. Following are the classifications:

(*a*) **Zone -0:** An area in which explosive gas/air mixture is continuously present (over 1000 hrs/year or more than 10% of time. No electrical apparatus is allowed in this area). However, if it is not practicable then intrinsically safe apparatus are used.

(*b*) **Zone-1:** Gas vapour or dust is present for long period of time under normal operation *i.e.* 10 – 1000 hrs/year or 0.1 to 10% of time. Electrical equipments to be used are flame-proof equipment, pressurised enclosure type, sand filled or oil immersed apparatus.

(*c*) **Zone-2:** Gas, vapour or dust would be present under abnormal conditions such as leaks. These may be present upto 10 hrs/year or 0 to 0.1% of time. Electrical equipments to be used are flame-proof equipment, pressurised enclosure type, sand or oil filled apparatus, non-sparking type.

The electrical apparatus is selected for use in hazardous area based on maximum surface temperature of apparatus. These classifications are given below (UK system):

Temperature Class	Maximum Surface Temperature (°C)
T_1	450
T_2	300
T_3	200
T_4	135
T_5	100
T_6	85

CASE STUDIES IN PLANT SAFETY

Case Study-1

This case study is of an urea plant—it had five transformers supplying power to the urea plant. One of the transformer (no.-3) was switched-off and was under repair. An electrician was asked to test the insulation of this transformer with the help of an instrument called Meggar. Testing was to be done at the electrical panel of respective transformer. So, electrician counted panel number from one end of the room and reached panel no.-3. He connected the instrument and immediately there was a flash. The electrician got severe burn injuries in hands and face.

Investigation revealed that the electrician had counted panel no.-3 from wrong side. So he ended up connecting the instrument to a live transformer resulting in an accident.

This was rectified by writing transformer number as each panel door (on front as well as on back side) so as to eliminate chances of error.

Case Study-2

While working inside a closed vessel, a breathing apparatus in the form of a face mask with an airline is provided to the technician working inside the vessel. This is done to avoid breathing problem inside a confined space.

One day, a fitter was asked to go inside a closed vessel for a repair work. A face mask with airline was provided to him. When technician tried to breath, he was flooded with water. A fellow worker saved him from chocking.

Investigation revealed that connection for breathing air was taken from the bottom of a pipeline instead of top of pipeline. The accumulation of moisture in air lead to flooding. So, as a preventive measure, the connection of breathing pipe was taken from top of the pipeline to prevent entry of condensate into breathing line.

Case Study-3

This case study relates to a demineralised water treatment plant. Hydro-chloric acid is used in this plant to recharge resin on a periodic basis by the plant operator.

One night while recharging resin with hydrochloric acid solution, the feeding pipeline got punctured and sprayed acid onto operator's hands and legs causing burn injuries.

Upon investigation, it was found that the ruptured pipeline was replaced recently. Instead of using rubberlined pipe, the engineer had used mild steel pipe. Mild steel pipe cannot withstand hydro-chloric acid of given concentration and hence had ruptured.

So inadequate knowledge about material of construction led to this accident.

Case Study-4

This case study refers to a transformer installed in a material handling yard (a place where raw materials such as coal, limestone *etc.* are unloaded and stored). The supply to the transformer was carried through cables from a power plant which was about 2km away.

So to switch on or switch-off power supply to the transformer, one has to traverse 2 km from material handling yard to power plant.

Contd...

One day, this transformer was taken for routine maintenance. The power supply fed to the transformer was through a circuit breaker installed in the power plant (Circuit breaker is a device used to switch-on or switch-off power supply). The supervisor incharge of this maintenance activity asked an electrician to go to material handling yard and clean its porcelain bushings. Meanwhile, supervisor went to power plant to switch-off supply to transformer. He assured that by the time electrician reaches material handling yard and begins maintenance on transformer he would have switched off power supply to this transformer. The supervisor while going towards power plant met with a colleague and talked to him. Thereafter, he went to power plant and switched-off supply.

When supervisor reached transformer site, he found the electrician dead due to electrocution.

Investigation into the accident revealed two things:

(1) Supervisor had sent electrician to work on transformer without first switching off its supply. He assumed that by the time electrician reaches site, the transformer would be switched off. This, however, did not happen due to his chance meeting with a colleague.

(2) The electrician blindly followed supervisor's orders without first cross-checking power supply status of transformer.

As a preventive measure, modified procedure required that both supervisor and electrician should together visit the plant to switch-off power supply and put-up "DO NOT SWITCH ON" board, secondly, while initiating maintenance, both have to be doubly sure that transformer is disconnected from power supply.

Summary

Safety is of utmost importance in any establishment. Millions of accidents take place in industry causing injury, disablement or even death. There are costs due to accidents such as direct cost and indirect cost. Direct cost include loss of life, loss of wages, loss of goods and services, cost of medical aid *etc.* while indirect cost of accident include time lost by other employees, loss due to fear and low morale, loss of profit, overhead cost, cost of rework *etc.*

Accidents primarily take place due to (i) unsafe conditions and (ii) unsafe acts. Unsafe conditions could be defective equipment, unguarded equipment, poor lighting, overloading *etc.* While unsafe

Contd...

acts are operation of equipment without proper safety gears, improper lifting of material, carelessness *etc.*

There are three ways to prevent occurrence of accidents *viz* eliminating unsafe conditions, eliminating unsafe acts and discovering causes of accidents and applying preventive measures.

Statistical measures are used to represent frequency and severity of accidents such as frequency rate, severity rate, frequency-severity rate *etc.*

Industrial safety is governed by a host of legislation such as Factories Act, 1948, Indian Explosive Act, 1884, Mines Act, 1952, Indian Boiler Act, 1923 *etc.*

Fires are major sources of death and destruction of properties. Fires are classified under four categories *viz* Class A, Class B, Class C and Class D. Class- A fires involve organic materials such as wood, paper, cloth *etc.* Class B fires involve flammable liquids *e.g.* petrol, diesel *etc.* Class-C fires involve gaseous substances and class D fires involve metals. Fire fighting systems consists of fire extinguishers, hydrant water system, sprinkler system *etc.* While fire detection system consists of smoke detectors, infra-red heat detectors, flame detectors, quartz bulb heat detectors *etc.*

Electricity poses a shock and fire hazard. Hence, precautions are needed while handling electricity.

Review Questions

1. Discuss various accident models.
2. Elaborate on various causes of accidents.
3. Discuss various accident prevention techniques.
4. How environmental factors affect accidents ?
5. What are various statistical tools used to measure accidents ?
6. Write a brief note on safety legislation in India. Discuss the implementation status of various Acts.
7. Explain various causes of fires and their prevention.
8. Write brief note on fire-fighting systems.
9. Explain various types of fire detection systems.
10. Discuss hazardous zone classifications.

Discussion Questions

1. Comment — "Environmental factors are one of biggest contributor of accidents in India".
2. Elaborate — "Majority of accidents can be prevented".
3. "Regular testing of fire prevention system is a must". Comment.
4. "Indirect cost of accidents are much more than direct cost of accident". Justify
5. "Accidents are caused by unsafe acts which can be improved upon through training". Comment.

Multiple Choice Questions

1. Direct cost of accident is:
   (*a*) Loss of wages.
   (*b*) Cost of rework.
   (*c*) Overhead cost.
   (*d*) Time lost by other employees.
2. Severity rate measures:
   (*a*) Extent of injury to a person.
   (*b*) Extent of demage to plant and machinery.
   (*c*) No. of days lost due to accidents.
   (*d*) None of these.
3. Fire insurance helps to
   (*a*) Reduce fire.
   (*b*) Reduce extent of damage.
   (*c*) Provide compensation.
   (*d*) None of these.
4. Fire on flammable liquid is:
   (*a*) Class A-fire.
   (*b*) Class B-fire.
   (*c*) Class C-fire.
   (*d*) Class D-fire.
5. Foam extinguishers are used on.
   (*a*) Class A-fire.
   (*b*) Class B-fire.
   (*c*) Class D-fire.
   (*d*) All of these.

True -False Questions

1. Behaviour model of accident theory relies on level of employee motivation and the ability to perform.
2. Indirect cost of accidents are difficult to estimate

3. Fire-fighting techniques are based on "Oxygen starvation" principle alone.
4. Flame-proof electrical equipments can be installed in zone - O hazardous area.
5. Purpose of accident investigation is to punish the guilty.

ANSWERS

Multiple Choice Questions

1. (*a*) **2.** (*c*) **3.** (*c*) **4.** (*b*) **5.** (*d*)

True-False Questions

1. True **2.** True **3.** False **4.** False **5.** False

❑❑❑

11

Industrial Disaster Management

Learning Objectives

After reading this chapter, you will be able to:

- Understand disaster and its effect on human beings.
- Understand risk and hazard management.
- Examine process of disaster preparation and mitigation.
- Understand regulatory concepts related to disaster management.

11.1 INTRODUCTION

A disaster is defined as unexpected disruption of normal life of human beings, and surroundings of a very high level causing immense suffering and losses.

It is the magnitude of loss and suffering that differentiates a disaster from an accident. When we think of disaster, pictures of Bhopal Gas Tragedy, Chernobyle-nuclear reactor explosion, Challenger space shuttle explosion *etc.* come to our mind. In all cases of disaster, there was loss of human lives and great damage to the surroundings. Even when no human lives are lost and only animal and plant lives are affected—that is also considered a disaster. Since human, animal and plant lives form an ecosystem and any disturbance to a particular element of ecosystem affects others as well. Hence, disaster is a major disruption and damage to any or all elements of ecosystem.

11.2 TYPES OF DISASTER

All disasters can be classified into two major categories *viz* natural disaster and man-made disaster.

(1) Natural Disaster: Natural disasters are—Hurricanes, cyclones, typoons, floods, droughts, earthquakes, volcanic erruptions, landslides, lightning strikes, extreme temperatures, famine, epidemics.

(*a*) *Cyclones/Hurricanes/Typhoons:* A large scale closed circulation system in the atmosphere which combines low pressure and strong winds that rotates counter-clockwise in the northern hemisphere and clockwise in the southern hemisphere. This is known as cyclone in Indian ocean and south Pacific, Hurrican in western Atlantic and eastern Pacific and Typhoon in western Pacific.

(*b*) *Floods:* Floods are sudden and extreme volume of water that flow rapidly and cause inundation. It results in destruction of housings, crops, cattle and people. Floods may initiate dam collapse, landslide *etc.*

(*c*) *Drought:* A drought is defined as deficient rainfull over extended period leading to inadequate water for plants, animals and human beings causing famine, epidemics and large displacement of population.

(*d*) *Earthquake:* Earthquake measured on Richter scale 6 and above are devastating. It leads to serious destruction of buildings and infra-structure. It may also trigger landslides or tidal waves (Tsunami).

(*e*) *Volcanic Erruption:* It happens when lava and gases are discharged from volcanic vent. Lava can travel as far as 40 km from original disaster site. It leads to destruction of human habitat.

(*f*) *Land slide:* These are sudden movement of large chunk of land.

(*g*) *Extreme Temperature:* very high and very low temperature in particular region may lead to loss of lives of cattle and plantations.

(*h*) *Famine:* It is a slow disaster building-up over several years of failed rain and lost harvests.

(*i*) *Epidemic:* Unusual increase in the number of cases of an infectious disease which already exists in a certain region or population or appearance of such cases in a region or population free from that diseases e.g., Cholera, Dengue, Malaria, Yellow fever, T.B., Menengities, Ebola *etc.*

(2) Man-made Disaster: These are caused by human beings either by accident or by negligence or by incompetence. For example, Chernobyle nuclear accident, Bhopal gas leak, Minamata case, B.P refinery explosion in texas, Mine colapse, Air/land/sea disasters *etc*. Man-made disasters can be further sub-divided as:

(*a*) *Long-term Disaster:* These are civil war/strife, International war, Food shortage, water shortage *etc*.

(*b*) *Technological Disaster:* These are industrial accidents, chemical explosion, nuclear explosion, mine explosion, pollution, fires, rail/road/train accidents *etc*.

Table. 11.1 World's Worst Disasters (List is indicative not exhaustive)

Disaster	Location	Year	Deaths
1. Earthquake	Near East and East Mediterranean	1201	11,00,000
2. Volcanic Erruption	Tambora, Sumbawa, Indonesia	1815	92,000
3. Land slide	Kansu Province, China	1920	1,20,000
4. Avalanche	Yungay, Juascaran, Peru	1970	18,000
5. Tornado	Shaturia, Bangladesh	1989	1300
6. Flood	Hwang-ho river, China	1887	9,00,000
7. Lightning strike	Hut in chinamasa Krael Near Umtali, Zimbabwe	1975	21
8. Smog	London fog, UK	1951	2850
9. Dam burst	Manchu river dam, Morvi Gujarat, India	1979	5000
10. Fire (single building)	The Theatre, Canton, China	1845	1670
11. Explosion	Halifax, Nova Scotia, Canada	1917	1963
12. Mining	Hankeiko Colliery, China	1942	1572
13. Industrial	Union carbide, Bhopal, India	1984	15000
14. Off shore platform	Piper Alfa, North sea, U.K.	1988	167
15. Nuclear Reactor	Chernobyle, Russia	1986	31
16. Air accident	KLM - Pan Am, Being 747 crash	1977	583
17. Rail Accident	Bagmati River, Bihar, India	1981	800
18. Road Accident	Petrol Tanker explosion inside salang Tunnel, Afghanistan	1981	1100
19. Atomic Bomb	Hiroshima, Japan	1945	1,41,000

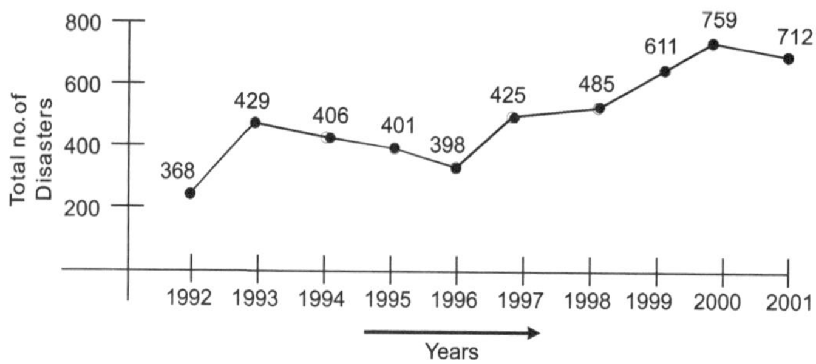

Fig. 11.1 Rising Trend of Disasters in the World (1992-2001)

Source: World Disaster report 2002, International federation of Red Cross

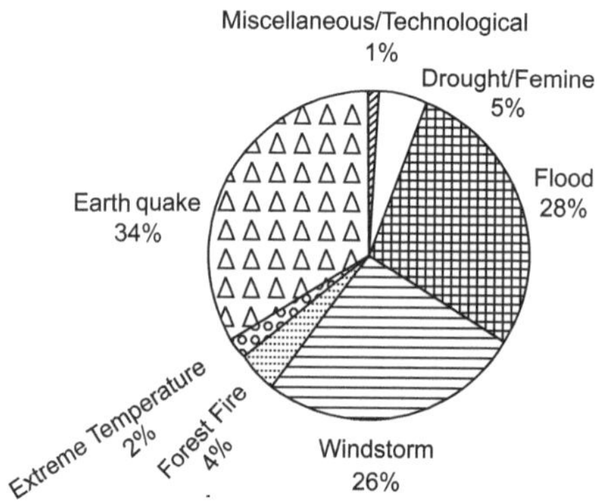

Fig. 11.2 Disaster Break up (1992-2001)

Source: World disaster report 2002, International federation of Red Cross

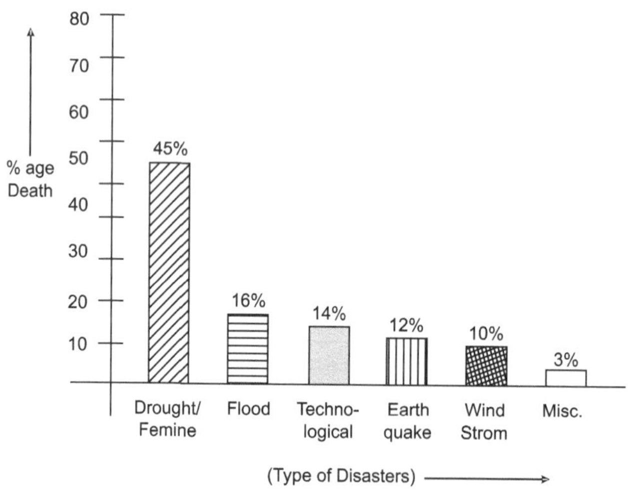

Fig. 11.3 Death from disasters (1992-2001)

Source: World disaster report 2002, International federation of Red Cross

11.3 SOME CASE HISTORIES OF DISASTERS AND THEIR CAUSES

(*i*) **Challenger Disaster:** On January 28,1986, just 73 seconds after launch, challenger exploded killing six astronauts and a school teacher Christa McAulif. It destroyed million dollar worth of equipment and tarnished NASA's reputation

Cause: 'O' rings which were part of sealing mechanism between the segments of booster rockets had failed. They has lost too much of resilience due to low temperature hence failed to seal. This resulted into escape of hot gases causing ignition of fuel in storage tank and a fatal explosion.

(*ii*) **Nypro, Flix-borough, U.K.:** On 1st June 1974, at about 4.53 PM, Flix borough works of Nypro (UK) Ltd was virtually demolished by an explosion. The explosion was in a reactor of caprolactum plant. Twenty-eight people were killed and 36 suffered injuries. Property damage was extended to wide area which included 1821 house, 167 shops and factories.

Cause: The plant had six reactors in series. Reactor no. 5 was bypassed two months ago for repair work. A temporary bypass assembly was laid between reactors 4 and 6. The bypass assembly was not properly engineered. The accident occurred due to rupture of 20 inch bypass assembly. Cyclohexane present in reactors formed a flammable vapour cloud with a source of ignition, leading into the explosion.

(*iii*) **Chernobyle Nuclear Reactor Explosion, Ukrain, 1986:** On 26 April, 1986, at around 1.20 am reactor unit no. 4 at the chernobyle power station exploded leading to severe damage to core and containment building. This resulted in escape of radiation into the atmosphere.

Cause: Following causes contributed to the disaster:

(*a*) Preplanned experiments carried out by the operators to check the possibility of using inertia of turbine generator (when isolated from steam supply and grid) to maintain the output of the unit station auxiliary machines.

(*b*) Poor reactor design.

(*c*) Scope for operator interference with safeguards.

(*d*) Inadequate and slow shutdown.

(*e*) Ability to run reactor with safeguards disabled.

(*iv*) **Texas city Disaster, Texas, USA:** On 16 April, 1947, at about 9.15 am, explosion aboard a docked ship ("Grand Camp") took place leading to subsequent fire. Five hundred seventy-eight people lost lives and 3500 were injured. Blast shattered window of houses as far as 40 kM away.

Cause: Origin of explosion was fire on ammonium nitrate fertilizer. Fire was initiated by a discarded cigarette.

(*v*) **Seveso, Italy:** On 10 July, 1978, bursting disc ruptured on a chemical reactor at works of Icmesa Chemical Company at Meda near Seveso, 15 miles from Milan. A white cloud was released drifted towards town. One of the toxic substance was TCDD (2,3,7,8 - tetra chlorodibenzo-p-dioxin). Over next few days, people felt ill and animals died in the contaminated area.

Cause:

(*a*) Rupture of bursting disc of chemical reactor leading to escape of toxic gases.

(*b*) No arrangement for containment of toxic fumes.

(*vi*) **Kader Toy Factory, Thailand:** On 10 May, 1993, fire started in a poorly built toy factory. The fire-exit doors were locked and staircase soon collapsed. One hundred eighty-eight workers, mostly young women died.

Cause:

(*a*) Poor building construction

(*b*) Locked fire-exit doors.

(*vii*) **Minamata Disaster, Japan:** Caused by Dumping of Mercury componds in Minamata bay, Japan by M/S Chisso Corporation, a fertiliser and Petrochemical company. The company was dumping toxic waste from 1932 to 1968. About 3000 people suffered various deformities, mercury poisoning and death.

Cause:

(*a*) Unsafe industrial waste disposal

(*viii*) **Gas Poisoning, Auburn, Indiana, USA:** On 28 June, 1988, workers at an electroplating plant in Indiana were poisoned by Hydro Cyanide gas (HCN) produced by mixing of hydrochloric acid with zinc cyanide during cleaning operation. Five workers died.

Cause:

(*a*) Carelessness

(*b*) Improper work instructions

Causes of technological disasters are poor system design, human error, system lethargy, unsafe practices, poor maintenance and safey practices, material failure, etc.

11.4 RISK AND HAZARD MANAGEMENT

In order to prevent occurrence of disasters, it is necessary that risk assessment and hazard identification is done.

The term risk is defined as the likelihood of occurrence of an incident of specific consequences.

Risk assessment can be defined as "the quantitative evaluation of the likelihood of undesired events and the likelihood of harm or damage being caused together with the value judgements made concerning the significance of the results". Purpose of risk assessment is to reduce/eliminate hazards.

The term hazard is used to describe a set of conditions which could lead to an accident with harmful consequenses.

$$\text{Risk} = \text{Hazard} \times \text{Vulnerability}$$

Risk control is carried out from the project stage itself. Following flowchart gives risk control steps:

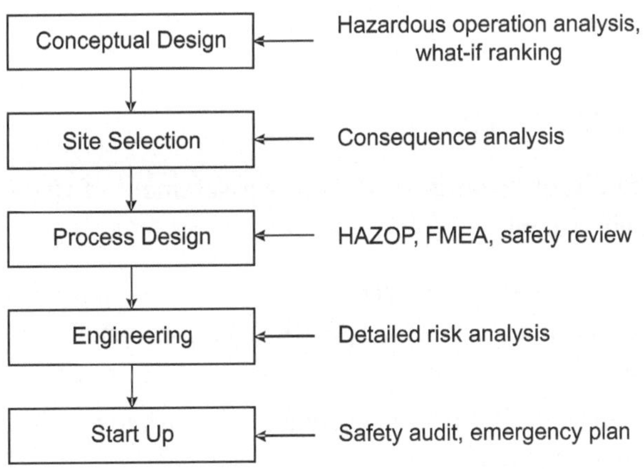

The detailed explanation is given below:

(*a*) **Conceptual design:** At conceptual design stage, Hazardous operation analysis of the plant process is carried out. Various hypothetical situations are considered in plant operation (What - if).

(*b*) **Site selection:** At site selection stage, various consequence of plant operations on surroundings are considered.

(*c*) **Process design:** To make process design foolproof, HAZOP (Hazardous Operation), FMEA (Failure Modes and Effects Analysis) and safety review are carried out.

(*d*) **Engineering:** At engineering stage, detailed risk analysis is carried out.

(*e*) **Start-up:** Before start up, safety audit is carried out and on-site and off-site emergency plans are devised.

11.5 HAZARD IDENTIFICATION

In a modern plant, there a three big hazards *viz* - fire, explosion and toxic release. In hazard evaluation, the first task is to identify the hazards that is inherent in a process or plant operation. Thereafter focus is on the evaluation of events and methods to eliminate them.

Once hazard has been identified, the hazard can then be evaluated in terms of risk it presents to the employees, neighbouring public property, environment and company.

Hazard analysis can be through reliability analysis, fault tree analysis, failure mode and effects analysis, what-if analysis, past experience, audit, hazard and operation study *etc*.

11.6 RISK ASSESSMENT

After identification of hazards, next step is assessment of risk to individuals, society and environment.

(*a*) **Individual risk:** Individual risk is defined as risk of death to an individual located permanently at a specific location, usually outside the site boundary.

It is used as a criterion for acceptability of risk to public around a major hazardous factory.

In Holland and Belgium, the maximum acceptable individual risk is based on the principle that a dangerous industrial activity shall not increase the risk to the individual member of the public by more than 1% of the natural death rate.

Table 11.2 shows individual risk of some human activities.

Table 11.2 Individual Risk of Some Human Activities.

Activity	Chances of death per million per year
Railway accidents (India)	40
Road accidents (India)	50
Lightning (UK)	0.1
Meteorite strike	0.00006
Aircraft crash (UK)	0.02
Rock climbing (UK)	40
Smoking (20 cigarettes/day)	5000

(*b*) **Social risk:** It indicates risk of deaths to a member of persons in anyone incident. Usually criteria is developed using Groningen FN Curve.

(*c*) **Environment:** It indicates damage to the environment posed by the hazard. It indicates ecological damage temporary or permanent in nature.

11.7 DISASTER PREPAREDNESS

After hazard identification and risk assessment, next step is preparedness for disaster.

Disaster Preparedness is the readiness to reduce the impact of disasters and wherever possible to predict or prevent disasters from occurring.

A disaster preparedness plan consists of the following steps:

(*a*) **Pre-Disaster Management:** This includes the following steps:

- (*i*) Hazard identification and analysis.
- (*ii*) Risk management.
- (*iii*) Consequence analysis.
- (*iv*) Safety review.
- (*v*) Safety audit.
- (*vi*) Emergency plan.

(*b*) **Disaster Management:** This includes the following steps:

- (*i*) Communication through telephones, sirens, public address system, mass media, *etc*.
- (*ii*) Relief and rescue operation.
- (*iii*) Medical aid.
- (*iv*) Mitigation of cause of disaster.

(*c*) **Post-Disaster Management:** This includes the following:

- (*i*) Rehabilitation of victims.
- (*ii*) Reconstruction of damaged property.

11.8 ON-SITE EMERGENCY PLAN

On-site emergency plan takes into account the area within the plant premises and people working in it.

It contains the following key elements:

(*a*) Effect of rescue and treatment of causalties.
(*b*) Safeguard of other personnel in the plant.
(*c*) Minimisation of damage to property and environment.
(*d*) Containing and bringing the incident under control.
(*e*) Identification of casualties.
(*f*) Dissemination of authoritative information to media.
(*g*) Securing safe rehabilitation of affected people.
(*h*) Preservation of relevant records and equipments for subsequent investigation for finding out the cause of the incident.

11.9 OFF-SITE EMERGENCY PLAN

Off-site emergency plan takes into account area outside the plant. Off-site emergency plan is to be prepared by the Government authorities. It is mandatory under Rule 14 of the manufacture, Storage and Hazardous Chemicals Rules, 1989 (amended in 1994) for district authorities to prepare an off-site emergency plan in respect of an industry storing/handling hazardous chemicals or at locations where accidents are likely to have an off-site adverse effect.

11.10 INDIAN REGULATORY FRAMEWORK FOR DISASTER MANAGEMENT

Indian regulations provide a fairly comprehensive framework for different phases of disaster management.

Following acts framed by Government of India, are generally applicable to Indian industries with regard to disaster management.

(*a*) The environment (Protection) Act, 1986.
(*b*) The manufacture storage and Import of Hazardous chemicals Rules, 1989.
(*c*) The Public Liability Insurance Act, 1991.
(*d*) The National Environmental Tribunal Act, 1995.
(*e*) The Chemical Accidents (Emergency planning, preparedness and Response) Rules, 1996.
(*f*) The National Environment Appellete Authority Act, 1997.
(*g*) The National Disaster Management Act, 2005.

After Gujarat earthquake in 2001, Government of India had set-up Disaster Management Authority in states and at centre with a view to prepare for emergency situations in a systematic and preplanned manner. This is kept under Ministry of Home Affairs, Government of India. Following has been done.

(*i*) Constitution of central, state and district crisis groups on chemical disasters.

(*ii*) Setting-up of crisis control room.

(*iii*) GIS (Geographical Information service) based emergency planning and response system for major industrial disasters.

Summary

Disaster is unexpected disruption of normal life of human beings and surroundings of a very high level causing immense suffering and losses. Disasters are broadly classified as natural disaster and man-made disaster.

Disasters can be prevented by proper risk assessment and hazard identification. Purpose of risk assessment is to reduce hazards. Risk control is initiated at the project stage so as to minimize hazards during plant operation. Hazards are analysed through many methods such as reliability analysis, fault-tree analysis, failure mode and effect analysis, what-if analysis, past experience, audit *etc.*

Disaster preparedness is the readiness to reduce impact of disasters. It consists of three phases *viz.* predisaster management, disaster management and post disaster management. Pre-disaster management consists of hazard identification and analysis, risk management, consequence analysis, safety review, safety audit and emergency plan. Disaster management includes—communication through telephone-sirens, public address system, TV *etc.*, relief and rescue operation, medical aid and mitigation of cause of disaster. While post disaster management includes rehabilitation of victims and reconstruction of damaged property.

Every industrial organisation prepares an on-site and off-site emergency plan based on percerved hazards. These are documents detailing process of disaster management and mitigation.

Government of India has framed various acts with regard to disaster management *viz* - The Environment (Protection) Act, 1986, The Manufacture, Storage and Import of Hazardous Chemicals Rules, 1989; The Chemical Accidents (emergency planning, preparedness and response) Rules, 1996 *etc.*

The National Disaster Management Act, 2005 has mandated that disaster management authority be set up in states and in centre to tackle emergency situations.

Review Questions

1. Differentiate accident from disaster.
2. What is technological disaster ?
3. Discuss the concept of risk and hazard management.
4. Discuss various steps required for disaster preparedness.
5. What is on-site and off-site emergency plan ? How do they help in mitigation of disaster?
6. Write a brief note on regulatory framework for disaster management.
7. Write a short note on risk assessment.
8. Why disaster preparedness is a must to control the after-effect?
9. What is the mandate of National Disaster Management Act, 2005.
10. Explain the process of hazard identification.

Discussion Questions

1. "Communication channel availability plays a vital role in controlling effect of disaster". Discuss.
2. Elaborate on "Disaster preparedness is a must to deal with any unexpected happenings''.
3. "Risk and hazard identifications is the first step in control of future disaster". Comment.
4. "Long-term disasters can be prevented by efficient planning and implementation". Justify.
5. "Most difficult part of post disaster management is psychological rehabilitation of victims". Comment.
6. "Some degree of risk of involved in every human activity". Discuss.
7. "Technological disaster are result of human error". Discuss.
8. "Environmental disasters leave a long-term footprint compared to other types of disasters". Justify
9. "Detailed analysis of past disasters, system modification and periodic training of personnel helps in prevention of recurrence of disaster". Comment.
10. "Existence of regulatory framework helps to protect the interest of disaster victims". Comment.

Multiple Choice Questions

1. Natural disaster is:
   (*a*) Toxic gas release. (*b*) Nuclear explosion.
   (*c*) Flood. (*d*) Train accident.
2. Site selection from risk and hazard point of view is based on
   (*a*) Nearness to sea shore. (*b*) Away from human habitation.
   (*c*) Consequence analysis. (*d*) None of these.
3. Off-site emergency plan is mandated by:
   (*a*) Factories Act, 1948.
   (*b*) Environment Protection Act 1986.
   (*c*) Manufacture, storage and hazardous chemical rules, 1989.
   (*d*) Public Liabilities Insurance Act 1991.
4. Risk is:
   (*a*) Hazard × severity. (*b*) Hazard × vulnerability.
   (*c*) Hazard × Frequency rate. (*d*) None of these.
5. Post disaster management involves:
   (*a*) Rehabilitation of victims.
   (*b*) Reconstruction of damaged property.
   (*c*) None of these.
   (*d*) Both (a) and (b).

True/False Questions

1. Hazard operation Analysis is applicable at design stage of a plant.
2. Major hazards in a modern plant are fire, explosion and toxic release.
3. Off-site emergency plan takes into account area inside the plant.
4. Man-made disasters are due to negligence or incompetence.
5. There are more death due to natural disasters than man-made disasters.

ANSWERS

Multiple Choice Questions

1. (*c*) **2.** (*c*) **3.** (*c*) **4.** (*b*) **5.** (*d*)

True/False Questions

1. True **2.** True **3.** False **4.** True **5.** True

❑❑❑

12

Asset Replacement Decisions

Learning Objectives

After reading this chapter, you will be able to:

- Understand the effect of failures on asset life cycle.
- Examine various factors affecting replacement decision.
- Determine philosophy of replacement policies.
- Comprehend the residual life extension strategies

12.1 INTRODUCTION

In any organisation, equipment replacement decisions are more common than new investment decisions. All machines undergo physical deterioration during their operational life which increases maintenance expenditure gradually. Also, new machines appear in the market with higher efficiency, lower operational and maintenance cost. These encourage machine replacement. However, the capital cost prevents machine replacement. So an optimum policy is to be found such that machine replacement is better than maintaining old machine. So replacement is necessary when operational and maintenance cost of existing machine is higher than average capital and operating cost of proposed machine.

In our daily life, some equipments are replaced when they fail to perform *e.g.* light bulb, hand tools *etc.* while some of the euipments are repaired such as automobile tyre-tubes, self starter, ceiling fan *etc*. While in some cases, a standby equipments is maintanined due to criticality of function such as standby tyre of a car. So asset replacement theory provides answer to optimal replacement period or when to replace?

12.2 ASSET LIFE CYCLE

Maintenance asset management essentially consists of three district stages during its useful economic life. Economic life is defined as the duration that an asset is expected to remain in service before it becomes cheaper to replace the equipment than maintain it. Asset management consists of three distinct stages—acquisition, operation and maintenance and disposal.

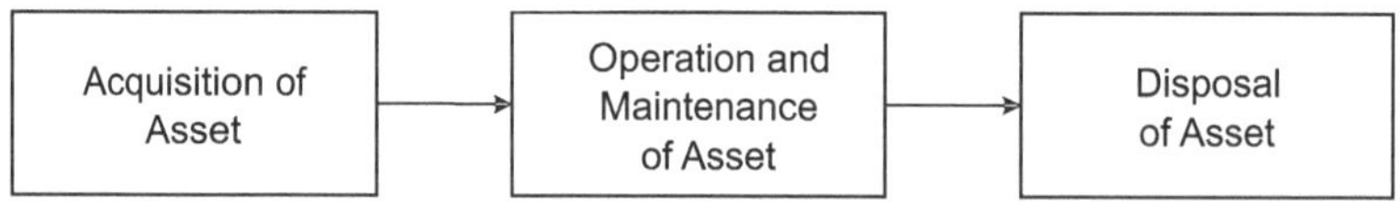

Fig. 12.1 Asset Managent Cycle.

Whenever a new equipment is to be purchased, life cycle costing of equipment is carried out. Life cycle cost of equipment is the estimation of total cost of ownership, operation, maintenance and disposal of an equipment during its economic life time.

It consists of following components:

(*i*) **Design and Development Cost:** It includes cost of preliminary studies, detailed design and development cost, fabrication assembly and testing cost, cost of documentation *etc*.

(*ii*) **Production Cost:** It includes cost of fabrication, assembly, testing, spares cost, operation and maintenance cost, cost of training, cost of failures *etc*.

(*iii*) **Operation and Maintenane Cost:** It includes cost of operation and maintenance manpower, energy cost, inventory cost, cost of test equipments, cost of modifications/overhauling/retrofitting *etc*.

(*iv*) **Disposal Cost:** It includes cost of obsolete inventory, cost of decommissioning, cost of transportation *etc*.

So life cycle cost = Acquisition cost + Operation and Maintenance cost + Disposal cost

12.3 REASONS FOR REPLACEMENT OF ASSET

Various reasons for replacement of asset are following:

(*i*) **Physical Deterioration:** Due to constant use over a period of time, an asset undergoes wear and tear resulting in higher maintenance cost, lesser equipment availability, poor product quality, higher energy consumption *etc*. These reasons lead to asset replacement.

(*ii*) **Obsolescence:** All assets become technically obsolete with the passage of time. This is due to newer and better product availability in the market. With every passing day, technologically superior products appear in the market. This results in loss in value of existing asset.

(*iii*) **Inadequacy:** The existing asset may be unable to meet current product demand. So asset may need replacement with a new asset having higher output capacity.

(*iv*) **Safety Problem:** The operation of existing asset may become unsafe due to its age and wear and tear.

For example, inner lining of reactors need periodic inspection and relining. Similarly tubes of water tube boiler need replacement.

(*v*) **End of Economic Life:** An asset might have completed its economic useful life as is indicated by frequent breakdowns, higher operational cost *etc*. This necessitates asset replacement.

(*vi*) **Decreased Reliability:** In critical applications, a fall in reliabiltiy is reason enough for replacement. For example, in aviation, millitary application, medical diagnostic applications *etc*.

12.4 TYPES OF ASSET DEPRECIATION

Depreciation is the gradual reduction of networth of an asset over a period of time. In depreciation accounting, a portion of earning derived from operation of asset is set aside every year in book-keeping accounts towards the cost of replacement of the asset,

The various types of depreciation are:

(*i*) **Physical Depreciation:** This is due to regular wear and tear of asset over its normal use. A good maintnance activity slows down this decay process.

(*ii*) **Functional Depreciation:** This is due to change in demand or service expected from the asset. Although asset may be still in good operational condition. *e.g.* an output requirement of 25 bags/hour of cement may be the current need. However, the bagging machine can provide only 15 bags/hr, so it is functional depreciation. Or parts made with 0.05 mm tolerance is the current need compared to machine capability of 0.5 mm.

(*iii*) **Technological Depreciation:** Development of latest technology makes existing asset uneconomical. This could be due to switchover to new type of raw material, improved product quality requirement *etc*. This makes equipment technologically obsolete.

(*iv*) **Monetary Depreciation:** A change in purchasing power of money causes monetary depreciation *e.g.* due to inflation. Hence money kept aside due to depreciation may not be sufficient to buy a new machine *i.e.* capital recovered during operation of asset is insufficient for replacement.

12.5. TYPES OF FAILURES

Types of failures also decide the replacement of an asset. Generally there are two types of failures — (i) gradual failure (ii) sudden failure.

(*i*) **Gradual Failures:** A gradual failure takes place over a period of time due to normal operational wear and tear. The asset follows typical "bath-tub" failure curve. *e.g.* gradual wear of bearings, bushes, automobiles tyres *etc*. Such failures can be predicted using condition-monitoring techniques. Hence, a replacement planning can be started well in advance.

(*ii*) **Sudden Failures:** Sudden failures take place without giving any prior warning. Such failures can be sub-categorised as:

(*a*) *Random Failures:* These failures take place due to random causes. So there is no pattern in their failures.

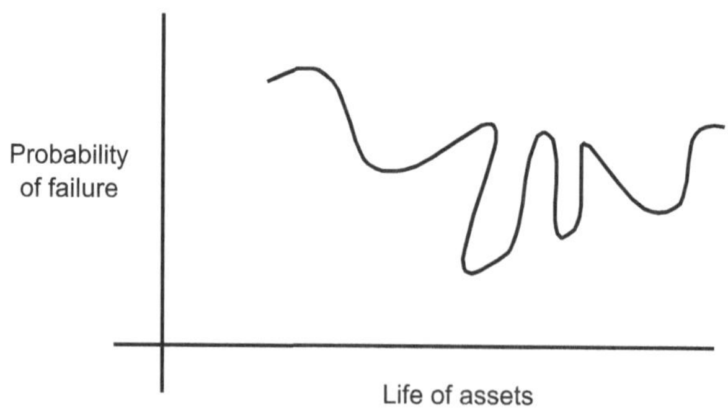

(*b*) *Progressive Failure:* Probability of such type of failures increases with the usage of asset *e.g.* Bearings, light bulb *etc.*

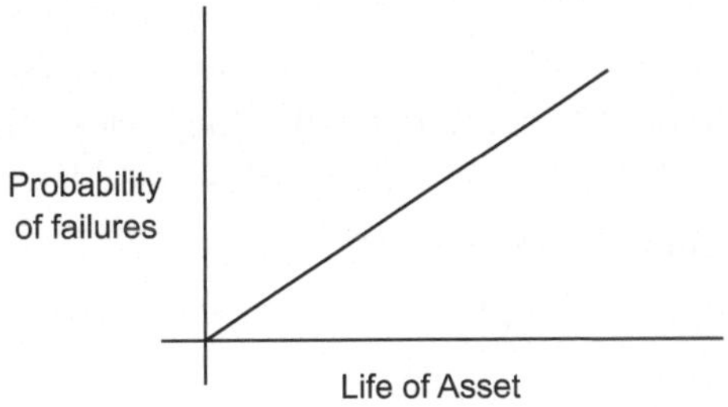

(*c*) *Retrogressive Failure:* Probability of such type of failures is more during initial stages and thereafter chances of failure decreases *e.g.* Medical diagnostic equipment, Air-craft, Sub-marine, Satellite, Missile *etc.*

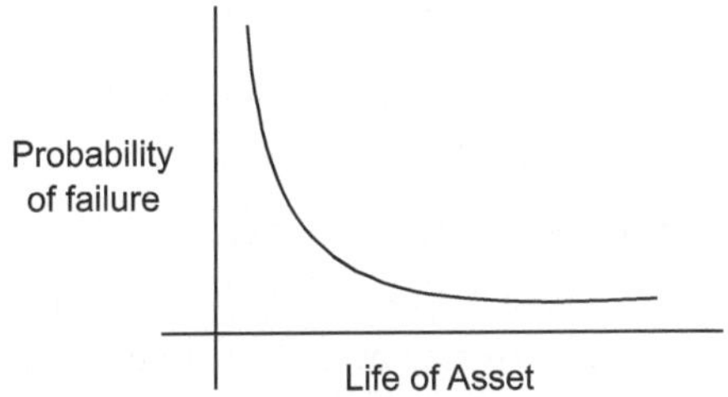

12.6. FACTORS AFFECTING REPLACEMENT DECISIONS

Following are the factors affecting replacement decisions:

(*i*) **Technological Factors:** While selecting an asset, technological factors play a vital role. For example, for manufacture of cables, the latest technology is to use XLPE (Cross-linked polyethylene) and paper-insulated cable technology is outdated. Similarly, super-critical boilers are being used for power plants and hence old technology is outdated and such equipments are never purchased. Also there is gradual shift from manual type equipments to semi-automatic to fully automatic to fully automatic and self diagnostic type.

(*ii*) **Functional Factors:** Functional requirement determines the type of the asset to be purchased. For example, if the product output requirement is 50 units/hour compared to existing 40 units/hour then equipment may be purchased with a capacity of 70 units/hour or for that matter if the requirement of tolerance level of product being manufactured is 0.05 mm from existing 0.25 mm. So replacement must meet the current requirement.

(*iii*) **Financial Factors:** It is the financial factor that decides whether the asset to be replaced or not. If the cost of operation and maintenance is far more than average acquisition cost then replacement is considered. There are number of ways of financial estimation *e.g.* Annual cost method, internal rate of return method, pay-back period method, present worth method, MAPI method *etc.* (discussed later in this chapter).

(*iv*) **Safety Factors:** Sometimes safety factor determines early replacement of an asset. For example, gas piplelines, corrosive chemical storage vessels *etc.*

(*v*) **Reliability Factors:** In critical application, reliability is of utmost importance, Hence, asset replacement is decided based on current reliability level *e.g.* air-crafts.

12.7 REPLACEMENT POLICY

Replacement policy is concerned with optimal replacement period. Following situations are considered for replacement policies:

(*i*) **Assets with Gradual Deterioration:** With assets deteriorating gradually over a period of time, there is a steady increase in operation and maintenance cost over a period of time while there is reduction in depreciation over time. This method considers replacement of asset at the original cost *i.e.* rise in cost due to inflation is not considered.

(*ii*) **Assets with Gradual Deterioration with Time Value of Money:** This is same as above except time value of money is considered.

(*iii*) **Assets with Sudden Failures:** The assets suddenly fail to perform and the failure is random in nature.

(*iv*) **MAPI Method:** (Machinery and Allied Product Institute, USA)— It is based on a method suggested by MAPI.

(*v*) **Assets with Gradual Deterioration with life Extension Strategy—**

Capital equipments such as boiler, gas turbine, steam turbine, alternator, transformer *etc.* which have very high cost of replacement have their life extended through residual life extension strategies.

(*i*) **Assets with gradual deterioration:**

Let, P = Price of asset.

S = Scrap value at the end of n years.

MC = Maintenance cost of asset in the year t.

Total ownership cost = Price of asset + Maintenance cost for 'n' years-scrap value.

or $$T(n) = P + \sum_{t=1}^{n} MC - S$$

Average cost $$A(n) = \frac{1}{n}\left[P + \sum_{t=1}^{n} MC - S\right]$$

so optimum replacement interval = minimum value of A(n)

(ii) **Assets with gradual deterioration with time value of money**: Time value of money or discounted cash flow method takes into account the fact that money invested in bank will accrue annual interest. In other words, Rs 1000 today is more valuable than Rs 1000 in 10 years' time.

Various techniques used in discounted cash flow method are:

(a) *Net Present Value Method:* This method equates yearly future cash flows to its current value today. This method determines present value of any future cash flow by assuming interest rate or discount rate to remain constant.

This can be illustrated as: Shri Ram invested Rs. 22,20,000 in a bank at an annual interest rate of 8% then the value of sum at the end of 1st year

$$= 22{,}20{,}000 + (0.08 \times 22{,}20{,}600)$$
$$= \text{Rs } 23{,}97{,}000$$

Similarly, the value of sum at the end of the 2nd year

$$= 23{,}97{,}000 + (0.08 \times 23{,}97{,}600)$$
$$= \text{Rs. } 25{,}89{,}408$$

So, the value of money keeps on increasing and at the end of 'n' years is given by

Future value $$F.V. = D\left(1 + \frac{IR}{100}\right)^n$$

Where, D = investment.

IR = interest rate.

So, Rs 22,20,000 yields Rs 25,89,408 at the end of 2nd year. In other words present valur of Rs 25,89,408 is Rs. 22,20,000.

Hence, present value $$P.V. = \frac{S}{\left(1 + \frac{IR}{100}\right)^n}$$

Where P.V. is the present value of S at an interest rate of IR for n years.

Mathematically,

$$\text{NPV} = \frac{CF_0}{(1+K)^0} + \frac{CF_1}{(1+K)^1} + \; + \frac{CF_n}{(1+K)^n} = \sum_{t=0}^{n} \frac{CF_t}{(1+K)^t}$$

Where CF_t = Cash flow occurring at the end of year 't'

(t = 0,1,2, ... n)

n = Life of asset,

K = discount rate

After calculation, if the net present value (NPV) comes out to be positive, asset is fit to be replaced.

(*b*) *Return on Investment (ROI):* This method expresses annual return or cash flows as a percentage of capital cost. Annual return or cash flow refers to savings in cost due to replacement of old asset. The savings may accrue due to lower maintenance cost, higher productivity, higher energy efficiency (*i.e.* lower energy consumption), *etc.*

So this method indicates annual return expected from capital investment (required for asset replacement).

$$\text{ROI} = \frac{\text{Annual Net Cash flow}}{\text{Capital cost of Asset}} \times 100$$

ROI must always be higher than interest rate. Greater the return on investment, better is the investment.

(*c*) *Internal Rate of Return (IRR):* This method calculates rate of return that an investment is expected to yield. The rate of return is usually calculated by a process of trial and error whereby the net cash flow is compiled for various discount rate until its value reaches zero.

So IRR is the discount rate which makes its NPV (Net Present Value) equal to zero.

or, $$O = \frac{CF_0}{(1+K)^0} + \frac{CF_1}{(1+K)^1} + \; \dots \; + \frac{CF_n}{(1+K)^n}$$

So an investment with positive value is preferred. In all the three methods discussed above, the main point of interest is the return a new asset provider after commissioning. The return provided by the asset may be in terms of savings due to lower energy consumption, higher productivity, high reliability, lower maintenance cost *etc*. Asset replacement is considered if chosen asset gives higher return than existing asset.

(*iii*) **Assets with sudden failures:** Certain assets such as light bulb fail suddenly without giving prior warning. Such assets need to be replaced. However, the replacement of individual unit costs higher than group replacement.

But, if the group replacements are done at short intervals, the cost/period is high. So if the group replacements are carried out at longer intervals, the cost/period is low but the cost of individual replacement increases. This is exhibited in a graph shown below.

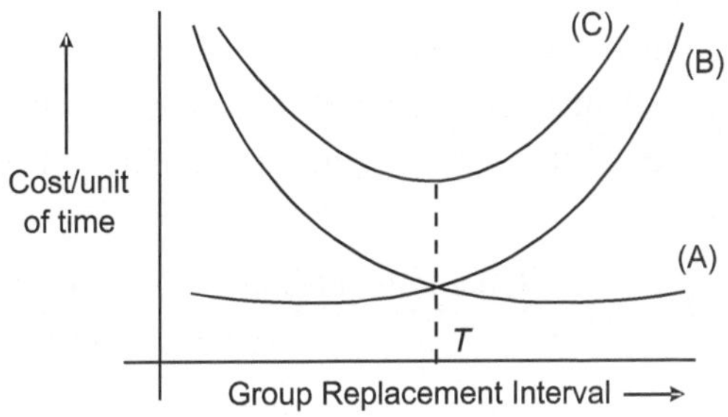

Curve A = Cost of group replacement per unit time.

Curve B = Cost of individual replacement per unit time.

Curve C = Total cost of group and individual replacement per per unit time.

Hence, point T indicates when the total cost of group and individual replacement per unit time is minimum. Hence, T is the optimum interval of group replacement.

Such policy can be used during preventive maintenance. For example, if during inspection of a machine, a component is seen to be likely to fail if left in service then it must be decided whether or not to replace this component only or all such similar components. Reason behind latter is that while some components do not yet show sign of impending failure, they are likely to fail before next scheduled maintenance.

Such replacement plan is applicable where large number of identical, low cost components increasingly fail with age.

(*iv*) **MAPI method:** Discounted cash flow techniques discussed above does not take into account following factors which may influence the replacement decision –

(*a*) inflation

(*b*) timings of cash flows in relation to opportunities for the use of money elsewhere.

(*c*) difficulty of raising money.

(*d*) risk *i.e.* uncertainity of factors such as repair cost, operating life, disposal value.

(*e*) possibility of development of technologically superior product.

However, due to difficulty in predicting future cash flows, MAPI method offers a better realistic solution.

MAPI system was developed by the "Machinery and Allied Products Institute", USA in 1949. This method calculates urgency rating which is a measure of urgency or making an investment compared to delaying it.

$$\text{Urgency Rating} = \frac{\text{Operating advantage} - \text{Capital consumption}}{\text{Net Investment}} \times 100$$

The calculcation of this after tax rate of return on investment is accomplished using MAPI forms and charts. Higher the after tax rate of return, more urgent is the need for asset replacement.

(*v*) **Assets with gradual deterioration using life extension strategies:** Life of all equipments and components are designed keeping in mind operational conditions (*e.g.* temperature, humidity, stress, corrosive atmosphere *etc*). However actual operational conditions may be different than design considerations and therefore put excessive stress on the equipment. This leads to reduced operational life. The equipment therefore, need to be replaced early. This accelerated replacement is necessitated by harsh woring conditions. So the equipment need to be replaced before its useful economic life is over.

Life extension strategies are applied to the following two classes:

(*i*) Useful economic life of equipment is over due to normal use

(*ii*) life of equipment is over before expected economic life due to change in operational and maintenance conditions.

Due to high cost of replacement of capital equipments, sometimes it is far more economical to go ahead with life extension rather than complete equipment replacement.

Factors determining life of equipment are:

(*i*) Design and specification standards of the equipment.

(*ii*) Manufacturing process, quality and tolerances.

(*iii*) Intensity of usage, working environment and nature of load variation.

(*iv*) Operational condition and maintenance quality.

(*v*) Natural or unforeseen factors (*e.g.* plants ravaged by flood, earthquake) and accidents.

Due to all above factors, life of equipment reduces. So assessment of residual life of equipment (RLA studies) is necessary to gauge the remaining life of equipment. Thereafter either renovation/overhauling/reconditioning/retrofitting or replacement decisions are taken.

12.8. FACTORS AFFECTING RENOVATION OF ASSET

Following factors affect renovation of asset:

(*i*) Recurring failures.

(*ii*) Performance deterioration.

(*iii*) Aging factor or increasing/equipment life cycle cost.

(*iv*) Progressively higher maintenance cost.

(*v*) Environmental norms.

(*vi*) Capacity enhancement.

(*vii*) Non-availability of spares due to model phase-out.

(*viii*) Higher energy consumption.

Examples of renovations are following:

(*i*) Boiler tube replacement.

(*ii*) Boiler burner panel replacement.

(*iii*) Coal mill logic, modifications.

(*iv*) Turbine blade replacement.

12.9 RESIDUAL LIFE EXTENSION STUDIES

Residual life extension studies involve the following:

(*i*) Physical inspection.

(*ii*) Diagnostic tests.

(*iii*) Analysis of operating data and design data.

(*iv*) Diagnosis for instrumental analysis:

(*a*) **Physical inspection:** Physical inspection involve visual inspection, dimensional measurement, thickness survey, hardness measurement *etc*.

(*b*) **Diagnostic test:** These are various non-destructive tests performed on the targeted components such as liquid penetrant test, ultrasonic tests, measurement of change of capacitance of high voltage windings, measurement of Ohm-Farad value, measurement of loss tangent of windings at 0.2 and 1.0 times line voltage *etc*.

(*c*) **Analysis of operating data and design data:** Operating data are studied and a trending is done. These are then compared with design parameters and deviations are noted and their effects are evaluated.

(*d*) **Diagnosis for instrumental analysis**: These tests are carried out when original material data is not available. So, a part of the component is subjected to destructive test to assess degradation. Thereafter, comparison is made to original generic nature of materials. The techniques commonly used are—Thermo-gravemetric analysis, Thermo-mechanical analysis and infra-red spectrography.

12.10 LIFE EXTENSION OR REPLACEMENT ?

Based on life extension studies, following decisions are taken:

(*i*) Whether to replace the whole machine or to replace the affected parts. The techno-economical factors are sole guiding factors for arriving at a decision.

(*ii*) For replacement of components, state-of-the art engineering technology is introduced into older machine. By this, reliability and the expected life of renovated machine would be similar to the original machine. Example of introduction of state-of-the art technology are—replacement of class- B bitumin insulation in windings of generators with class-F insulation, improvement in generator protection system with digital electronic relays *etc*.

CASE STUDY ON RENOVATION AND MODERNISATION OF A BOILER

This case study pertains to a boiler of a power utility company. The original equipment was supplied by Russians. The rated capacity of unit was 50 MW, 220 Tons/hour of steam at 100 kg/cm^2 pressure and 540°c temperature. The boiler used to feed steam to a steam turbine for power generation. The unit was running at reduced load due to its age (having clocked more than 1,50,000 hours). The main reasons were repeated failures of superheater and economiser tubes. The flue gas temperature had risen to 200°C resulting in low efficiency.

The residual life assessment of the unit indicated that tubes had failed due to creep cracks indicating expiry of life. So it was recommended to refurbish the boiler by replacing pressure part components. RLA studies indicated the following replacements.

Pressure Parts

(*i*)	Water walls.	65% replacement
(*ii*)	Down comers.	100% replacement
(*iii*)	Super-heaters.	100% replacement
(*iv*)	Main steam line.	100% replacement
(*v*)	Feed lines.	50% replacement

Non-Pressure Parts

(*i*)	Mill ducting.	100% replacement
(*ii*)	ID fan ducting.	100% replacement
(*iii*)	Furnace bottom hopper.	100% replacement
(*iv*)	Expansion joints.	100% replacement
(*v*)	Valves (main feed control value, main steam stop value, safety values).	100% replacement
(*vi*)	Refractory and insulation lining.	95% replacement

After renovation, following improved performance was observed—

(*i*) Boiler steam output was 235 tons/hour against rated capacity of 220 Tons/hour.

(*ii*) Flue gas temperature reduced to 145°C leading to higher thermal efficiency.

(*iii*) Higher availability and reliability.

(*iv*) Lost capacity restored from 35 MW to 50 MW.

This shows that a unit which had expired economic life was restored to original operating conditions without replacing the existing old unit. The total cost of renovation and the time taken for restart was much less than the cost of similar new unit.

Summary

Industrial assets need replacement due to variety of reasons such as aging, technological obsolescence, higher operating cost, decreased reliability *etc.* Since the capital cost of new equipment is very high hence replacement is considered necessary only when operational and maintenance cost of existing machine is higher than average capital and operating cost of proposed machine.

Asset replacement decisions are also influenced by failures of asset. In case of gradual failures, asset replacement planning can be done. These assets follow a bath-tub failure curve. However, assets which fail suddenly, no such replacement planning can be done. Replacement decisions are also affected by a host of factors such as technological factors, functional factors, financial factors, safety factor and reliability factors.

Replacement policies determine optimal replacement period of an asset. It is based on type of deterioration of the asset.For assets with gradual deterioration, the asset need to be replaced when total average cost of ownership is minimum. However, when time value of money is considered then a number of discounted cash flow techniques are used to determine optimum replacement interval. However, for assets with sudden failure (without giving prior warning), a group replacement policy could be adopted. This could be built into preventive maintenance plan. In such cases. group asset replacement need to be done when the total cost of group and individual replacement per unit time is minimum. MAPI method of asset replacement is useful when it is difficult to predict future cash flows. Life extension strategies could be used with gradually deteriorating assets.

Review Questions

1. Explain various components of asset life cycle.
2. Explain with example—Why asset replacement decisions are very complex?
3. Enumerate various reasons for replacement of asset.
4. What is depreciation? What are various types of asset depreciation ?
5. What are different types of failures? How do they affect replacement decisions?
6. Discuss different types of replacement policies.
7. Explain in detail MAPI method of replacement policy.

8. Write brief note on (a) Return on Investment (b) Internal rate of return.
9. What do you mean by life extension strategy ? Why is it preferable to go for life extension strategy rather than complete asset replacement.
10. Write briefly on residual life extension studies.

Discussion Questions

1. "Overall ownership cost is most important rather than purchase price or operating cost of an asset". Comment.
2. "It is advisable to increase economic life of asset through life extension strategies rather than total asset replacement". Justify
3. "Sometimes technological obsolescence determines asset replacement rather than functional performance impairment". Elaborate
4. "Techno-economic factors are more important than purely technical factor for asset replacement". Explain
5. "Financial factors do not matter in case reliability and safety of asset is involved". Justify
6. "Probability of progressive failures can be modelled by normal probability distribution". Comment
7. "Random failures such as failure of insurance spares can be modelled by poisson distribution". Explain
8. "For assets with random failures characteristic such as light bulb, it is economical to go for group replacement rather than individual replacement". Justify
9. "Maintenance policy affects asset economic life". Justify
10. "Working environment and intensity of usage affects functional life of an asset". Elaborate.

Multiple Choice Questions

1. Acquisition cost of asset consists of :
   (*a*) Purchase price.
   (*b*) purchase price + Insurance + Tranportation.
   (*c*) Purchase Price + Taxes and duties + Insurance + Transportation.
   (*d*) None of these.

2. Asset is replaced due to:
   (*a*) Physical deterioration.
   (*b*) Functional deterioration.
   (*c*) Financial deterioration.
   (*d*) All of these.
3. Asset replacement or life extension is affected by:
   (*a*) Technical factors.
   (*b*) Economic factors.
   (*c*) Techno-economic factors.
   (*d*) None of these.
4. Gradual failures follow a pattern of:
   (*a*) Normal distribution.
   (*b*) Position distribution.
   (*c*) Bath-tub curve.
   (*d*) Exponential distribution.
5. Residual life extension studies require:
   (*a*) Diagnostic tests.
   (*b*) Past failure history.
   (*c*) Physical inspection.
   (*d*) All of these.

True-False Questions

1. Purpose of asset replacement analysis is to decide when to replace.
2. Asset failures are more during initial and final stages of life of an asset.
3. Monetary depreciation is due to inflation.
4. MAPI method takes into account inflation and risk factors.
5. Change in working environment enhances useful asset life.

ANSWERS

Multiple Choice Questions

1. (*c*) **2.** (*d*) **3.** (*c*) **4.** (*c*) **5.** (*d*)

True/False Questions

1. True **2.** True **3.** True **4.** False **5.** False

❑❑❑

13

Shutdown Maintenance

Learning Objectives

After reading this chapter, you will be able to:

- Understand need and advantages of shutdown maintenance.
- Comprehend planning and scheduling of shutdown.
- Understand the importance of safety during maintenance.

13.1 INTRODUCTION

Among all other types of maintenance *viz* preventive, breakdown, predictive, design out etc., this type of maintenance requires planning, scheduling and controlling skills of very high order.

Shutdown refers to complete stoppage of plant and utilities for maintenance for major overhauling, retrofitting, major modification, addition and alteration of plant and machinery. Shutdown maintenance is undertaken when it is not feasible to carry out intended jobs during running condition of plant.

Also inspection and condition monitoring of major plant equipments such as reactor lining inspection in a urea plant, blade inspection in steam/gas turbine, boiler refractory inspection and repair *etc*. cannot be carried out during normal running of plant. This requires stoppage of plant to carry out necessary checks and repairs.

Generally, continuous process plants such as fertiliser, petrochemical, thermal power plants, cement *etc*. which runs 24 hours a day, provides little opportunity for preventive maintenance. Hence, periodically either entire plant or part of the plant is stopped for maintenance. The periodicity of stoppage may be once in six months or once in a year or once in two years *etc*. Generally, time selected for plant stoppage co-incides with overflowing finished goods, raw material shortage, poor product demand, stoppage of fuel (*e.g.* natural gas from GAIL), plant expansion *etc*.

To carry out such a major activity in a given time span and budget, requires meticulous planning of spares, manpower (both skilled and unskilled), tools and tackles, specialised contractual support from original equipment manufacturer (OEM) *etc*.

So the entire planning and scheduling is carried out by maintenance planning section (in a large organisation) or by planning engineer (in a medium or small organisation).

In the following sections, various aspects of shutdown maintenance *viz* planning procedure, arrangements of spares and manpower, method of progress review, sequence of shutdown and startup of plant are discussed.

13.2 NEED FOR SHUTDOWN MAINTENANCE

Following are the need for shutdown maintenance—

(*i*) To carry out prevention maintenance, inspection, repair *etc* in a continuous process plant (operating 24 × 7) which cannot be accessed during normal operation.

(*ii*) To carry out major inspection, overhauling, retrofitting *etc*. of major equipments such as turbine, boiler, process equipments such as urea reactor, primary reformer (in an ammonia plant) or digester (in a phosphoric acid plant) *etc*.

(*iii*) To carry out modification, addition and alteration in plant system such as water pipelines, fuel pipelines, steam piplelines, gas pipelines *etc*.

(*iv*) To carry out major modification/addition in plant electrical distribution system.

(*v*) To integrate new plant with the existing plant

(*vi*) To carry out safety related periodic inspection such as thickness measurement of ammonia storage tank walls, monitoring of oxidizer or reactor lining and internal condition (*e.g.* in caprolactum plant), boiler tube inspection and hydraulic testing, refractory lining of furnaces *etc*.

(*vii*) To stop production due to poor product demand, raw material shortage, fuel/feedstock stoppage *etc* and utilize this time for maintenance (opportunity maintenance).

(*viii*) Emergency plant condition requiring immediate stoppage of plant for repair.

(*ix*) Statutory compliance *e.g.* annual boiler test.

13.3 ADVANTAGES AND DISADVANTAGES OF SHUTDOWN MAINTENANCE

Advantages

Following are the advantages:

(*i*) Due to sufficient availability of time, manpower and spares, good quality of job can be done.

(*ii*) Hiring of specialised technical personnel allows complex jobs to be carried out which would not have been possible without shutdown.

(*iii*) Due to meticulous planning, scheduling and stoppage of plant, operational safety is ensured *e.g.* danger from live wire, pressurised pipelines, fire and explosion hazard *etc* is absent.

(*iv*) Since routine and breakdown maintenance activities are absent hence entire resources can be fruitfully utilised.

(*v*) Proper and detailed inspection and repair prevents unexpected and incipient failures.

(*vi*) Ensure higher operational efficiency.

(*vii*) Ensure higher yield.

Disadvantages

(*i*) Stoppage of plant leads to revenue loss due to loss of production.

(*ii*) Energy loss is associated with plant startup and shutdown.

(*iii*) Minimum fixed charges for power, natural gas *etc* have to be paid irrespective of consumption.

(*iv*) After startup up, it requires time to finetune the process to original settings resulting in material, manpower and energy wastages.

13.4 DETERMINATION OF SHUTDOWN INTERVAL

Shutdown maintenance is carried periodically. The interval varies from industry to industry and can be once in a year to once in a 3 year or so. Various factors affecting choice of shutdown interval are:

(*i*) Market demand for product undergoes cyclical changes. So shutdown is chosen during weak product demand period. For example, car manufacturing industry chooses month of December for shutdown as the demand is low. Since customers wait to buy new car next year. Similarly refrigerator manufacturers prefer July/August due to poor product demand in monsoon months and non-wedding season.

(*ii*) Stoppage of feed stock may decide time for shutdown. For example, stoppage of natural gas from GAIL may decide plant shutdown.

(*iii*) Periodic inspection of critical equipments may decide shutdown *e.g.* boiler, turbine, major process equipment such as reactor, oxidizer *etc.*

(*iv*) An overflowing inventory may prompt plant shutdown.

(*v*) Old plant needs frequent inspection so apart from technical consideration, inteval between two shutdowns is also decided by commercial interests. It varies from organisation to organisation and is based on their operational philosophy.

13.5 SHUTDOWN PLANNING

Planning is the most critical component of entire shutdown maintenance exercise which decides the success or failure of the complete operation. Seventy per cent effort should be on meticulous planning and remaining 30% should be on execution.

Following data are required for initiating shutdown planning:

(*i*) Clear number of days available for maintenance.

(*ii*) Total skilled/semiskilled /unskilled manpower available.

(*iii*) Necessary spares and their availability.

(*iv*) Budget sanctions.

(*v*) List of critical job to be carried out.

(*vi*) Availability of external experts.

(*vii*) Availability of contractual manpower with tools and tackles.

(*viii*) Availability of standby power source, water source, medical facilities, canteen facilities and other peripheral facilities.

So there should be a proper match between available resources and quantum of work to be completed. The major constraints are— resource availability, time limit, budget constraint and quantum of jobs at hand.

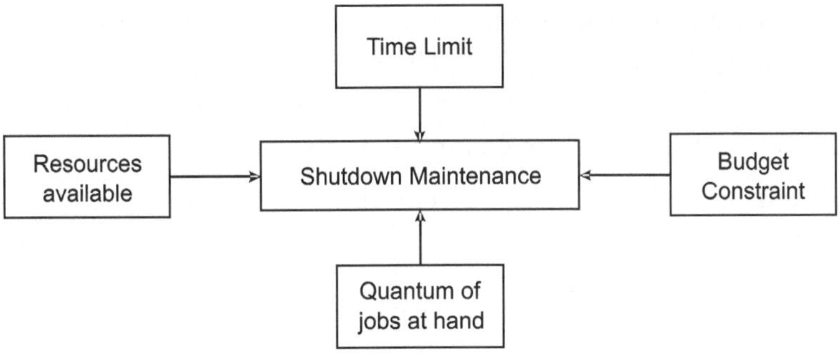

Fig. 13.1 Major Constraints on Shutdown Maintenance

Major components to be taken care-off are following:

(*i*) **Job outsourcing:** Looking at the quantum of jobs available, one has to identify jobs to be outsourced. Reason for outsourcing could be monetary, lack of manpower availability, risk involved, specialised nature of job *etc*. Outsourcing saves lot of time, money and effort and allows to concentrate on critical jobs

Next step is allocation of jobs to different contractors depending upon their relevant expertise. This requires clear identification of jobs with time limit and performance guarantee.

(*ii*) **Manpower requirement:** After segregating contractual jobs the available manpower is to be deployed on the list of jobs at hand. This may also include visit of external experts.

(*iii*) **Spares requirement:** No shutdown maintenance can be done unless all necessary spares are available. This requires stocktaking and chasing the venders.

Spares may include insurance spares, ordinary spares, consumables, fabricated parts *etc.*

(*iv*) **Estimation of peripheral services:** Peripheral services may include standby power source, water source, additional canteen facilities,

facilities for boarding and lodgging of experts, about hiring of tools and tackles *etc.*

(*v*) **Budget sanction:** Next step involves total estimation of expenditure on maintenance activities. This requires prior approval from competent authority in the organisation.

(*vi*) **Monitoring and review team:** Lastly, a monitoring and review team need to be formed to take shutdown activities from beginning to its logical end meeting all the constraints.

The purpose of team is not only to monitor and review but also to do course correction if any deviation is noticed. These deviations could be minor or major changes in plant, unexpected non-availability of manpower, spares or accidents, fire *etc*. So team has to plan for contingency situations

Planning process flow chart is given below:

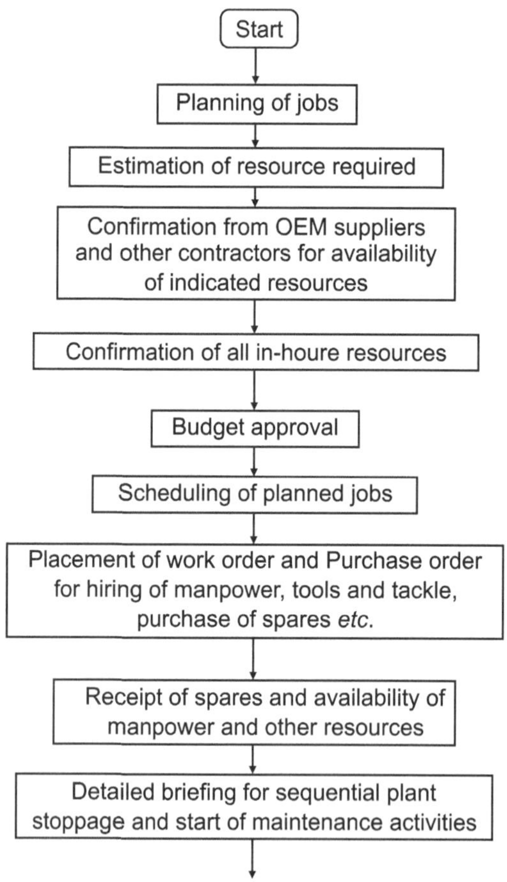

Contd...

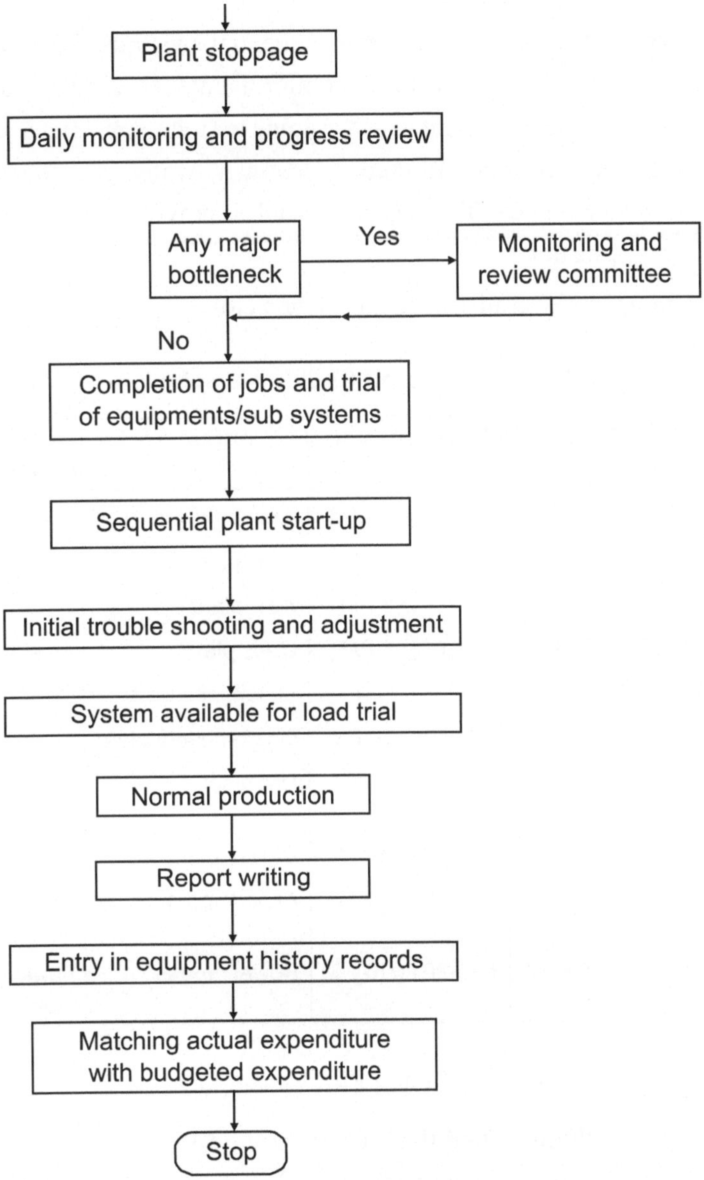

Fig. 13.2 Flowchart for Shutdown Planning

13.6 SCHEDULING OF SHUTDOWN MAINTENANCE

Scheduling involves allotment of available resources and sequencing of jobs, with start and stop time.

Schedules are of following types:

(*i*) **Overall Schedule:** It indicates sequential switching off of plant and sub systems and restart of plant after maintenance. It indicates time, date and section of plant which is going to stop for maintenance. It also mentions services affected *e.g.* power, water, air conditioning, compressed air *etc*. This type of schedule is provided to top management for information.

Example of overall schedule is given below—

Table: 13.1 Overall Shutdown Schedule.

S. No	Date/Time	Date/Start-time	Area	Services affected
1.	18-6-2011/10 AM	30-6-2011/ 6 AM	Ammonia plant	—
		1-7-2011/ 8 PM	DAP plant	
		1-7-2011/ 8 PM	ASP plant	
2.	18-6-2011/12 Noon	1-7-2011/ 5 PM	Urea plant	—
3.	18-6-2011/2 PM	1-7-2011/ 10 PM	Melamine plant	—
		1-7-2011/ 10 PM	Gas recovery plant	
4.	19-6-2011/6 AM	28-6-2011/ 6 AM	DM plant	DM water, soft water
5.	19-6-2011/6 AM	28-6-2011/10 PM	Chiller plant	Air conditioning
6.	19-6-2011/6 AM	28-6-2011/6 AM	Air compressor	Compressed air
7.	19-6-2011/6.30 AM	27-6-2011/10 PM	Power house	Power (stand by power available)

Other method is to use CPM and PERT techniques.

(*ii*) **Specific Plant Schedule:** Plant specific maintenance schedule is prepared for maintenance personnel. It contains details of maintenance activities, date, time and names of personnel doing it.

Apart from above schedule, an intimation is given to canteen, medical room, security *etc*. for extra personnel working in the plant.

An example of specific plant schedule is given below:

Table. 13.2 Specific Plant Schedule

S. No.	Area Engineer	Activity	Plant Area and Equipment No.	Start date and Time	Finish Date and Time	Technicians	Supervisor
1.	Sanjeev Kumar	Impeller and Bearing replacement of boiler feed pump.	Power House Boiler No. # 6 feed pump No. # 2	26-4-2011 8 AM	27-4-2011 10 AM	Ravi, Mohan, Suresh.	Kapil.
2.	Mukesh Patel	Air circuit breaker fixed and moving Contact replacement	Ball Mill PCC # 2	26-4-2011 8.30 AM	26-4-2011 9 PM	Naresh, Vipin, + One Contract Labour	Suresh.
3.	Hari Ram	H. T. Panel No # 2 maintenance	Power House, material handling yard HT panel	27-4-2011 9AM	27-4-2011 6 PM	Mahesh, Sohan +4 contract labour	Kapil
4.	Rakesh Desai	Air Compressor No-5 maint.	Utilities, Compressor room	29-4-2011 10AM	31-4-2011 12 noon	Salil + 3 contract labour	Kalmadi
5.	Vineet	Coal crusher No # 2, Maint	Material Handling Yard	30-4-2011 3 PM	31-4-2011 10AM	Kuruvilla + 4 labour	Suresh
6.	Jai Singh	Inspection of digestor lining, pipeline replacement	Phosphoric acid plant digester No # 2	2-5-2011 6 Am	2-5-2011 10 Pm	Verghese + 6 labour	Sandeep

13.7 SAFETY PRECAUTION

When maintenance activities are carried out on a massive scale, there are chances of accident, fire *etc*. So, safety engineer has to take extra precautions so as to prevent any untoward incident.

This requires proper safety measures to be used by working personnel. Also necessary premits have to be used, say-working in confined area, hazardous area, electrical systems *etc*.

Even contractual labourers need to be sensitised regarding use of safety gears such as safety shoe, safety helmet, safety belts *etc*.

13.8 PLANT SHUTDOWN AND START-UP

For plant shutdown, raw material fed to the system are stopped and sequentially all other plants are stopped one after another. At the end, utility services are stopped.

During shutdown, alternate arrangement for power, water, compressed air, flood lighting *etc*. is made. A skeleton staff is maintained to look after these services.

While start up requires careful operational planning. When an equipment or machinery is ready after maintenance then its trial is taken and associated defects if any are rectified. Thereafter, all sub-systems are boxed-up and operational trial is taken.

Majority of accidents take place at the time of trial. Many a times, different maintenance crews are working on a same equipment and while trial is being taken by one of the crew, there are chances of accidents while other teams are still working. Hence, it requires well co-ordination at the site level among various maintenance crews working on a given equipments or sub-system.

Summary

Shutdown maintenance is carried out by stopping all plant machinery and utilities. Such type maintenance is useful for major overhauling, major modification, plant extension, plant addition and alteration *etc*. Such type of maintenance is carried out annually or biannually in continuous process plants such as fertilizers, petro-chemicals, refinery, cement, thermal power plant *etc*.

Shutdown maintenance offers excellent opportunity for undertaking complex jobs. Shutdown interval is affected by market demand for

Contd...

product, stoppage of feed stock, periodic inspection of capital equipments, overflowing inventory *etc.*

Shutdown maintenance requires meticulous planning. The major constraints are time limit, resource availability, budget constraint and quantum of jobs at hand.

Major components of shutdown jobs are outsourcing manpower requirement, spare requirement, budget sanction, and monitoring and review.

Shutdown planning requires planning of jobs, estimation of resources, sanction of budget, scheduling of planned jobs, procurement of spares and consumables.

Schedule is the allotment of available resources and sequencing of jobs. It is the next important job after planning. Schedules are of two types—Startup/shutdown schedule and job schedule.Startup/shutdown schedules gives details of startup and shutdown of various sections of plants and utilities while job schedule describes the various activities to be taken up during shutdown. It shows job, manpower, start and finish time *etc.*

After completion of shutdown, plant start-up is takenup. It takes time due to various issues that crop-up during trial. After sucessful completion of trials, plant is brought back into operating state.

Review Questions

1. How shutdown maintenance is different than other types of maintenance?
2. Discuss the need for shutdown maintenance.
3. Compare advantages and disadvantages of shutdown maintenance.
4. How do you decide shutdown interval for a complex plant ?
5. Discuss effect of resource availability, time limit, and budgetary constraints in successful execution of shutdown maintenance.
6. How proper planning of plant shutdown and startup plays an important role in quick plant restart ?
7. What is the role played by review and monitoring team during shutdown maintenance ?
8. Explain the importance of job outsourcing in shutdown.
9. Elaborate on safety precautions to be taken during shutdown maintenance.
10. Discuss factors affecting time interval of shutdown maintenance.

Discussion Questions

1. "Shutdown maintenance requires meticulous planning and execution". Elaborate.
2. "Shutdown maintenance is mainly dependent on production department". Justify.
3. "Overhauling and retrofitting of major equipments can be carried out only during shutdown". Comment.
4. "Record of preventive and breakdown maintenance acts as feed for shutdown maintenance". Elaborate.
5. "Over maintenance is as bad as no maintenance". Justify.
6. Do you agree "a tight financial control is not possible for satisfactory completion of a job.
7. "Energy loss during shutdown and restart can be reduced by efficient planning and scheduling" Discuss.
8. "CPM and PERT scheduling techniques can play a vital role in shutdown maintenance". Comment.
9. "Daily briefing and review of jobs in progress is essential for safe and timely completion of a project". Justify.
10. "Recording of all activities in equipment history card is the next vital job in post shutdown period". Comment.

Multiple Choice Questions

1. Shutdown maintenance requires:
   (*a*) Money. (*b*) Manpower.
   (*c*) Spares. (*d*) All of these.
2. Major constraint in shutdown maintenance is
   (*a*) Budget. (*b*) Time limit.
   (*c*) Availability of resources. (*d*) All of these.
3. Safety hazards could be minimised by
   (*a*) Safety review.
   (*b*) More drills.
   (*c*) Use of personnel productive gears.
   (*d*) All of above.

4. Periodicity of shutdown maintenance depends on :
   (*a*) Shortage of raw material.
   (*b*) Low product demand.
   (*c*) Major overhauling of capital equipments.
   (*d*) All of these.

5. Most critical part of the shutdown maintenance is
   (*a*) Planning and scheduling. (*b*) Execution.
   (*c*) Review and monitoring. (*d*) None of these.

True-False Questions

1. Shutdown maintenance can be replaced by prevention maintenance.
2. Cost of shutdown maintenance outweighs its benefits.
3. Shutdown maintenance must include emergency evacuation procedures.
4. Continuous monitoring can reduce maintenance bottlenecks.
5. Purpose of review commitee is to monitor expendiure against the budget.

ANSWERS

Multiple Choice Questions

1. (*d*) **2.** (*d*) **3.** (*d*) **4.** (*d*) **5.** (*c*)

True/False Questions

1. False **2.** False **3.** True **4.** True **5.** True

❑❑❑

14

Computerised Maintenance Management System

Learning Objectives

After reading this chapter, you will be able to:

- Appreciate importance of computerisation of maintenance.
- Understand various modules of CMMS.
- Examine choice and selection of various software available in the market.
- Understand process of CMMS implementation.

14.1 INTRODUCTION

Computers have invaded all domains—personal, commercial and industrial. This process started in late 1980s in India. Like electricity, one cannot think of life without computers. This is quite evident in various business applications such as e-business, e-commerce, enterprise resourse planning (ERP), customer relationship management (CRM) to educational institutions, where presentations are made through laptop. Today, even small kids surf internet and chat with their friends on-line.

Despite all these progress, it took quite sometime for computers to find acceptance in maintenance field. There was non-availability of right kind of software to users as software need to be user application specific. But late

1990s saw rapid use of computers in industrial organisations not only in maintenance but also in accounts, finance, human resource, production, quality assurance, stores and purchase functions. This helped organisations to integrate different functions and generate a common data base. This tremendously improved speed, efficiency and data retrieval.

In this chapter, the basic concept of CMMS is developed. Its aim is to familiarise readers with tremendous opportunities that computer offers to maintenance management field.

14.2 BENEFITS OF COMPUTERISED MAINTENANCE MANAGEMENT SYSTEM (CMMS)

A computerised Maintenance Management System or CMMS is a type of software based system that performs functions in the area of maintenance management. The CMMS system has many modules such as work order generation and tracking, preventive maintenance module, spare parts module etc. All these are inter-linked.

The various benefit provided by CMMs over manual system of maintenance operation are following:

(*i*) Proper planning, control and monitoring of preventive, predictive and breakdown maintenance.

(*ii*) Efficient tracking and control of pending work.

(*iii*) Better utilisation of resources such as men, machine, material and logistics.

(*iv*) Monitoring of plant/equipment performance through trend analysis, Pareto–analysis, A-B-C analysis etc.

(*v*) Efficient documentation and reporting system.

(*vi*) Better interface with internal customers such as production, materials, human resource, quality, safety etc.

(*vii*) Improved diagnosis of productivity and profitability.

(*viii*) Improved diagnosis of maintenance problem.

(*ix*) Better decisions due to availability of sufficient data.

(*x*) Maintenance cost reduction due to efficient utilization of available resources.

(*xi*) Increased plant availability.

(*xii*) Availability of maintenance information for top management decision making.

(*xiii*) Timely statutory inspection and record keeping.

(*xiv*) Timely planning of major inspections/overhauls or residual life assessment.

14.3 BASIC COMPONENTS OF CMMS

A CMMS system is composed of various sub-systems called modules such as inventory management module, preductive maintenance module, work order and breakdown maintenance module etc. All these sub-systems or modules are inter-linked so as to provide a comprehensive report. The various maintenance modules are given below:

(*i*) Machinery data and preventive maintenance module.
(*ii*) Work order and breakdown maintenance module.
(*iii*) Predictive maintenance module.
(*iv*) Inventory management module.
(*v*) Residual life assessment, overhaul and statutory compliance module.
(*vi*) Modification and safety module.
(*vii*) Maintenance information module.
(*viii*) Finance module.
(*ix*) Job outsourcing module and so on:

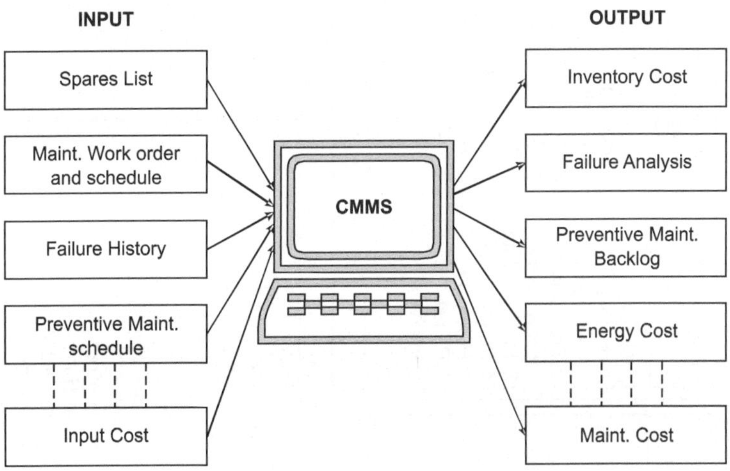

Fig. 14.1 CMMS System Overview.

CMMS system takes input in terms of cost of spares, spares list of preventive maintenance, breakdown hours, energy consumption etc. and provides useful output such as cost of current inventory, failure analysis, cost of energy consumption, overall maintenance cost, etc.

The output is taken as bar chart, pie-chart, histogram *etc.* for performance review and improvement.

The data is fed to various sub-systems or modules on a fixed interval basis. The reports are taken out daily, monthly or yearly basis. The sub-systems or modules are expandable and depends upon user's current and future need.

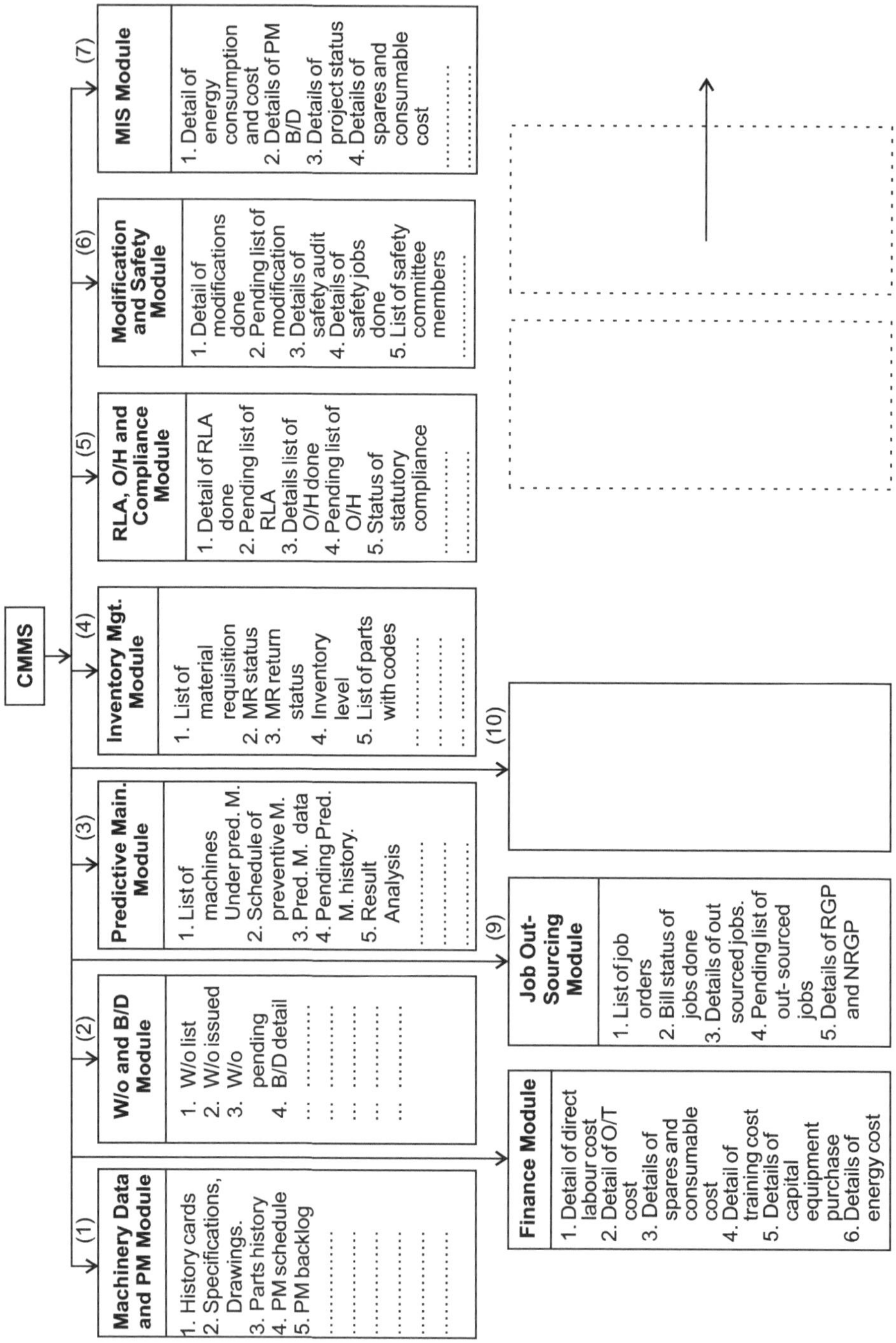

Fig. 14.2 CMMS Modules

Abbreviations—

(1) PM—Preventive Maintenance.

(2) W/O— Work Order.

(3) B/D—Break Down.

(4) RLA—Residual Life Assessment.

(5) O/H—Overhauling.

(6) RGP—Returnable Gate Pass.

(7) NRGP—Non-Returnable Gate Pass.

The various sub-areas of CMMS are following (these are indicative and not exhaustive and will depend upon user need):

(*i*) Machinery data and PM module.

(*ii*) Work order and breakdown maintenance module.

(*iii*) Predictive maintenance module.

(*iv*) Inventory management module.

(*v*) Residual life assessment, overhauling and statutory compliance modules.

(*vi*) Modification and safety module.

(*vii*) Maintenance information system module.

(*viii*) Finance module.

(*ix*) Job outsourcing module and so on.

(*i*) **Machinery Data and PM Module**: The section has data related to specification of machines/equipments, drawings (schematics foundation plan, process and Instrumentation diagram, assembly diagram etc.) List of spare parts for all critical machines with code number, schedule of preventive maintenance of machines, list of PM back-log etc.

(*ii*) **Work order and Breakdown Maintenance Module:** This module contains list and content of work orders issued to carry out jobs on day-to-day basis, list and details of breakdown maintenance carried out *etc*.

(*iii*) **Predictive Maintenance Module:** This module has list of all equipments under predictive maintenance, data obtained from predictive maintenance, analysis of data (barchart, pareto analysis, trend analysis etc.).

(*iv*) **Inventory Management Module:** This module contains list and detail of all indented material (spare parts, consumables), status of material recieved/rejected, monthly consumption and cost *etc*.

(*v*) **Residual life Assessment, Overhauling and Statutory Compliance Module:** This module has details of RLA studies and jobs carried out, details of overhauling done and pending, list and status of statutory compliance (e.g. related to boilers, electrical, pollution, storage of petroleum products *etc*.)

(*vi*) **Modification and Safety Module:** This section has list and detail of all modifications done on equipments and sub-systems, details of safety audits, list of safety committee members, list of safety related jobs carried out, details of past accidents and data *etc*.

(*vii*) **Maintenance Information System Module**: This module contains all consolidated information on entire maintenance department activities such as energy consumption and cost, cost of spares and consumables, details of PM and breakdown, status of present and future projects, labour overtime cost *etc*.

(*viii*) **Finance Module:** This module contains all costing related to various maintenance activities such as direct labour cost, overtime cost, spares and consumable cost, cost of training, cost of capital equipment purchase, capital and revenue budget, energy cost *etc*.

(*ix*) **Job Outsourcing Module:** This contains list of all jobs sent to vendors/fabricators/motor rewinders etc., approved list of vendors, details of work orders issued, status of pending jobs, status of bills, details of NRGP(non-returnable gate pass) and RGP (returnable gate pass) *etc*.

These are various modules depending upon user need and there can be other user specific modules. All these modules can generate relevant reports regarding jobs completed, backlog, maintenance cost *etc*. All these are done on monthly basis and at the end of financial year, a final report is generated. Maintenance information system module generates a format specific monthly and yearly report on major maintenance activities and their expenditure for submission to top management. This MIS contains details such as energy consumption and cost (monthly), spares and consumable cost, labour cost, cost of capital equipment purchase, various maintenance indices and efficiencies etc.

14.4 SELECTION OF CMMS SOFTWARE

Maintenance management softwares are available as standard package or can be specifically developed. The standard softwares have advantage of lower cost, greater acceptability and ease of availability. While it may have disadvantages such as difficulty in upgradation, difficulty in linking up with other integrated software system such as ERP (Enterprise Resource Planning), may not meet user specific need. However, internally developed software has many advantages but requires more time and is expensive. The choice has to be made by the maintenance head considering all pros and cons.

Following features need to be considered while selecting a software:

(*i*) **Easy to use:** The software chosen should be menu driven and easy to use so that even a supervisor is able use it. The various modules of software should be properly interlinked.

(*ii*) **Hardware need**: The hardware required for data storage, speed etc. need to be understood before selection. Also, the operating platform for software should be such that it can be easily hooked up with plant software such as ERP (Entreprise Resource Planning).

(*iii*) **Ease of operation and maintenance:** The software should be easy to install, operate and troubleshoot. It should have provisions for future upgradation and uplinking with other software operating in the organisation. The sale agreement should include software service and modifications.

(*iv*) **Training to employees:** There should be provision to provide training to employees after installation at no extra cost.

14.5 READILY AVAILABLE SOFTWARES

There are variety of maintenance software that are available in the market with wide variety of applications and features. Some of these are listed below:

Table 14.1 List of Software and Suppliers

	Name of Software	Supplier
(i)	Scope.	ABB Service.
(ii)	Maint. Scope.	Advanced Software Design.
(iii)	Maintenance Complete 2000.	Cybercode.
(iv)	CMMS Plus.	Fluor Daniel.
(v)	Prism.	Avantis.
(vi)	Advanced Maintenance. Management System.	Micro West Software Systems.
(vii)	SMART/MMS.	PM Systems.
(viii)	PMXpert 65.	PMXpert Software.
(ix)	AIMS-CMMS.	Phoenix Data Systems Inc.
(x)	Maintenance Manager.	Symbiotic Systems.

14.6 CMMS IMPLEMENTATION

Generally, functions other than maintenance such as accounts, HR, materials etc are the first to be computerised. The maintenance functions run manually. This is due to some reasons such as reluctance of top management to invest, lack of information about CMMS softwares, indifference to computerisation etc.

However, once the top management is made to understand the cost advantage, the computerisation of maintenance becomes lot more easier. Once it is decided then implementation can be done by the following steps:

(*i*) Raising awareness and motivation of maintenance staff

(*ii*) Installation of software

(*iii*) Training of maintenance personnel.

(*iv*) Adoption by various maintenance sections.

(*v*) Trouble shooting and modifications

(*vi*) Creation of back up files.

14.7 CAUSES OF FAILURES OF CMMS

Even after careful selection of a maintenance software and its proper planning, a CMMS can fail to produce desired result due to the following reasons:

(*i*) **Poor training of users:** Poorly trained users try to revert back to manual system thereby defeating very purpose of computerisation.

(*ii*) **Fear of computers:** Learning abilities decline with age so old maintenance staff who are comfortable with manual system try to avoid computers due to inability to learn.

(*iii*) **Poorly selected system software:** A poorly selected system software may create problem such as inability to provide new required functions, synchronisation problem, with the newly introduced software in other departments, inability to provide data in the desired format etc.

(*iv*) **Software glitches:** These are due to programming errors.

(*v*) **Very high expectation:** A CMMS can provide result as per its programmed features and cannot provide services which are not programmed.

Summary

Computerised maintenance management system (CMMS) is an indispensable part of today's maintenance. Various maintenance softwares are available that can be devised as per specific customer need. Advantages of CMMS is manifold such as tracking and control of pending jobs, cost analysis, trend analysis of equipment performance, cost reduction, improved productivity, timely planning of major inspections, overhauls etc.

A CMMS system consists of number of interactive sub-systems. These sub-systems are based on need of customers. These are machinery data and PM module, work order and breakdown maintenance module, inventory management module, overhaul and statutory compliance module, finance module etc. Data need to be entered into these modules and processed information can be taken out such as inventory cost, failure analysis, energy cost, maintenance cost, down time detail etc. These modules are expandable for future need.

The most important part of CMMS is the selection of appropriate software. Readymade softwares are available in the market. However, a tailor made software is more appropriate. The chosen software should be easily integrate with other plant softwares such as ERP.

After selection of required software, next step is raising awareness of maintenance staff. Training of maintenance personnel forms an important component of CMMS.

Review Questions

1. Discuss the role of computers in maintenance management .
2. Compare manual and computerised system of maintenance.
3. Discuss the basic components of CMMS.
4. Explain how will you prepare preventive maintenance management module of CMMS?
5. How would you select appropriate software for a maintenance department?
6. Discuss the importance of inter-connections of maintenance software and ERP system.
7. Discuss various causes of failure of computerised maintenance management system.
8. Explain steps involved in implementation of CMMS system.

9. Discuss desirable features that a maintenance software should possess.
10. Why awareness and proper training plays an important role in sucessful implementation of CMMS?

Discussion Questions

1. "Computerised Maintenance System is a necessity". Justify
2. "Older organisations with older people find it difficult to adopt to computerised environment". Comment.
3. Elaborate on – "Adaptability with existing software systems is a key requirement for purchase of a new software".
4. "Cost of software depends upon added features which may or may not be useful". Discuss.
5. "The main barrier in successul implementation of CMMS is not computer illiterate manpower but a committed top management". Comment.
6. "Cost of computerisation is much less than the benefit derived".Discuss.
7. "Out of all barriers to computerisation, it is the human attitude that is most important". Comment
8. "Data analysis and their representation is easier in computerised environment". Justify
9. "It is possible to maintain a paperless maintenance management system in CMMS". Comment.
10. "Manipulation of daily data is difficult in CMMs system". Comment.

Multiple Choice Questions

1. Purpose of CMMS is:

   (*a*) Easy data analysis. (*b*) Efficient tracking of jobs.

   (*c*) Better utilisation of resources. (*d*) All of these.

2. Basic components of CMMS system are divided by:

   (*a*) Software provider. (*b*) Customer.

   (*c*) Top management. (*d*) None of these.

3. Major factor to be considered in selection of a software is

   (*a*) Cost. (*b*) Hardware requirement.

   (*c*) Training need of employees. (*d*) All of the above.

4. A trial run of software can eliminate:
   (*a*) Software errors
   (*b*) Problem in module accessibility
   (*c*) Mismatch
   (*d*) All of these

5. CMMS system can help to reduce downtime by
   (*a*) Recording and monitoring
   (*b*) Allowing analysis
   (*c*) Providing a database
   (*d*) All of these.

True/False Questions

1. For a small plant, CMMS is not a necessity.
2. As the system expands, software glitches are bound to happen.
3. CMMS system proves to be expensive proposition in the long run.
4. A good CMMS system is one which has maximum number of features.
5. Periodic updation of software is necessary for trouble free operation.

ANSWERS

Multiple Choice Questions

1. (*d*) **2.** (*b*) **3.** (*d*) **4.** (*c*) **5.** (*d*)

True/False Questions

1. True **2.** False **3.** False **4.** False **5.** True

❑❑❑

15

Maintenance Audit

Learning Objectives

After reading this chapter, you will be able to:

- Understand purpose and types of audit.
- Understand various areas of maintenance audit.
- Understand methodology of audit.

15.1 INTRODUCTION

Purpose of maintenance audit is to critically assess and evaluate the working of maintenance function for the intended result and suggest suitable measures to make existing systems/procedures/techniques more effective.

Maintenance audit is defined as a periodic review of various sub-systems of maintenance department in order to gauge the current health of the system and point out areas of neglect, weak spots and obsolete practices.

This can be compared to a periodic health check-up of a person. The purpose is to check the function of critical organs of the body and detect any abnormality so as to apply corrective action in time.

A maintenance audit is carried out by a team of experienced professionals. Experienced maintenance audit professional are available like certified management system professionals such as ISO-9000, OHSAS-18000 etc. so that job is carried out in professional manner.

15.2 TYPES OF AUDIT

Before we delve into details of maintenance audit, first we'll know about the various types of audit:

(*i*) **Management Audit:** This type of audit is carried out to check maintenance of system and precedures in an organisation. Examples of such audit are ISO 9000, TS 16000, OHSAS 18000 etc.

(*ii*) **Statutory Audit:** This type of audit is necessitated due to the prevailing laws of the country. Example of such audit are annual boiler inspection.

(*iii*) **Technical Audit:** Such audits are carried out to inspect and assess a technical component of a plant. Maintenance audit is an example of technical audit.

(*iv*) **External and Internal Audit:** External audit is carried out by a team of experts called specially for the task, while internal audit is carried out by department personnel.

(*v*) **Preliminary and Detailed Audit:** Preliminary audit is conducted to identify thrust area. While detailed audit is conducted to completely focus on a given area.

(*vi*) **Third Party Audit:** It is an audit carried out by an independent agency.

(*vii*) **Second Party Audit:** It is an audit carried out by one organisation on another e.g. customer audit on supplier.

(*viii*) **First Party Audit:** It is an internal audit.

15.3 PURPOSE OF MAINTENANCE AUDIT

Following are the purpose of maintenance audit:

(*i*) To increase awareness in the organisation about the best maintenance practices.

(*ii*) To create a learning opportunity for the development of people.

(*iii*) To identify area of neglect and provide suitable focus to improve upon it.

(*iv*) To identify opportunities for improvement and encourage suitable actions.

(*v*) To improve effectiveness of man, material, process and technology.

(*vi*) To eliminate wastage and pillferage.

(*vii*) To benchmark with the best practices in the industry.

(*viii*) To get quality system certification.

15.4 AREAS OF MAINTENANCE AUDIT

Following are the several areas of maintenance audit:

(*i*) Maintenance strategies and targets.
(*ii*) Maintenance cost.
(*iii*) Maintenance organisation performance.
(*iv*) Outsourcing policy.
(*v*) Equipment performance.
(*vi*) Methodology of maintenance.
(*vii*) Maintenance information system.
(*viii*) Maintenance facilities.
(*ix*) Maintenance material management performance.
(*x*) Preventive and predictive maintenance system.

Here, we can elaborate one by one:

(*i*) **Maintenance strategies and targets:** Review of maintenance strategies of organisation vis-a-vis target selected is carried out. For the purpose is to check effectiveness of maintenance strategy w.r.t. to the organisation.

(*ii*) **Maintenance cost:** Maintenance cost related to spares consumption, manpower, inventory turnover, energy, overtime, energy to production ratio, total employee to production ratio etc. are evaluated.

(*iii*) **Maintenance organisation performance:** It is measured by ratio of maintenance manpower to asset ratio, planned maintenance work to unplanned maintenance work ratio, line to support staff ratio *etc.*

(*iv*) **Outsourcing policy:** Purpose is to check reasons, validity and necessity of outsourcing. Also to check loopholes or gaps in policies.

(*v*) **Equipment performance:** It is intended to evaluate equipment performance in terms of availability, reliability, efficiency, maintenance cost *etc.*

(*vi*) **Methodology of maintenance:** Purpose is to evaluate effectiveness of the maintenance methodology adopted by the organisation.

(*vii*) **Maintenance information system:** Audit measures and evaluates maintenance informetics system w.r.t. to sufficiency of data, presentation style, usefulness *etc.*

(*viii*) **Maintenance facilities:** The audit evaluates maintenance facilities such as testing, fabrication, rewinding, machining etc. in terms of quality of output, necessity, future need, adequacy *etc.*

(*ix*) **Maintenance material management performance:** Various indices/ ratios are calculated to judge performance. These performance indices could be inventory turnover, purchase value to issue value, cost of spares procured to total maintenance cost etc.

(*x*) **Preventive and predictive maintenance system:** Audit checks for optimum deployment of resources, achievement of target in terms of equipment availability. For predictive maintenance system, audit checks use of modern condition monitoring systems, interface between preventive and predictive maintenance, adequacy of predictive maintenance techniques deployed vis-a-vis plant need and resulting improvement in performance in terms of downtime and failures.

15.5 METHODOLOGY OF MAINTENANCE AUDIT

It consists of the following distinct steps:

(*i*) Initial briefing.

(*ii*) Preliminary Preparation.

(*iii*) Information collection through questionnaries , survey and interview.

(*iv*) Plant visit.

(*v*) Data analysis.

(*vi*) Conclusions and recommendations.

(*vii*) Report submission.

(*i*) **Initial Briefing:** The audit team holds meeting with representatives of the company regarding scope, coverage, goal, schedule and information requirement. Purpose is to understand the help required by the audit team and to develop a rapport.

(*ii*) **Preliminary Preparation:** Next, audit team chalks out plan to proceed with the audit. It prepares details of questionnair, survey plan, interview methodology and plant visit.

(*iii*) **Information Collection:** Information is collected through a detailed questionnaire, survey and interviews. Purpose is to elicit required information for the audit. The information so collected are compiled and target areas are chosen.

(*iv*) **Plant Visit:** Next step is to visit the plant and collect firsthand information from supervisions, technicians, engineers etc. All data related to operation practices, plant performance bottlenecks *etc.* are collected. So the purpose is to understand available organisation resources, manpower, budget, utilisation effectiveness, sources of wastages etc.

(*v*) **Data Analysis:** Data collected through qustionnaires, surveys, interviews and plant visits are summarised and analysed. Criteria/ indices are used to compare present and past performance and compared with other industries. The indices could be in terms of maintenance expenditure, man hours utilisation, energy consumption, equipment availability, safety practices *etc.*

(*vi*) **Conclusions and Recommendations:** After detailed data analysis, conclusions are drawn about efficiency and effectiveness of maintenance management system. Strength of the systems are highlighted and weaknesses are identified. Based on conclusions, recommendations are made to improve not only weaknesses but also bring the system at par with the benchmarked practices. All these outcomes of audit are discussed with the top management representatives of the company.

(*vii*) **Report Submission:** Last step is preparation and submission of detailed report indicating scope of audit, methodology adopted, data gathered, conclusion drawn and recommendations suggested.

The report is presented directly to the top management.

CASE STUDY OF MAINTENANCE AUDIT

A maintenance management audit was conducted in a rubber plant. Details of the audit are presented below:

1. **Organisation:** The company manufactures rubber parts for leading car manufacturers. Its turnover is 45 crore. The company purchases raw material (synthetic rubber or EPDM) and produces required grade of rubber using process oil and various filler chemicals. These are used in an extruder to produce strips of required shape and cross section. Thereafter, these strips are joined in fixed lengths in various hydraulic presses. It has employee strength of 350 and runs in continuous three shifts.
2. **Scope and coverage of audit:**
   - Maintenance organisation.
   - Maintenance policy and strategies.
   - Planning and scheduling of maintenance.
   - Energy conservation.

- Monitoring and control of maintenance performance parameters.
- Spares management.
- Shutdown maintenance.

3. **Methodology of audit:** Following methodology was adopted:
    - Meeting with heads of department to aprise them of the audit.
    - A questionnaire was used to collect required information.
    - Necessary data was collected such as down time reports, daily reports, maintenance information system, spares consumption details, equipment history cards etc.
    - Plant visits and interviews were conducted to collect more detail about mechanical, electrical and instrumentation maintenance.
    - Training details were collected from HRD
    - Details of statutory compliance and safety records were collected.
    - Recommendations were based on audit.
    - Findings were presented before senior management.
    - A report was sumitted to the managing director.
4. **Details of existing maintenance system:** The rubber plant has three main sections:
    - Mixing section, producing required grade of rubber as raw material.
    - Extrusion section, producing required cross-section and length of rubber beading through extruders.
    - Press section, joining rubber beading of required length through heat sealing.

    Critical process machinery are—Mixing Mill, Extruders and presses (hydraulic and pneumatic).

    The maintenance department did not have any maintenance policy or manual. Majority of time was spent on breakdown maintenance. There was no record of money spent on maintenance i.e. spares, consumable etc.

    The utility section had slow speed marine DG sets, chiller, air compressors, water softner and 2000 KVA transformer connected to electricity grid.

    Production schedule decided the maintenance of equipment. There was rarely any annual plant shutdown.

There is hardly any condition monitoring techniques used.

Due to rising breakdown, GM (Works) had decided to provide 10 days in a year for annual shutdown.

Maintenance and production heads were at loggerheads due to production loss and wastages.

Electrical and mechanical maintenance are looked after by separate maintenance cells with little co-ordination.

5. **Recommendations**: Following were suggestions:

(*i*) *Maintenance organisation:* Maintenance organisation need to reorganised with clear focus on preventive maintenance. All engineers looking after electrical, mechanical and instrumentation maintenance including shift foreman should report to a section engineer who inturn will report to maintenance head.

(*ii*) *Maintenance policy and strategy:* Maintenance department need to develop policy and strategies. It needs to set objectives such as plant availability = 90%, maintenance cost reduction /year = 5%.

Preventive maintenance schedule need to be implemented so as to bring down high rate of breakdowns. This requires recruitment of additional technicians.

(*iii*) *Planning and scheduling of maintenance:* It is recommended to introduce CMMS — Computerised maintenance management system. This will enable logging all maintenance activities. It is recommended that a fixed number of hours of preventive maintenance be allowed during model changes. Without doing preventive maintenance, rate of breakdown would not come down. Condition monitoring of critical equipments must be started and data logged.

(*iv*) *Energy conservation:* State of energy consumption is alarming and is increasing every year without increase in production.

One wasteful practice was noted in extrusion section where a heater used for joining polyster conveyor belt was kept on 24 hours a day in anticipation of belt breakage. Such wasteful practices can be avoided by using instant heaters of higher wattage which can be switched on when required.

During long periods of model change or line breakdown, process heaters were kept on leading to energy wastage Heaters used in presses need design change so as to reduce specific energy consumption.

Plant lighting and ventilation is highly inadequate and needs redesigning.

Instead of using chilled water for process cooling, cooling water from cooling tower can be suitably used. Power distribution system is highly inflexible leading to poor leading on gensets.

(v) *Monitoring and control of maintenance performance parameters:* A detailed monthly report need to be generated regarding sectionwise breakdown hours and analysis, monthly power and fuel consumption, spares consumption, total maintenance cost, preventive maintenance hours etc. The data be displayed as bar/pie charts etc. in the organisation notice board.

(vi) *Spares management*: Proper record of spares and consumables be kept and monitored on monthly basis. Inventory level need to be checked for dead/obsolete spares and their disposal.

(vii) *Shutdown maintenance:* As mentioned earlier, a general shutdown of 10 dyas in year has been planned henceforth. This is a welcome step. All critical equipments need to be checked thoroughly to avoid repeated problems

A register can be kept for shutdown jobs. Required jobs be entered on regular basis so that these can be taken up at annual shutdown.

Lastly, training of technicians and engineers was found to be poor. Although it was not in scope of audit but it is worth mentioning that engineers and technicians were not aware of the latest condition monitoring techniques and development in maintenance field. Hence, it is recommended that periodic updation of technical knowledge is required. Human resource department should organise this by making an annual training calender.

Summary

Purpose of maintenance audit is to critical assess and evaluate the working of maintenance function for the intended result and suggest suitable measures to make it more effective.

Maintenance audit can be internal or external audit. The various areas of maintenance audit are maintenance strategies and targets, maintenance cost, maintenance organisation performance, outsourcing policy, equipment performance etc.

Methodology of maintenance audit consists of—initial briefing, preliminary preparation, information collection through questionnaires, surveys and interviews, plant visit, data analysis and conclusion and recommendation.

Initial briefing has meeting of audit team with company representatives. They discuss scope and coverage of audit. Preliminary information about the organisation is provided to the audit team.

Preliminary preparation includes preparation of questionnaire, survey plan, interview methodology and plant visit.

Information collection is done through questionnaire, survey, interviews and plant visit.

Plant visit provides audit team with field data. Thereafter, data analysis is carried out. This forms the basis for conclusions and recommendations. Finally, all conclusions and recommendations are presented in the form of a report.

The recommendations of report are used to drive improvement in the plant.

For any audit to succeed, methodology selection by the audit team and whole hearted participation of plant personnel is a must.

Review Questions

1. What is maintenance audit? Why is it required?
2. Discuss various types of audit ?
3. Elaborate on various areas of maintenance audit.
4. Discuss methodology to be adopted for carrying out maintenance audit.
5. Discuss the need for experienced and certified maintenance audit professionals.
6. Why is it difficult ot audit maintenance cost area ?

7. What are the barriers in obtaining information about maintenance system?
8. Describe in detail the preparation required for collection of data during plant visit.
9. Describe in detail the contents of audit report.
10. How do you compare maintenance performance using various maintenance indices?

Discussion Questions

1. "Maintenance audit is not a fault finding exercise". Justify.
2. "Collection of information is the most vital part of maintenance audit". Elaborate.
3. "Methodology of maintenance audit is decided by the desired outcome". Discuss.
4. "Maintenance audit recommendations are to be based on techno-financial analysis". Comment.
5. "Human bias is most important barrier in successful implementation of audit recommendations". Discuss.
6. "Maintenance Management Information System provides insight into current health of existing plant". Discuss.
7. Elaborate on "It is difficult to develop maintenance indices benchmarks due to difficulty in data acquisition".
8. Comment – "Maintenance being expenditure function, audit becomes a fault finding exercise".
9. "Current condition of plant is the joint responsibility of maintenance and production". Justify.
10. "A good MIS acts as a check and balance on maintenance performance". Discuss

Multiple Choice Questions

1. Purpose of audit is to:

   (*a*) Find faults. (*b*) Fix responsibility.

   (*c*) To make improvement. (*d*) Increase budget allocation.

2. Various methods of information collection are:

   (*a*) Through questionnaire. (*b*) Interview.

   (*c*) Survey. (*d*) All of the above.

3. Maintenance audit is carried out by:
   (*a*) Chartered accountants. (*b*) Maintenance professionals.
   (*c*) Cost accountants. (*d*) None of these.
4. Maintenance audit areas are:
   (*a*) Equipment performance. (*b*) Manpower training.
   (*c*) Maintenance facilities. (*d*) All of these.
5. Scope of maintenance audit does not include:
   (*a*) Maintenance cost. (*b*) Outsourcing policy.
   (*c*) Equipment performance. (*d*) Maintenance manpower salary.

True/False Questions

1. Maintenance audit provides scope for bench marking.
2. Maintenance Information System acts as a base for maintenance audit.
3. Plant visit does not serve any purpose in audit.
4. Information gathered from foreman is more useful than senior manager.
5. Most difficult part of audit is the implementation of audit recommendations.

ANSWERS

Multiple Choice Questions

1. (*c*) **2.** (*d*) **3.** (*b*) **4.** (*d*) **5.** (*d*)

True/False Questions

1. True **2.** True **3.** False **4.** True **5.** True

❑❑❑

16

Financials of Maintenance Management

Learning Objectives

After reading this chapter, you will be able to:

- Understand classification of cost and costing process.
- Understand maintenance budget and various methods of cost control.
- Examine various performance indices.
- Explore how to develope maintenance information system.

16.1 INTRODUCTION

The process of costing consists of analysing, recording, classifying and appropriate allocation of expenditure for determination of cost of product or services. The cost so calculated are compared against standard cost and variance is calculated. The purpose of process of costing is to provide information to the management for planning, control and decision making.

Process of costing consists of following steps:

(*i*) Collection of maintenance cost.

(*ii*) Classification of cost.

(*iii*) Apportionment of cost to concerned area/department.

Whenever any material is withdrawn from stores, a cost centre or cost code is mentioned in the material issue slip. The cost centre is the cost code alloted by the finance department to a particular department. Any expenditure incurred on a particular department are booked under that cost centre. This helps in calculating the total expenditure incurred on a particular department at the end of the month.

Hence all maintenance expenses both labour and material are booked on particular department where the maintenance services are utilised.

16.2 NEED FOR COSTING

Need for costing are following:

(*i*) It provides comparative cost data for different types of repair and for different periods incurred on different departments.

(*ii*) It makes it possible to compare with standard cost.

(*iii*) It provides a measure of maintenance workload.

(*iv*) It helps in improving maintenance labour productivity.

(*v*) It helps to reduce wasteful expenditure.

(*vi*) It helps in material usage control.

(*vii*) It helps in setting bench-mark in terms of maintenance cost and energy cost.

(*viii*) It helps to decide in house/outsource decisions.

(*ix*) It helps to decide spare capacity utilisation.

(*x*) It helps to decide automation.

16.3 CLASSIFICATION OF COSTS

Costs can be classified under the following heads:

(*i*) **Classification based on nature of cost:** For example, material cost, labour cost, overheads etc.

(*ii*) **Classification based on function:** This could be cost incurred on production, on administration, on selling and distribution etc.

(*iii*) **Classification based on variability:** This could be fixed cost, variable cost and semi-variable cost.

(*iv*) **Classification based on type of cost:** These are direct cost and indirect cost.

(*v*) **Classifications based on recurring and non-recurring nature of cost:** These are revenue and capital cost.

(*vi*) **Classifications based on normality:** For example, normal cost and abnormal cost.

(*vii*) **Classification based on managerial decisions:** For example, opportunity cost, replacement cost, conversion cost, sunk cost, marginal cost, differential cost etc.

16.4 DEFINITIONS

(*i*) **Material cost:** It includes cost of all materials such as spare parts, nut-bolts, lubricating oil, cotton waste etc.

(*ii*) **Labour cost:** It includes wages and salaries of maintenance workmen and staff, annual benefits such as provident fund, ESI, gratuity etc. overtime payment, incentive/bonus, cost of contract labour etc.

(*iii*) **Overheads:** It includes proportionate cost of administration, stores, rent of equipments, conveyance for outdoor jobs, cost of statutory inspection and compliance (as per Air Act, Water Act, Boiler Act, Factories Act etc), Cost of training etc.

(*iv*) **Revenue cost:** These are recurring operational costs such as cost of consumables, salary and wages, water and electricity bill for maintenance department, telephone bill etc.

(*v*) **Capital cost:** These are one time cost such as cost of buying new equipment (boiler, generator, turbine etc.) or cost of life extension of major equipments e.g. major overhauling or retrofitting.

(*vi*) **Indirect cost:** It includes cost of energy loss due to start-up and shutdown, wages paid to idle workers during material shortage or breakdown, energy loss during idle period, loss due to scrap and rework etc.

(*vii*) **Fixed cost:** Costs which do not vary with volume of production e.g. salary and wages, depreciation etc.

(*viii*) **Variable cost:** Cost, which vary in direct proportion to the production e.g. raw material, direct labour etc.

(*ix*) **Semi variable cost:** Cost which partly vary with volume of production and partly remain constant.

16.5 COSTING SHEET

It is a consolidated statement which contains maintenance cost under different sub-heads. An example of cost sheet is given below

Table 16.1 Format of Costing Sheet

Area	Revenue			Capital		
	Material	Labour	Overhead	Material	Labour	Overhead
(A) Production (i) Plant and Machinery (ii) Mould and tools (iii) Material handling equipment (iv) Auxiliary systems–piping, cabling, lighting (v) Building						
(B) Services (i) Plant and Machinery e.g. generators, compressers, Pumps, boilers, etc. (ii) Piping, cabling, Ducting, insulation, support systems. (iii) Tank farm (iv) Effluent Treatment plant (v) Building						
(C) Miscellaneous (i) Roads (ii) Toilets (iii) Canteen (iv) Fire Fighting System						

16.6 MAINTENANCE BUDGET

A budget is an estimated future cost of operation. It contains allocation of financial resources under various sub-heads.

All activities require financial resources and budget provides detailed allocation to various such anticipated future needs.Maintenance budget is of two types (i) Revenue budget and (ii) Capital budget. Revenue budget takes into account recurring expenditure such as cost of spares, consumables, salary and wages, training cost etc. Capital budget takes into account new asset addition such as cost of purchasing a new boiler or turbine or a new building for a powerhouse etc.

Process of preparing revenue budget starts with previous year's budget as reference. After considering inflation and anticipating future need, new budget is prepared. In case, past data is not available or is not reliable then what is prepared is called "zero based budget".

Capital budget is prepared for the following:

(*i*) Purchase of new equipment for future expansion.
(*ii*) Purchase of new equipment for replacement of existing equipment due to old age, obsolescencce, poor efficiency *etc*.
(*iii*) Expenditure on accessories: Cabling, piping, supports *etc*.
(*iv*) Civil works: Foundation, building, road *etc*.
(*v*) Cost of installation, testing, commissioning of new equipment *etc*.

Budgeting exercise is carried out before the beginning of each financial year. After preparation of budget, it is approved by the finance department.

Implementation and adherence to budget is monitored every quarter and deviations, if any is set right.

16.7 COST CONTROL

Cost control is the process of achieving targets, comparing actual cost with target cost and taking remedial measures. Two important techniques of cost control are:

(*i*) Budgetory control.
(*ii*) Standard costing.

(*i*) **Budgetory Control:** In this method, actual cost is compared with budgeted cost and variances, if any are rectified. Hence, budgeted value is taken as reference point and actual cost is comapared with this. This is preferred form of cost control

(*ii*) **Standard Costing:** In this technique, standard cost is calculated for various activities/processes and actual cost is compared with it. Remedial action is initiated for any variance.

16.8 ADVANTAGES OF COST CONTROL

Following are the advantages of cost control—

(*i*) It helps to achieve expected return on investment.
(*ii*) It improves productivity.
(*iii*) It ensures reasonable price to customers.
(*iv*) It ensures economic utilisation of scarce resources.

16.9 STEPS INVOLVED IN COST CONTROL

Following are the steps involved in cost control:

(*i*) Formulation of plan in the form of budgets, standard cost or estimates.

(*ii*) Communications of plans and policies down the line.

(*iii*) Computation of variances and analysis.

(*iv*) Corrective action for variances.

16.10 CONTROL OVER WASTE, SCRAP AND SPOILAGE

Following are the steps involved in cost control—

(*i*) **Waste:** It is a discarded substance having no value such as smoke, dust, gases, slag etc. It arises in course of manufacturing process.

Waste is due to the following reasons:

(*a*) Evaporation of oil, paints, solvents *etc.*

(*b*) Absorption of moisture *e.g.* in lime.

(*c*) Natural deterioration *e.g.* rusting, deterioration of rubber items etc.

(*d*) Losses due to breakage *e.g.* glass sheets.

Waste represents a significants loss and can be minimised by adequate care and storage.

(*ii*) **Scrap:** It represents discarded material having some value. Metal scrap, metal chippings, sheet metal pieces are examples of scrap having some value.

Scrap is due to following reasons:

(*a*) Legitimate scrap due to manufacturing and project process.

(*b*) Administrative scrap due to design obsolescence.

(*c*) Defective scrap due to sub-standard raw material and poor workmanship.

In order to control scrap, one must control quality of raw material and workmanship. Periodic design review can minimise design obsolescence.

(*iii*) **Spoilage:** It represents material damaged in manufacturing process, Spoiled work cannot be repaired or reconditioned, spoilage involves loss of material, labour and manufacturing overheads.

Spoilage is of two types:

(*a*) **Normal spoilage:** Spoilage which are within the limits. The limits being decided through study of men, material and process.

(*b*) **Abnormal spoilage:** It is due to abnormal reasons such as power failure, strike, flood, accident, material shortage etc.

16.11. MANAGEMENT INFORMATION SYSTEM (MIS)

It is a structured system for providing information to management on regular basis for the purpose of planning and control.

Following are the steps in the preparation of MIS—

(*i*) Collection of data.

(*ii*) Processing of data.

(*iii*) Storage and retrieval of data.

(*iv*) Evaluation of data.

(*v*) Dissemination.

Format of MIS for maintenance department varies from organisation to organisation. Generally, a monthly MIS is prepared by the maintenance department and submitted to maintenance head/operations head. On the basis of these monthly MIS, annual MIS is prepared. A typical format of monthly maintenance MIS is given below:

Table. 16.2 Format for MIS

ABC Co. Ltd

Maintenance -MIS

Month________ Unit No. ________

Year _________

1. Power and Fuel cost = Rs.............
   - (i) Grid Power = Rs..........(...Units)
   - (ii) Captive Power = Rs ....(...Units)

     Total = Rs...........
   - (iii) Cost of oil/coal/gas = Rs ....

     (... KL/Tons/Sm^3)
   - (iv) Steam consumption = ... Tons
   - (v) Overall plant power factor = ...

Contd...

2. Spares Cost
   (i) Imported = Rs ....
   (ii) Indigenous = Rs ....
   Total = Rs ....
   (iii) Cost of Consummables = Rs ....
3. Contracts Job work cost = Rs ....
4. Plant-wise downtime hours
   Plant A = ... hrs
   Plant B = ...hrs
   Plant C = ...hrs
5. Worker overtime hours = ...hrs
   Cost = Rs ....
6. Total preventive maintenance hours = ...
7. Maintenance manpower training cost = Rs ....
8. New asset addition—
   (i) Boiler = Rs ....
   (ii) Transformer = Rs............
   (iii) DG Set = Rs ....

16.12 MAINTENANCE PERFORMANCE INDICES

For effective monitoring and analysis of maintenance function, certain performance indices are devised which provide an indication about the health of maintenance function in an organisation.

Major parameters to be considered are:

(*i*) Downtime.

(*ii*) Various maintenance related expenses.

(*iii*) Various functional ratios.

(*i*) **Downtime**: Downtime refers to the period during which the equipment was out of order. This indicates production loss, energy wastage, man hour wastage and has ripple effect on downstream production lines.

This parameter is either expressed in hours or in percentage of total operation hours.

$$\% \text{ age downtime} = \frac{\text{Total Breakdown hours}}{\text{Total Operation hours}} \times 100$$

(*ii*) **Maintenance Expenses:** Various maintenance expense indices are following :

(*a*) $\% \text{ age maintenance cost} = \frac{\text{Total Maintenance Cost}}{\text{Total Sales}} \times 100$

(*b*) % age manpower cost

$$= \frac{\text{Cost of Maintenance Personnel (salary + benefits)}}{\text{Cost of total workforce}} \times 100$$

(*c*) % age Spares cost $= \dfrac{\text{Cost of spares procured}}{\text{Total Maintenance cost}} \times 100$

(*d*) % age Consumable cost $= \dfrac{\text{Cost of Consummables}}{\text{Total maintenance cost}} \times 100$

(*e*) % age Energy cost

$$= \frac{\text{Cost of Energy used}}{\text{Total Plant Operational expenditure}} \times 100$$

(*f*) % age Overtime cost

$$= \frac{\text{Total Maintenance overtime cost}}{\text{Total Salary of Maintenance Personnel}} \times 100$$

(*iii*) **Functional Ratios:** These ratios act as a parameter for comparison amongst similar industries (Bench Marking). For example, two car manufacturers employing almost similar technology and having same capacity can compare number of persons employed/cars produced as a benchmark.

Various functional ratios are:

(*a*) Maintenance to Production ratio

$$= \frac{\text{No. of Maintenance Persons}}{\text{No. of Production Persons}}$$

(*b*) Total employees to Production ratio

$$= \frac{\text{Total Production (in numbers)}}{\text{Total Employees}}$$

(*c*) Energy to Production ratio $= \dfrac{\text{Total energy consumed (KWH)}}{\text{Total production (in numbers)}}$

(*d*) Steam Consumption to Production ratio

$$= \frac{\text{Total steam consumed (in tons)}}{\text{Total Production (in numbers)}}$$

(*e*) Water Consumption to Production ratio

$$= \frac{\text{Total water consumed (in litres)}}{\text{Total production (in numbers)}}$$

$$\% \text{ Availability} = \frac{\text{Running hrs}}{\text{Ref. hrs}} \times 100$$

$$\% \text{ Plant Load Factor} = \frac{\text{Actual } O/P}{\text{Possible } O/P} \times 100$$

$$\text{Operating Index} = \frac{\text{Production rate}}{\text{Std target rate}} \times 100$$

$$\text{Performance Index} = \text{Availability} \times \text{operating index}$$

Now-a-days, a new parameter is used called "overall equipment efficiency" (OEE). Purpose of OEE is to incorporate primary causes of performance loss.

The Primary Causes are:

(*a*) Loss of equipment availability (due to breakdown time, set up time, non-availability of manpower, non-availability of material etc.)

(*b*) Loss due to rework.

(*c*) Loss due to operational inefficiency, startup/shutdown etc.

(*d*) Loss due to minor delays, slow running, procedures etc.

OEE combines people and machinery performance. It provides a common basis for objective setting, reward etc.

$$\text{O.E.E.} = \% \text{ uptime} \times \% \text{ efficiency} \times \% \text{ quality}$$

Where % uptime

$$= \frac{\text{Planned Operating time} - (\text{Downtime} + \text{setup time})}{\text{Planned Operating time}} \times 100$$

$$\% \text{ Efficiency} = \frac{\text{Amount Produced}}{\text{Runtime} \times \text{Standard Production rate}} \times 100$$

$$\% \text{ Quality} = \frac{\text{Amount Produced} - \text{Amount of defect/rework}}{\text{Amount Produced}} \times 100$$

Summary

Financials of maintenance is collection of maintenance cost and control. Maintenance department like othe functional area has various cost centres such as spares cost, consumable cost, energy cost, labour cost, cost of preventive maintenance, cost of capital equipment etc. Purpose of costing is to appropriate allocation of expenditure so as to monitor and control cost. This information is used by the management for planning, control and decision making.

A maintenance budget contains allocation of financial resources under various sub-heads. Maintenance budget falls under two distinct groups, viz. revenue budget and capital budget. Revenue budget deals with recurring expenditure such as consummables, regular spares, salary and wages, training cost etc. Capital budget deals with purchase of new capital equipment such as boiler, turbine, lathe machine, new building etc. The budget preparation is an annual exercise and is based on previous year's budget. However, if the budget is made without reference to previous year's budget then it is known as zero based budget. Preparation of revenue budget is a routine exercise but capital budget is prepared based on new asset addition planning.

Cost control is exercised through budgetory control and standard costing. In budgetory control, actual cost is compared with the budgeted cost and variances if any are rectified. Standard costing calculates cost of various activities and compares with actual cost.

Management Information System (MIS) forms the basis for providing information on maintenance expenditure to top management for future planning and control.

To gauge the performance level and also to bench mark, various maintenance indices are calculated such as percentage downtime, percentage maintenance cost, energy to production ratio, O.E.E etc. These are calculated on monthly or yearly basis and trending is done.

Review Questions

1. What are the various sources of collection of maintenance cost ?
2. Explain the need for maintenance costing.
3. Differentiate between fixed and variable cost and direct and indirect cost.
4. What are overheads? Explain with an example.
5. What is a maintenance budget? What are various components?
6. Explain terms – (*i*) Budgetory control (*ii*) Standard costing.
7. What is meant by cost control? Enumerate its advantages.
8. What is MIS ?Why is it required? Explain with an example.
9. What do you mean by performance indices ?
10. What is O.E.E ?

Discussion Questions

1. "Maintenance function is an expense account". Comment.
2. "Costing of maintenance activity is extremely difficult". Do you agree?
3. Discuss format of a Maintenance Information System.
4. Discuss "Cost Centre" concept as applied to maintenance.
5. Explain capital and Revenue budget and their scope.

Multiple Choice Questions

1. Purpose of costing is
   (*a*) To improve labour productivity (*b*) Reduce wastage
   (*c*) Benchmark maintenance cost (*d*) All of these
2. Salary of workmen is
   (*a*) Fixed cost (*b*) Variable cost
   (*c*) Semi variable cost (*d*) None of these
3. Energy loss due to start up is
   (*a*) Indirect cost (*b*) Direct cost
   (*c*) Fixed cost (*d*) Variable cost
4. Spares and consumables are included in
   (*a*) Capital budget (*b*) Revenue budget
   (*c*) Annual budget (*d*) None of these
5. Maintenance indices are used to
   (*a*) Benchmark performance (*b*) Reduce wastage
   (*c*) Improve efficiency (*d*) All of these

True/False Questions

1. All maintenance costs are variable costs.
2. It is difficult to estimate overheads accurately.
3. Process of preparation of capital budget and revenue budget is same.
4. Cost control is a technique to reduce maintenance work.
5. Abnormal spoilage can result from strikes and floods.

ANSWERS

Multiple Choice Questions

1. (*d*) **2.** (*a*) **3.** (*a*) **4.** (*b*) **5.** (*d*)

True/False Questions

1. False **2.** True **3.** False **4.** False **5.** True

❑❑❑

17

Advances in Maintenance Management

Learning Objectives

After reading this chapter, you will be able to:

- Understand goals, characteristics and process of implementation of TPM.
- Understand benefits and goals of RCM.
- Examine benchmarking benefits, types and process.
- Understand benefits and process of value analysis.
- Comprehend the purpose and process of reverse engineering.
- Explore AFD (Anticipatory failure elimination process).
- Understand purpose and reasons for outsourcing.

17.1 INTRODUCTION

In this chapter, recent topics such as TPM, RCM, bench marking, value analysis, reverse engineering, anticipatory failure determination and outsouring of maintenance are discussed. We will begin with first topic *i.e.* TPM.

17.2 TPM

The concept of TPM originated from USA and was adopted by Japanese. In 1961, Nippon Denso adopted TPM. Earlier, all equipment maintenance was looked after by maintenance workers. However, due to adoption of preventive maintenance, the workload got enhanced. So management decided to give minor maintenance jobs such as cleaning, lubrication, bolt tightening etc. to operators. This allowed maintenance workers to concentrate on major maintenance jobs e.g. preventive maintenance, maintenability improvement etc. For this development, Japan Institute of Plant Engineers (JIPE) awarded distinguished plant prize (PM prize) to Nippon Denso in 1971.

TPM can be defined as a productive maintenance carried out by all employees with a view to maximize plant and personnel effectiveness and achieving optimum life cycle of equipment.

17.2.1 Goals of TPM

Following are the goals of TPM:

(*i*) To improve equipment effectiveness by reducing breakdown.

(*ii*) To achieve autonomous maintenance by production and maintenance staff sharing maintenance responsibility.

(*iii*) To ensure clean, accident free and pollution free environment.

(*iv*) To prolong life span of equipment.

(*v*) To involve all employees in achieving zero breakdown, zero accident and zero defect.

17.2.2 Characteristics of TPM

Takahashi, Japan has identified three specific characteristics of TPM:

(*i*) Improvement of overall performance through life cycle approach.

(*ii*) Improvement of productivity of work force through participation of all.

(*iii*) Improvement of equipment availability through small group activities by identification and removal of causes of failure of equipments.

TPM addresses losses due to the following causes:

(*i*) Setup and adjustment losses.

(*ii*) No/Poor maintenance.

(*iii*) Idling and minor stoppage of equipment.

(*iv*) Capacity losses due to reduced speed.

(*v*) Losses due to defective product.

17.2.3 Steps in TPM Implementation

Following are the steps:

(*i*) Create awareness among employees about TPM.

(*ii*) Identify operators for routine maintenance.

(*iii*) On-the-job training of operates to carry out routine maintenance.

(*iv*) Provide display boards for visual control on machine problems

(*v*) Develop system for breakdown analysis and integration with routine maintenance.

(*vi*) identify spares (short and long lead time) through vendor development. Progressively converting long lead time spares to short lead time spares.

(*vii*) Maintain harmony between production and maintenance through day-to-day interaction.

17.3 RELIABILITY CENTERED MAINTENANCE (RCM)

The concept of RCM has its origin in airline industry, wayback in 1960s. The US department of defence used RCM concept in 1978. Nowlan and heap (1978) wrote a comprehensive document on relationship among maintenance, reliability and safety which created RCM methodology. Thereafter, it spread through industries in 1980s and 1990s.

RCM is a systematic approach to maintaining the internal reliability of any system. The guiding principle of RCM is that no preventive maintenance tasks should be performed unless it is for safety, availability, reliability and cost effectiveness in preventing a failure. RCM approach is to optimize maintenance to achieve highest reliability at the lowest cost.

17.3.1 Benefits of RCM

Benefits of RCM are following:

(*i*) Longer useful life of physical assets.

(*ii*) Improved safety of personnel and equipment.

(*iii*) Improved operating performance.

(*iv*) Cost effective maintenance plan.

(*v*) Improved equipment reliability.

17.3.2 Goals of RCM

RCM system has three major goals viz:

(*i*) Enhanced safety and reliability of system by focussing on the most important functions. Purpose is that the equipment performs intended task.

(*ii*) Prevent the consequences of failure. Purpose is not to prevent failures themselves but the consequence. Failures depend on the location and method of operation and installation.

(*iii*) Reduction of maintenance cost by avoiding unnecessary maintenance actions.

Steps in RCM Process:

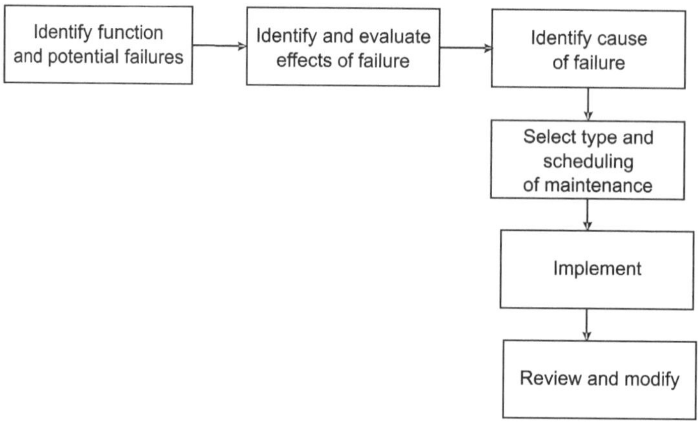

Fig. 17.1 RCM Process.

RCM process can be explained with the help of an example. Suppose a pump fails to operate. Next step is to analyse the cause of failure. After identification of cause of failure, rectify the error and include in the maintenance schedule.

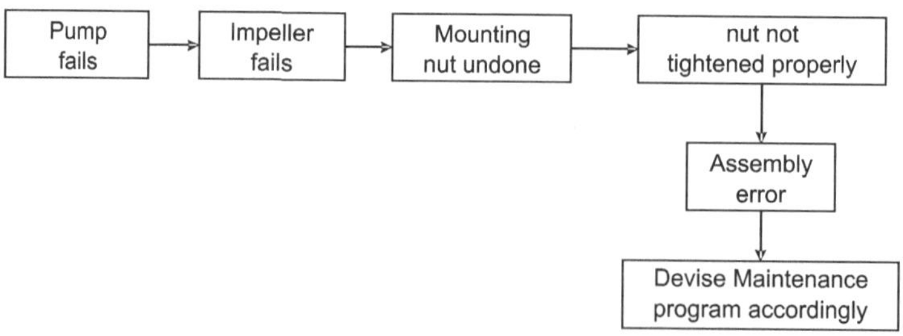

Fig. 17.2 RCM Process Applied to Pump Failure.

17.4 BENCHMARKING

Benchmarking is the process of comparing business metrics such as productivity, sales, quality, operations cost, power consumption etc. with the best practices. This is the process of comparing operational and /or financial parameters with the best available. That may be a competitor in a region, country or world. The purpose is to upgrade the existing performance parameter to a higher level by taking best practices as a reference standard. The process does not stop after achieving the intended performance as it is a continuous process leading to improvement.

This is something very much similar to a boy who compares his grades with that of first boy in the class and sets a time bound target either to achieve or to surpass first boy's grade. Similarly in business, organisations compare their performance with other organisations and identify key elements that contribute to higher level of performance.

M/s Xerox corporation defines benchmarking as — "The search for industry best practices which lead to superior performance". The metrics could be revenue per employee, asset utilization, customer satisfaction, return on assets, maintenance cost, preventive maintenance to total maintenance ratio etc.

17.4.1 Need for Benchmarking

The need for benchmarking are following:

(*i*) To identify areas of improvement.

(*ii*) To identify risks.

(*iii*) To improve maintenance efficiency.

(*iv*) To introduce change in maintenance practices.

(*v*) To make continuous improvement.

(*i*) **To identify area of improvement:** Area of improvement could be preventive maintenance, inventory and procurement, training, workflow control, usage of computers etc. After identification of area, specific metrics are compared with best practices and time bound targets are set for their achievement.

(*ii*) **To identify risk:** Benchmarking process helps to identify certain risks such as competition, legal environment, technology, environmental pollution *etc*.

(*iii*) **To improve maintenance efficiency:** Benchmarking helps to improve maintenance such as return-on-investment, customers satisfaction, % percentage preventive maintenance etc.

(*iv*) **To introduce change in maintenance practices:** Benchmarking allows employees to understand their performance vis-a-vis competitor's thereby compelling them to achieve superior performance. It changes the employee mind set from being complacent to competitive.

(*v*) **To make continuous improvement:** Benchmarking is not a one time process and does not stop after achieving milestone. It is a continuous journey of improvement.

17.4.2 Benefits of Benchmarking

Benefits of Benchmarking are following:

(*i*) **Faster pace of improvement:** Benchmarking helps to improve the current performance to a much higher level. Instead of improvement in smaller steps say 5% better improvement/year, the competitor may be well ahead of this figure. Hence, setting a bigger target helps to improve efficiency at a much higher speed than achievable without benchmarking.

(*ii*) **Mindset change:** Benchmarking allows employees to first look towards the competitor and then look within to find and analyse the gap. This feeling of "lagging behind" in numerical terms helps to initiate a change in the organisation culture. From being a "status quo" situation to a "pro-active" situation. This change helps to discard organisational lethargy by infusing enthusiasm to achieve a higher level of performance and fame.

(*iii*) **Change agent:** Benchmarking acts a change agent for human resource. The employees do gap analysis and prepare themselves to bridge the gap. Training could be one of the "enabler" that could reduce the gap between existing and desired level of performance.

(*iv*) **Source of higher return and better brand image**: A higher target improves business performance and provides better return on-investment. It could be reduction in waste, reduction in customer rejection, improvement in energy consumption, increase in recycle of waste, improvement in customer feedback etc. This ultimately improves profit margin and builds a better brand image.

17.4.3 Types of Benchmarking

Following are the types of Benchmarking undertaken by the organisations:

(*i*) Internal benchmarking.

(*ii*) Competitive benchmarking.

(*iii*) Non-competitive benchmarking.

(*iv*) Best practice benchmarking.

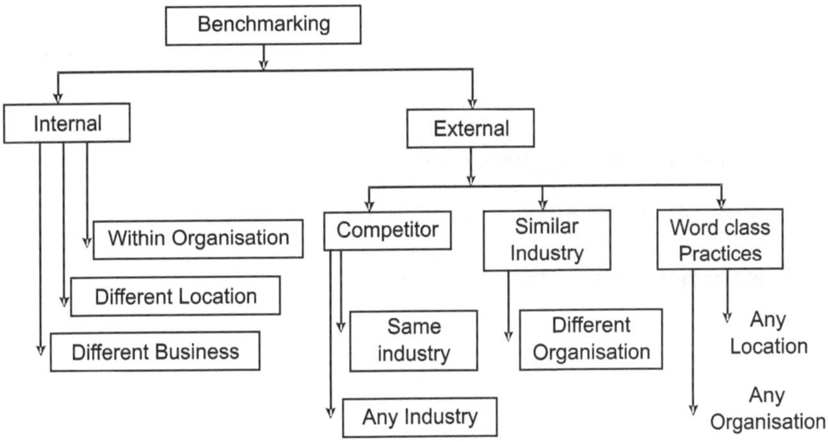

Fig. 17.3 Types of Benchmarking

(*i*) **Internal benchmarking:** It involves comparing business metrics within an organisation such as one production plant with another or one branch office with another. It helps to create a two-way communication channel by sharing information and data. It helps to build trust and create transparency.

(*ii*) **Competitive benchmarking:** It involves comparing business data with direct competitors. Sometimes getting business data from competitor is difficult.

(*iii*) **Non-competitive benchmarking:** It involves comparing business process with a non-competitive organisation or different type of industry. For example, comparing percentage preventive maintenance between a car manufacturer and a bike manufacturer or customer feedback between telecom operator and mobile hand set manufacturer.

(*iv*) **Best practice benchmarking**: This approach involves comparison with world class organisation.

17.4.4 Failures in Benchmarking

Benchmarking process is used to create an atmosphere of continuous improvement. However, not all such initiatives result in improvements.

The reasons are following:

(*i*) Lack of committment from top management.

(*ii*) Lack of focus.

(*iii*) Lack of perseverance.

(*iv*) Non-involvement of right staff.

(*v*) De-motivated staff.

(*vi*) Conflicting benchmarking and organisation objectives.

(*vii*) Loss of interest after initial euphoria.

17.4.5 Benchmarking Process

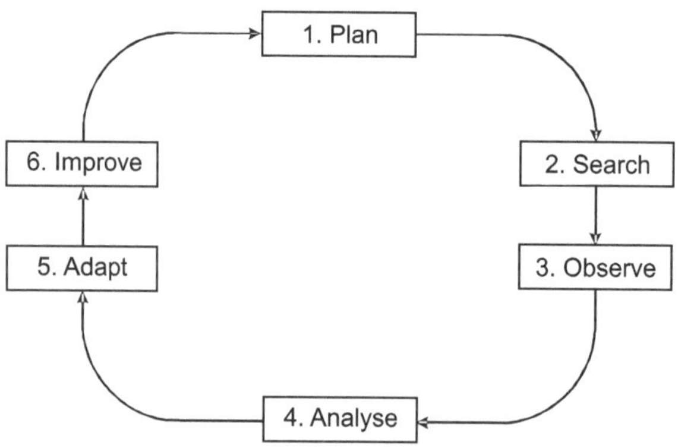

Fig. 17.4 Benchmarking Process

(*i*) **Plan**: First step in Benchmarking is to plan for "What to Benchmark"?

(*ii*) **Search:** Collect data from organisations on agreed business or process metrics.

(*iii*) **Observe:** Observe process details of target organisations i.e. how do they do it?

(*iv*) **Analyse:** Analyse data collected and compare with current organisational data. Identify the gaps.

(*v*) **Adapt**: Plan and implement actions according to the existing systems so that there are no disruptions in day-to-day activities.

(*vi*) **Improve:** Last step is to monitor progress and improve upon it.

17.5 VALUE ANALYSIS

Value analysis is a method of enhancing value of a product or service by improving performance, reliability, quality, safety and life cycle cost. It is an excellent tool for cost reduction by eliminating unnecessary parts or steps that do not contribute to the value of a product or service.

Value analysis began at General Electric Co., USA during World War-II due to shortage of skilled manpower and raw material. M/s Lawrence Miles, Jerry Lefttow and Harry Erlicher looked for acceptable substitutes to tide over with the crisis. These substitution not only reduced cost but also improved performance or functionality of product. Later on, this concept gained popularity and is being widely used today.

17.5.1 Benefits of Value Analysis

Following are the benefits of value analysis—

(*i*) Cost saving due to elimination/simplification of a part/item.

(*ii*) Product design improvement.

(*iii*) Risk reduction

(*iv*) Improvement in product performance.

(*v*) Import substitution

(*vi*) Lesser product manfacturing time (reduced cycle time of Product)

(*vii*) Lower Product cost.

17.5.2 Process of Value Analysis

It consists of the following steps:

(*i*) Identification of product for value analysis.

(*ii*) Defining functions.

(*iii*) Identifying alternatives.

(*iv*) Evaluating alternatives.

(*v*) Adopting best alternative

(*vi*) Implementing alternative.

17.5.3 Example of Value Analysis

Consider a bottle of home cleaning agent such as phenyle with a glass bottle having cylindrical body and top end opening with threaded cap. It is cumbersum to clean inaccessible parts in a toilet/bathroom. But if polymer bottle has a gripping facility with a slant opening then it is lot more easier to pour cleaning agent on an in accessible surface. So, the new design has added value to the product.

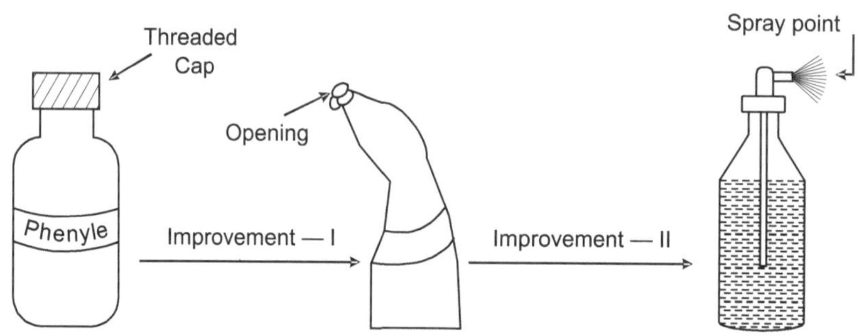

Fig. 17.5 Domestic Cleaning Agent (Glass Body)

Fig. 17.6 Tapered Mouth (Opaque Polymer Body)

Fig. 17.7 Bottle with Spray point (Transparent Polymer Body)

The functionality is improved further if a transparent bottle with spray point is used. The transparent polymer body is unbreakable and shows level of the liquid cleaning agent. The spray allows to clean any inaccessible surface. So, this is improvement in functionality due to value analysis.

17.6 REVERSE ENGINEERING

Reverse Engineering is the process of learning technical and operational principle of a product. It follows reverse process of manufacturing. It involves taking apart a product and analyse working and fitment of various parts in order to manufacture it.

17.6.1 Purpose of Reverse Engineering

Following are the purpose of reverse engineering—

(*i*) Product analysis.

(*ii*) Import substitution.

(*iii*) Commercial espionage.

(*iv*) Product improvement.

(*v*) Lost product manufacturing drawings.

17.6.2 Process of Reverse Engineering

Reverse engineering begins with study of shape, structure, functions of item. Thereafter, the item under study is dismantled part by part. Drawings of these dismantled parts are made along with material of construction. Then functionality and fitment of each part is studied. Individual parts are fabricated and complete assembly is done. Lastly, trial and correction is done.

17.6.3 Examples of Reverse Engineering

During World War-II, gasolene fuel cans used by Germans were of the very good design. USA and UK millitary adopted this design by reverse engineering.

Other example is of B - 29 bomber of U.S. Airforce which was forced to land in erstwhile USSR. The Russians then reverse engineered the bomber design and produced what was known as "Tupelev Tu-4" aircraft.

17.7 ANTICIPATORY FAILURE DETERMINATION

Anticipatory failure Determination (AFD) originated in erstwhile U.S.S.R. The process was used for identification of potential failures of machines and to improve reliability.

There are various methods/techniques used for reliability improvement such as Root cause Analysis (RCA), Task Analysis (TA), Change Analysis (CA), Barrier Analysis (BA), Fault Tree Analysis (FTA), Human Performance Enhancement System (HPES), Accident Evaluation and Barrier Function Analysis (AEB) etc. Approach to Anticipatory Failure Determination is different and novel.

Traditional methods of failure prevention ask question "What can go wrong"? While in AFD process, question asked is "How to make the system fail"? This is similar to making an atomic/millitary installation safe from terrorist attack. So the process of AFD explores the unexplored areas not considered till now by traditional methods.

Let us consider an example to illustrate application of AFD thought process. Consider a system having motor driven blower that supplies air to furnace. Traditional process starts with a question — "What can stop furnace from operating"? The probable reasons are listed below:

(*i*) Fuel supply to furnace fails. (*ii*) Combustion air to furnace fails. (*iii*) Exhaust gas exit damper is blocked. (*iv*) Fuel fails to ignite. (*v*) Power supply to blower motor fails. (*vi*) Motor burns out. etc.	Potential failure modes. (by traditional methods)

Traditional fault proofing methods will look at reasons mentioned above. However, there are other causes that may stop a furnace e.g.:

(*i*) Total opening of fuel bypass value, intentionally. (*ii*) Forcing a nut/bolt to go into blower suction. (*iii*) Closing exhaust gas damper intentionally. (*iv*) Switchng off fuel pump and blower from site. (*v*) Change of combustion parameters.	Failure modes (by AFD method)

Here, we have listed reasons which have not been considered earlier. These causes are by sub-versive intention i.e. with an intention to cause failure. Hence AFD method takes into account all modes of failure as if the sole intention is to make system fail. So all such possible failure modes are anticipated and measures are built into the system to make it completely sabotage proof.

17.7.1 AFD Process

Following are the steps in Anticipatory Failure Determination (AFD) process

(*i*) Select a part of the system.

(*ii*) Analyse previous failures to understand how to create failures.

(*iii*) Identify potential failure scenario.

(*iv*) Analyse causes of failures.

(*v*) Find ways to cause failures.

(*vi*) Find remedial measures to avoid failures in Sr no (iv) and (v).

AFD process is extremely useful where failure of a system may cause extremely harmful effects. Examples are –

(*i*) Possibility of leakage of radio-active material into environment

(*ii*) Possibility of causing explosion in a fuel storage yard.

(*iii*) Possibility of breach of security system in a atomic/millitary installation.

(*iv*) Possibility of cutting off oxygen/power supply to an operation theatre.

(*v*) Possibility of causing escape of toxic gases/chemicals into environment. (Bhopal Gas Tragedy). *etc*.

17.7.2 Possible Ways to Prevent Failures

(*i*) Making item/system/place inaccessible.

(*ii*) Making handling such situations risky.

(*iii*) Making less reliance on human based system.

(*iv*) Making two or three layers of defence mechanism.

(*v*) Making part/item temper/error proof.

(*vi*) Taking system trial before actual operation.

(*vii*) Making repair/removal possible through special tools.

(*viii*) Making system with fewer spares.

(*ix*) Devising system with error-fees - diagnosis.

17.8 OUTSOURCING OF MAINTENANCE

Outsourcing is the "buzz-word" in manufacturing and service sector, today. Be it Business Process Outsourcing (BPO), Knowledge Process Outsourcing or Product Outsourcing, it has emerged an viable business alternative. In maintenance domain, outsourcing plays as important role. So depending upon organisational philosophy, various outsourcing options are exercised.

17.8.1 Purpose of Outsourcing

Following are the purpose of outsourcing:

(*i*) Eliminating non-core activities.

(*ii*) Saving cost.

(*iii*) Better quality of work.

(*iv*) Reducing inventory.

(*v*) Super-specialised service.

17.8.2 Decision for Outsourcing

Following matrix provides basis for deciding outsourcing.

	Strategic	Non-strategic
Competitive	Do not outsource	Requires deliberation
Non-competitive	Re-engineer	Outsource

Fig. 17.8 Outsourcing Matrix

(*i*) Functions that are strategic and competitive requires to be retained.

(*ii*) Functions that are strategic but non-competitive need to be re-engineered so that these are performed at a competitive cost.

(*iii*) Functions that are non-strategic but competitive requires further deliberations to decide whether to outsource or not.

(*iv*) Functions that are non-strategic and non-competitive requires to be outsourced.

Difference between strategic and competitive parameter can be explained as — In an oil exploration industry, maintenance cost is of strategic importance and hence need not be outsourced but in a small batch process plant or a residential colony, maintenance is not strategic and can be outsourced. Similarly function being performed at a competitive cost compared to a market place need not be outsourced. While functions that are expensive to perform compared to a market place must be outsourced (non-competitive)

17.8.3 Areas of Maintenance Outsourcing

Maintenance outsourcing, mainly comprises maintenance staff for preventive/ breakdown maintenance, staff for fabrication work, staff for tooling/die/ lathe work, semi-skilled and unskilled labour, specialised services such as errection/commissioning/installation of critical/capital equipments and staff for routine maintenance jobs such as water coolers/filters, computers, photocopy machines, telephones *etc*.

Outsourcing process consists of two steps *viz*:

(*i*) **Complete Outsourcing:** All maintenance jobs are completely outsourced. For example, all low cost carrier airlines outsource complete maintenance function of their aircrafts.

(*ii*) **Partial Outsourcing:** All non-critical functions such as fabrication work, errection-commissioning, requirement of unskilled labour or specialised services are outsourced while other maintenance services such as preventive maintenance, breakdown maintenance *etc*. are carried out by inhouse staff.

Depending upon organisational philosophy, degree of outsourcing is decided. For example, depending upon criticality and volume of work, all tooling and die making function may be outsourced.

Summary

This chapter is a collection of assorted topics of relevance. These are TPM (Total Productive Maintenance), RCM (Reliability Centred Maintenance), Benchmarking, Value analysis, Reverse engineering, Anticipatory Failure Determination and outsourcing of maintenance.

Purpose of TPM is to involve all employees in achieving optimum plant output. This philosophy helps to reduce breakdown and improve productivity by collective sharing of responsibility by maintenance and production staff. It is an autonomous maintenance thereby maximising plant and personel effectiveness.

Philosophy of RCM is to discard those maintenance activities which do not enhance safety, availability, cost and effectiveness. Approach is to optimize maintenance to achieve highest reliability at lowest cost.

Benchmarking is the process of comparison of business metrics with the best practices. Purpose is to upgrade the existing performance parameter to a higher level by emulating best practices.

Value analysis is a process of enhancing the functional usuability, reliability, safety and life cycle cost of a product or service. The process looks at a product with a view to simplify its design, reduce cost and improve functionality.

Reverse engineering is the opposite of manufacturing of product. It is the process fo dismantling a product, understanding the working and re-producing the product.

Anticipatory failure determination is the process of identification of potential failures of machines with a view to improve reliability. It is different from various other techniques of reliability improvement. It looks at a product with a view to sabotage its working and works backward to eliminate that possibility.

Maintenance outsourcing is process of getting jobs done by a third party. Such decisions are taken for reducing cost, improving quality, reducing inventory and eliminating non-core activities.

Review Questions

1. What are the major goals of T.P.M ?
2. Discuss the steps used in implementation of TPM. What are obstacles in implementation ?
3. Explain the concept of Reliability Centred Maintenance.
4. Enumerate the benefits of RCM.
5. What is Benchmarking ? Why is it needed?
6. Discuss types of Benchmarking and Benchmarking process.
7. What is value analysis ? What are the benefits ?
8. Explain Reverse Engineering. Give suitable example to illustrate the concept.
9. Explain the concept behind Anticipatory Failure Determination. How is this different from other methods of failure analysis ?
10. Explain the importance of outsourcing in maintenance especially in the era of globalisation, privatisation and liberalisation.

Discussion Questions

1. "TPM is the process of improvement of man and his surrounding".
2. "It is a easier to implement TPM if bottom up approach is followed". Justify
3. "RCM is not a straight forward method rather a philosophy of reliability improvement". Comment.
4. Elaborate "Value analysis is the best way to reduce product cost".
5. Comment – "Reverse Engineering can result in product duplication".

Multiple Choice Questions

1. Benefits of RCM are:
   (*a*) Longer equipment life. (*b*) Reduced maintenance cost.
   (*c*) Improved reliability. (*d*) All of these.
2. Purpose of Outsourcing is to:
   (*a*) Develop new product. (*b*) Reduce cost.
   (*c*) Reduce inventory. (*d*) All of these.

3. Outsourcing decision is taken for:
   (*a*) Competitives strategic items.
   (*b*) Competitive and non-strategic items.
   (*c*) Non-competitive and strategic items.
   (*d*) Non-competitive and non-strategic items.
4. Benchmarking can be applied to:
   (*a*) Competitor. (*b*) Supplier.
   (*c*) Customer. (*d*) All of these.
5. Anticipatory Failure Determination process is far superior to other failure analysis process due to:
   (*a*) Analysis of all possible failure modes.
   (*b*) Analysis of all possible failure modes including sabotage.
   (*c*) None of these.
   (*d*) All of these.

True-False Questions

1. Outsourcing reduces maintenance cost.
2. Benchmarking helps to improve product-image.
3. Value analysis is based on functional utility concept.
4. Revese engineering is a way of producing new products.
5. RCM increases cost of Maintenance.

ANSWERS

Multiple Choice Questions

1. (*d*) **2.** (*d*) **3.** (*d*) **4.** (*d*) **5.** (*b*)

True/False Questions

1. True **2.** False **3.** True **4.** False **5.** False

❐❐❐

PART IV

Human Resource for Maintenance Management

Strategic Outline

This part deals with the management of maintenance manpower especially planning of manpower and their training in a world of fast changing technology. It explains the need, objectives and methods of maintenance manpower planning. Explores various sources of manpower recruitment and their training including redeployment due to economic recession, organisational re-structuring, change in organisational business strategy *etc.* Lastly this section also deals with design and evaluation of a training program.

Human factor plays an important role in maintenance. This section deals with human errors, their prevention and ethical aspects.

Chapters in this section

18. Maintenance Manpower Planning, Training and Re-deployment.
19. Human Factors in Maintenance.

18

Maintenance Manpower Planning, Training and Re-deployment

Learning Objectives

After reading this chapter, you will be able to:

- Understand need, objectives and methods of maintenance manpower training.
- Explore various sources of manpower recruitment and training.
- Analyse reasons and understand ways of redeployment.
- Understand design of a training program and its evaluation.

18.1 INTRODUCTION

Maintenance manpower is an important organizational asset. Hence, it is imperative that proper planning is done so as to provide required quantity and quality of manpower to the organization on demand.

As with any profession, maintenance manpower is also not available off the shelf. Also the need varies from organization to organization. Hence, a gap analysis is required to be carried out to identify the position and numbers required. Based on the projected demand, recruitment process is initiated. After selection of suitable candidate, induction/training is imparted to new joinee. This enables the new joinee to get acquainted with organizational structure, systems, culture *etc.* which helps to bond with the organization easily.

In this chapter, we shall study human resource planning, training and redeployment from maintenance management perspective.

18.2 MAINTENANCE MANPOWER PLANNING

Maintenance manpower planning is the preparation of forecast of maintenance personnel requirement. It is a continuous on going process necessitated by retirement, resignation, retrenchment, death *etc.* in existing setup. It may also be due to creation of new department/plant/shop, installation of new machinery, automation, expansion *etc.*

Keeping in view of the future requirements, maintenance head has to project such requirements to Human Resource Department well in advance.

18.3 OBJECTIVE OF MAINTENANCE MANPOWER PLANNING

The objective of maintenance manpower planning is to estimate future requirement of employees (at workmen, junior management and senior management levels) with required attitude, experience and skill sets. The planning need to be carried out for short term (upto one year), medium term (1 year to 3 years) and long term (5 year and above). The need for maintenance manpower planning arises due to the following:

(*i*) **Technological changes:** Due to changes in technology of the maintenance/production system, new manpower is required *e.g.* due to plant automation, new skilled manpower may be required.

With adoption of latest automated production systems as well as advanced maintenance systems, there is urgent need to maintain these sophisticated machines. This requires, hiring of new employees possessing the necessary skills set and experience.

(*ii*) **Future expansion:** An organization may plan for future expansion of similar or different products based on current or anticipated market demand. For example, an auto ancillary company manufacturing radiators for cars may go for manufacture of radiators of heavy commercial vehicles such as busses and trucks or may decide to go for manufacturing of car air-conditioning kits. So, for such future expansion, manpower is required with new skill sets.

(*iii*) **Periodic replacement:** With periodic depletion of manpower due to retirement, transfer, promotion, resignation, death *etc*., the existing posts need to be filled up.

(*iv*) **Changes in market or government regulation:** Maintenance manpower planning has to keep in mind changes in market conditions or changes in government regulation. For example, with greater emphasis on pollution control, organisations need to have well manned effluent treatment plant. This will require an experienced operator. With enactment of Energy Conservation Act. 2001, certain categories of industries need to recruit Energy Managers to lookafter energy management aspect of the organization as envisaged in the act. This requires fresh recruitment.

18.4 METHODS OF MAINTENANCE MANPOWER PLANNING

There are various methods of manpower planning. These are following:

(*i*) **Trend projections:** It involves analyzing past and present trends of employee requirement and accordingly project future anticipated need.

(*ii*) **Econometric modeling:** It involves building complex software based models to simulate the future requirements based on certain assumptions.

(*iii*) **Experts' opinion:** A panel of experts give opinion about the future trends in various sectors of industry. Sometime various business associations such as FICCI, ASSOCHAM, NASSCOM etc. provide forecast about various sectors of economy.

Human Resource and maintenance department can analyse and estimate their future need keeping in view expert's opinion.

(*iv*) **In-house estimate:** The maintenance department can estimate its future need based on current and future level of activities such as plant capacity addition, plant automation, new project *etc*. So based on projected activities maintenance head can estimate future requirement.

(*v*) **Regression analysis:** Regression analysis is statistical technique that finds relationship between past indicators and staffing levels. For example, relationship between current sales level and sales to employees ratio. Such analysis relies on a strong relationship between two variables.

(*vi*) **Productivity ratio:** It is ratio of workload to number of employees. So increase in demand shows the increase in number of employees. Similarly, decrease in demand shows reduction in number of employees.

A maintenance function being a specialized area, needs manpower in various areas such as skilled technicians (electricians, turners, fitters, plumbers, lathe operators, die fitters *etc.*), as junior staff (Supervisor, junior engineer, foreman, assistant engineer *etc.*) as middle management executives (senior engineer, assistant manager, deputy manager, manager) and as senior management executives (Deputy General Manager, General Manager *etc.*) The skill set required for each category is different and depends on type of industrial sector. For example, type of experience required in a tyre industry will be totally different than those required in semi-conductor industry or fertilizer industry or glass manufacturing industry. Hence, maintenance man-power planning requires not only numbers and positions but also experience in similar industry.

18.5 RECRUITMENT

Recruitment is the process of searching, selecting and attracting right talent for the right job. Recruitment is the single most important factor that has a large impact on the organizational growth.

18.6 SOURCES OF RECRUITMENT

Various sources of recruitment are following:

(*A*) **Internal Sources:** Internal sources means selection of employees from within the organization. This method of selection cuts down hiring cost and time, improves employee's morale and leads to higher employee commitment. Various sources of internal recruitment are.

(*i*) **Serving employee:** Serving employees provide a large talent pool whose data base and past performance is already with the concerned department. Hence, a well informed and better selection can be made to fill up vacancies due to transfer, promotion, VRS (voluntary retirement scheme), death retirement or new position.

(*ii*) **Past employees:** All past employees who were laid off, and resigned, can serve as good sources for recruitment. The major advantage being known past performance.

(*iii*) **Employees referrals:** All existing employees can refer their friends, relatives, former collegues *etc.*

(*iv*) **Customer and business partner referrals:** Customers and business partners such as venders, contractors, suppliers *etc.* can provide suitable candidates.

(*B*) **External Sources:** Some organizations have policies as not to recruit relatives of existing employees. Also entry of outside candidate infuses new experience and competency that is necessary for growth of the organization.

The various sources of external recruitment are:

(*i*) **Advertisement:** This is one of the most widely used methods of recruitment. Advertisements appear in print media such as newspaper, magazines, specialized career magazines, business magazines and in electronic media such as T.V., internet (job portals), radio *etc.*

(*ii*) **Placement consultants:** Placement consultants or head hunters provide a large talent pool starting from entry level freshers to top management executives. They can provide sector specific manpower such as for information technology area, bio-technology area or FMCG (Fast Moving Consumer Goods) area, Business Process Outsourcing (BPO) area *etc.*

(*iii*) **Campus recruitment:** Campus recruitment is the route followed for filling up entry level vacancies. Generally, large reputed organizations visit Management Institutes, Engineering colleges, Information Technology Institutes, Fashion Design Institutes *etc.* to hold campus recruitment. Visiting organizations deliver pre-placement talks followed by written test, group discussion and interviews.

(*iv*) **Employment exchange:** Job seekers register themselves in employment exchanges. These employment exchanges forward bio-data of eligible candidates to prospective organizations. Now this has become outdated in the era of job portals.

(*v*) **Job fairs:** Various educational institutes organize job fairs in a centralized locations (Usually metro city or state capital) and invite prospective organizations for recruitment.

(*vi*) **Forced applications:** Organizations receive job applications for which there may or may not be any demand. Such applications are sent by job seeker on their own. Such applications are kept in a data bank for future use.

(*vii*) **Poaching from competitors:** Organizations poach candidates from their competitors. For example, coke and Pepsi provide ideal platform for poaching good candidates from each other.

After recruitment, the next step is training. Training is carried out for fresh recruits as well as for experienced personnel.

18.7 TRAINING

Training is a planned activity designed to improve performance of the employee by imparting skills and techniques necessary for the job. A trained employee, thus become an asset for the organization.

Organization plans training of targeted employees by identifying their training needs. Thereafter, training in required field is imparted to the employee. After conclusion of training program, its evaluation is carried out to judge its effectiveness.

Hence, training acts as a protection against human obsolescence.

18.8 NEED FOR TRAINING

Training serves the following purposes:

(*i*) It helps fresh joinees out of college/institute to learn the basics so as to perform their tasks. It acts as a source of coaching so as to begin their career.

(*ii*) It helps an employee for taking up higher responsibilities after promotion.

(*iii*) It helps to learn latest techniques in maintenance management and engineering.

(*iv*) It helps an employee to move from one job to another. For example, from maintenance planning to technical cell.

(*v*) Maintenance technology is changing very fast, so there is need to know the latest development in the field.

(*vi*) It helps to face uncertain and unpredictable situations encountered in maintenance e.g. due to automation, downsizing, change in organisational business policy.

(*vii*) To develop skills in fields other than maintenance. For example, in finance, material management, statutory compliances *etc.*

18.9 TARGET EMPLOYEES FOR TRAINING

Following are employees who need training:

(*i*) **Fresh joinees:** These include gradate engineers, diploma engineers, ITI workmen. They go through induction training, orientation, safety training and on-the-job training.

(*ii*) **Working executives:** These include middle level engineers and managers. They need to broaden their knowledge horizon so as to take up higher responsibility. They undergo training in behavioural and management aspects.

(*iii*) **Workmen/supervisors:** These include junior level employees. They need to improve upon their basic skill sets. They undergo refresher courses in respective trades and basic human psychology.

(*iv*) **Senior management executives:** These include top level employees in maintenance department hierarchy. They undergo training in business strategy, behavioral traits and cross-functional areas (finance, operations, materials *etc.*)

18.10 WHY MAINTENANCE TRAINING IS SPECIAL?

This is due to following reasons:

(*i*) Unlike production or accounts, maintenance is not a repetitive job. Every situation is unique.

(*ii*) It requires knowledge of diverse areas such as materials, project, finance, human psychology *etc.*

(*iii*) Maintenance technology is a fast changing field. It requires maintenance personnel to be in touch with the latest development.

18.11 TYPES OF TRAINING IN MAINTENANCE

Various types of maintenance training are following:

(*i*) **Technical Training:** This type of training is mainly imparted to fresh joinees and junior level employees (workmen/supervisor/engineer). Purpose is to enhance and consolidate technical skills necessary to carry out day-to-day activities. For example, how to repair electrical faults in a furnace or how to replace bearings in a pump or how to lap/grind valve seats or how to rectify faults in hydraulic/pneumatic systems *etc.*

(*ii*) **Management Training:** This type of training is imparted to middle and senior level maintenance executives. Purpose of training is to broaden their horizon and shift focus from merely technical aspect to overall management aspect. It teaches people management, financial aspects, production and operation, cross-functional co-ordination *etc*.

(*iii*) **Behavioural Training:** This type of training is imparted to middle and senior level employees. Purpose of training is to develop right attitude for the job, correct certain behavioural patterns *etc*.

(*iv*) **Induction and Orientation Training:** All new entrants to the organization irrespective of the level, undergo Induction and Orientation training. Purpose of this training is to introduce the organizational culture, systems and procedures, introducing employees at key places, about the organization (*e.g.* employee strength, turnovers, different plants and office locations, product details etc.) and also to allow interaction of new employees with existing employees of other departments.

(*v*) **Safety and Environment Training:** This type of training is imparted to all levels of employees. Purpose of this training is to inculcate a right attitude towards safety and environment, teach and make them aware of basic safety gears and safety procedures and environmental aspects.

Safety training includes mock fire drills, demonstration of use of safety appliances and gears, impart knowledge of accidents/disasters management procedures, basic course in first-aid *etc*. It includes disseminating basic information about toxic chemicals—their handling and safe disposal, handling of chemical accidents, environmental impact of chemicals *etc*.

(*vi*) **Organizational System Training:** This type of training is imparted to all levels of employees in the organization. Organisational system pertains to ISO-9002/OHSAS/TS-16000/ISO 14000 *etc*. Purpose of this training is to enable all employees understand various clauses of the system and their impact on their working. This training ultimately helps organization to acquire certification. This certification not only helps to boost organization image but also is a necessity in certain business environment.

18.12 TRAINING DELIVERY SYSTEMS

Following are various modes of training delivery in maintenance field:

(*i*) **Class Room Lecture:** This is one of the most common methods of training. This method is used for fresh joinees as well as experienced engineers/ managers. It consists of lecture through slides, power point presentations, black-boards, posters, electronic moving slides, cut-models of equipments *etc*.

(*ii*) **On-the-job Training:** This method is mostly used for fresh graduate engineers, diploma engineers and ITI technicians. The new recruit is put on the job with a senior person. The new joinee learns by watching his senior and also by doing job himself. He always remains under the watchful eye of the senior who teaches various techniques of the trade on-the-job.

(*iii*) **Audio Visual Presentation:** This technique uses a short film that demonstrates technical concept or management idea. It may use graphical representation or use presenter or use cartoons. Certain situations can be enacted by actors and can be explained by narrator.

(*iv*) **Broadcast:** It is another medium that is used for information dissemination purpose.

(*v*) **Industrial Visit:** Industrial visits are arranged for a group of trainees. Trainees see and interact with the plant personnels at the site and gather information.

(*vi*) **Seminars/ Symposiums/ Exhibitions:** Seminars are organized by either by the organization or allow trainees to participate seminars. The lectures are delivered by industry experts. For example, various business associations organize industrial exhibition focussing on a particular sector of industry such as lighting industry, machine tool industry, food machinery industry *etc*. Trainees collect information through brochures/leaflets of products and interact with company personnel to clarify their doubts. Such exhibitions are called trade fairs and are held in metro cities such as Delhi, Mumbai, Chennai *etc*.

(*vii*) **Management Workshops:** This technique is used for middle and senior level executives usually in a hotel or in a resort, away from plant so as to free the mind of the participants from plant problems. Certain management games are conducted in small groups. This helps to learn intricacies of management principles. This may include role-play in which participants are given a crisis situation and are asked to find solutions. So various training methods are enumerated above which can be used judiciously based on need of target employee.

18.13 SOURCES OF TRAINING

Having identified employees with training need, the next step is to identify sources of training. Various sources available for maintenance training are following:

(*i*) **Hiring of Experts:** Subject experts are hired who can deliver lectures, demonstrate to target audience in-house. Such experts have in-depth knowledge and experience in the identified field and help participants to gather requisite information/knowledge.

(*ii*) **Training Institutes:** There are training institutes that specialize in their chosen field of expertise and advertise their training calendar in advance through their web-site, training brochures, advertisements in newspapers and magazines *etc*. For example, administrative staff college of India, National Institute of Industrial Engineering. National Power Training Institutes, Indian Institute of Management, Indian Institute of Technology, National Institute of Technical Teachers Training and various other public and private training institutes. These institutes also conduct tailor-made program for a specific organization. For technicians training, various vocational institutes offer training programs.

(*iii*) **In-house Training:** Human Resource department can organize in-house training programs by inviting plant personnel for lecture/ demonstration. For example, such method is used for fire-safety demonstration, use of various types of fire extinguishes are demonstrated by fire fighting personnel. Public sectors undertakings (PSU) organize in house training programs for gradate engineer trainees by inviting plant personnel as faculties.

(*iv*) **Vender Training:** After installation of new equipments/machinery the vender organizes a training program for the operating staff. This training helps to familiarize the operating and maintenance staff with technical details. Such trainings are either held at the plant or at the vendors' premises.

For example, M/S Thermax organizes training program for technicians and engineers at its Pune plant. It is a well structured training that includes a plant visit. Similarly, turbine, boiler, furnace manufacturers provides training to operation and maintenance staff at site. Similarity, pneumatic manufacturer – 'FESTO' organizes training programs for technicians and engineers. Larsen and toubro Ltd organizes training program on switch gears. So all major equipment manufacturer/supplier provide such tailor made training to operation and maintenance staff.

(*v*) **University/R & D Centers/Training Institutes:** Various Universities and Research and Development centers involved in specific areas can provide training. For example, IITs undertakes industry related consultancy in specific areas and can help to disseminate knowledge to plant staff.

Even universities engaged in state-of-the-art technologies can provide valuable input to industries engaged in similar field. There are reputed institutes such as Indian Institute of Industrial Engineering, Administrative Staff College of India, National Power Training Institute, who provide such training facilities.

(*vi*) **Continuing Education Program:** A person can enhance his knowledge or skill sets through continuing education programs. Such programs are — various associatedship examinations such as AMIE, IETE, AMAeSI, IIIE, etc., part time or evening courses in specific areas can be pursued at ITIs, polytechnics and engineering colleges. Even correspondence courses are offered by various universities such as diploma in Maintenance Engineering and Management, P.G. Diploma in Drives etc. by Annamalai University. Employers encourage by giving financial incentives or time-off so that such courses can be pursued by the employees.

18.14 IDENTIFICATION OF TRAINING NEED

It is necessary to first identify a training need of an employee before the requisite training can be imparted to him. The sources of identification of training needs are following.

(*i*) **Through Performance Appraisals:** Every employee undergoes performance appraisal on a yearly basics. The superior officer records his observations about employee's strengths and weaknesses. Accordingly, identifies areas of weaknesses for improvement. So the area of weakness can be converted into area of strength through a combined effort of employee's own actions and through area specific training. Performance appraisal not only records observations of superior officer but also records observations of employee (see appraisal). So, it identifies area of training for employees.

(*ii*) **Through Supervisors/ Superior Officer:** HR or training department can get valuable input about employees though their respective supervisors or superior officers. Their observations can provide necessary input for employee improvement.

(*iii*) **Through Maintenance Head:** A maintenance head provides valuable input to training department. This may be necessitated by expansion of plant, modification/automation in the plant, job rotation/ re-deployment, installation of news machinery, management policy of multi-tasking etc.

(*iv*) **Through H.R. Department:** Human Resource Department (HRD) can initiate training programs in specific areas such as Organization Systems (ISO), Fire-fighting, Plant safety, First-Aid, Employee conflicts management *etc*.

18.15 DESIGN OF TRAINING PROGRAM

After identification of target employee and his training need, next comes design of training program that meets needs of the trainee. Following points need to be kept in mind for designing training program:

(*i*) After identification of type of training program such as technical, behavioural, management *etc*, the designed program must consider age, qualification, experience and desired outcome for the design of training curriculum.

(*ii*) Internal or external training program should be decided based on availability of program (internally or externally), cost and availability of employee for the duration of training.

(*iii*) The training program for workmen level should have less lectures and more practice. The lectures must preferably be in local language understood by the majority.

(*iv*) For executive level program, curriculum must have ideal mix of lectures and group based activities including presentations.

(*v*) The program must contain basics and latest developments in the focussed area.

(*vi*) Main subject must be divided into small modules. Each module must have sub-areas. Each sub-area must follow logically one after another. They key ideas can be highlighted by posters, cartoons, jokes *etc*.

(*vii*) For designing content of a sub-area, especially concerned with operation and maintenance, India's as well as world's best practices must be highlighted (bench marking)

(*viii*) A site visit is highly useful along with interaction with the field staff.

(*ix*) Feedback obtained at the end of training program must be used to modify the course contents.

18.16 EVALUATION OF TRAINING PROGRAM

Evaluation of training program is required to assess the effectiveness and suitability of curriculum. This, in turn helps to modify the content and characteristics so as to make training more relevant, useful and result oriented.

Generally, following methods are used for evaluation of training program.

(*i*) **Through questionnaire:** A well designed questionnaire seeks information about various aspects of training such as course content, its depth, coverage area, quality of lectures, expectations of participant, suggested changes *etc*. Answers to these probing questions provide scope for improvement.

(*ii*) **Through a test:** A test, at the end of the training program is conducted to assess the level of knowledge acquired. The test can be objective type based on the topics covered. The test gives an idea of individual's achievement.

(*iii*) **Through an interview:** The participants are interviewed to judge the effectiveness of the training program.

18.17 MAINTENANCE MANPOWER RE-DEPLOYMENT

In the era of global competition, maintenance manpower is an organization asset. However, due to economic recession, business cycle swings, govt policy changes, labour unrest, take overs, mergers and acquisitions, top management has to take certain decision such as downsizing (lay-off), re-deployment in other areas, multi-tasking, lean manufacturing systems *etc*. Possibility of such situations requires maintenance department to be able to handle events with ease. Focus of discussions will primarily be confined to re-deployment and multi-tasking.

18.18 OBJECTIVES OF MAINTENANCE MANPOWER RE-DEPLOYMENT

Manpower re-deployment means disengaging existing manpower from current position to re-engaging to new environment. This not only means moving man-power from one type of jobs but also includes attachment to altogether to a new system of operation. Objectives of re-deployment are following:

(*i*) To introduce new system of manufacturing and maintenance.

(*ii*) To enable maintenance manpower to do a wide variety of tasks under new and challenging environment.

(*iii*) To take care of man-power shortage.

(*iv*) To economise on manpower cost.

(*v*) To utilize capabilities of manpower so as improve system efficientcy.

(*vi*) To benchmark best practices.

(*vii*) To develop over-all perspective.

(*viii*) To minimize inter-area conflicts.

(*ix*) To improve employability.

(*x*) To "right size" employee strength

(*xi*) To create a pool of "all-rounders" to take care of any exigencies *e.g.* resignation, death, transfer *etc.*

18.19 METHOD OF MAINTENANCE MANPOWER RE-DEPLOYMENT

Following are the various method of re-deployment of maintenance manpower:

(1) **Training for the new job:** The person to be transferred is trained for the new job by allowing him to operate in parallel with existing person in-charge.

(2) **Direct transfer to the job:** The person is directly given the charge of new section/area/plant. So he learns on the job.

(3) **Allowing to operate in parallel for short duration:** The person to be transferred is allowed to operate in parallel during his free time *e.g.* two hours everyday or for a particular day in a week.

This process in applicable to skilled workmen as well as to senior engineering staff where the content of job is very technical in nature.

CASE STUDY OF MAINTENANCE MANPOWER RE-DEPLOYMENT

Case Study-1

(*i*) **Re-deployment of skilled workmen:** This case study pertains to utility section of a continuous process plant. The utility section had recently installed a gas turbine for captive power consumption. Due to automated and sophisticated gas turbine controls, very few manpower was needed at workmen level.

Before the commissioning of gas turbine, the captive power plant consisted of 6NOS. of 1000 KVA Diesel generating sets which required four persons in the shift operation. But due to operation of gas turbine, the DG-sets were used as stand-by power source. Hence, DG set operators were free till gas turbine tripped on fault or was under maintenance.

The De-mineralized water plant (DM-plant) was manually operated skill-intensive, critical plant which had no reserve manpower. The DM plant operators were smart enough to go on leave in cyclical manner so that they are able to take overtime. To overcome this vicious circle and to develop required reserve manpower, it was decided to allow DG set operators to operate in parallel with DM plant operators.

Initially, there was resistance from DM plant operators. So, it was decided to transfer one DM plant operator to ETP (Effluent treatment plant) and ETP operator to DM plant. This strategy worked due to various reasons. First, ETP job was less technical in nature and was taken as Punishment posting. Secondly, the job was not very critical compared to DM plant. So during the period of stay of DM plant operator in ETP, the ETP operator was trained in DM plant operations. Initially in parallel for few days, thereafter in independent shift. This newly trained ETP operator trained DG-SET operator.

In this way, all DG set operators were trained for DM plant operation over a period of time and a reserve pool of skilled manpower was developed through re-deployment.

Case Study-2

(*ii*) **Re-deployment of senior maintenance Engineer:** This case study pertains to a refrigerator manufacturing plant. The entire plant was sub-divided into different sections such as press shop, paint shop, foaming shop, evaporator manufacturing section, assembly lines and utilities. A group of such plants were under senior engineer/deputy manager level person in maintenance.

It was observed that foaming, paint shop and press shops were poorly maintained and there was neglect and collusion between workmen and Deputy Manager for excess overtime. So it was decided to re-deploy maintenance manpower at supervisory level to improve system efficiency, carry out much needed renovation and improve maintenance response time. So there was mutual transfer of two senior engineers. The utility senior engineer was asked to shift to foaming and paint shop section. Similarly, foaming and paint shop senior engineers was shifted to a well managed utility section.

The utility senior engineer's task was well defined — identify trouble makers and confine them to less important jobs, take efficient people to form a core team to carry out renovation and improve maintenance response time.

This whole process took more than a year to accomplish the job. So by deploying maintenance supervisory staff, system improvement was carried out apart from building a trained manpower pool.

Summary

With fast changing technology, latest equipments with sophisticated technology have become obsolete. Moreover, there are variety of products available with advanced features, both indigenous and imported. Under such scenario, it is absolutely necessary to update maintenance personnel with the latest development. Hence, manpower planning and training assumes a greater importance in this context.

Purpose of maintenance manpower planning is to replenish the existing stock of maintenance personnel depleted by various factors such as technological changes, requirement for future expansion, regular replacement due to retirement, death, transfer, promotion *etc.* The planning process helps to estimate future need. Various method used are trend projections, econometric modelling, expert's opinion, regression analysis *etc.* After estimating future requirement, the next step is recruitment. Recruitment is carried out through several sources. The sources could be employee referral, internal sources, advertisements in print and electronic media, placement consultant,

Contd...

job fairs, campus recruitment etc. These new recruitees could be fresh candidates out of college or experienced personnel.

The next step is training of maintenance personnel. Training is imparted to new recruits as well as to serving employees at all levels. Before training is imparted, it is necessary to gauge training needs of the person so that training need matches with the employee's need. The purpose of training is improve knowledge and skill sets necessary to perform the existing work satisfactorily. Training is also necessary for vertical movement in hierarchical ladder.

Training need for maintenance manpower could be technical, management, behavioural, orientation, safety and environmental etc. Once training need and type of training is identified, the next step is selection of training delivery system. This could be class room lecturers, on-the-job training, audio-visual presentation, seminars, industrial management workshops *etc.*

Having chosen the delivery mode, the immediate step is identification of source of training. Several choices are available such as in-house training, vender training, lectures by invited experts, sponsorship to reputed training institutes such as Administrative Staff College of India (Hyderabad), Indian Institute of Industrial Engineering (Mumbai). National Power Training Institute, National Productivity Council *etc.*

Training needs of an employee are identified from various sources such as through annual performance appraisals, from immediate superior, from Human Resource Department *etc.*

Final step is the design of training curriculum followed by evaluation of training program.

Re-deployment of maintenance manpower is necessitated by manpower shortage, downsizing, expansion, *etc.* The process involves training for new job, transfer, and allowing new person to work with existing staff *etc.* In this way, a person is allowed to take up responsibility in a new job. Re-deployment helps in building new skill sets in the employees so that person can do multi-tasking.

Review Questions

1. What is maintenance manpower planning? Why is it required?
2. Discuss objectives of maintenance manpower planning.
3. Explain in detail various method of recruitment.
4. Define recruitment. Discuss sources of recruitment.

5. Why sometimes internal sources of recruitment are preferred over external sources?
6. What is training? Why is it important for maintenance personnel?
7. Discuss various types of training for maintenance personnel.
8. How training can help to build technical and managerial competence in an organization?
9. Discuss maintenance manpower redeployment.
10. How training program is evaluated? What are various criteria for evaluation?

Discussion Questions

1. "Trained and experienced maintenance manpower are scare resource". Justify.
2. "Training cannot act as a replacement for experience" Do you agree?
3. "Continuous learning is a hedge against obsolescence in maintenance". Comment.
4. "In the era of globalization, maintenance manpower redeployment on periodic basis is a must". Discuss.
5. "Recruitment of talent is easier than retaining". Do you agree?
6. "Multi-tasking is must for maintenance personnel in lean maintenance system". Comment.

Multiple Choice Questions

1. Maintenance manpower planning is required due to
   (*a*) Periodic replacement (*b*) Future expansion
   (*c*) Technological changes (*d*) All of these
2. Internal sources of recruitment are
   (*a*) Customer and business partner referrals
   (*b*) Placement consultants
   (*c*) Job fairs
   (*d*) Advertisements
3. Behavioural training is imparted to
   (*a*) Junior level employees
   (*b*) Fresh joinees
   (*c*) Middle and senior level employees
   (*d*) All of these

4. Training needs are identified through/from
   (*a*) Superior officer (*b*) H.R. department
   (*c*) Performance appraisals (*d*) All of these
5. Objective of maintenance manpower re-deployment is
   (*a*) Reduce manpower cost (*b*) Retrenchment
   (*c*) Multi-tasking (*d*) All of these

True/False Questions

1. Maintenance manpower planning by trend-projection method involves analysis of past and present trends.
2. Maintenance manpower training is given to newly recruited employees only.
3. Middle level executives need only managerial training.
4. Maintenance manpower redeployment can be a tool for lean manufacturing systems.
5. Design of training curriculum should have more technical aspects for senior management level.
6. Training is a hedge against technical obsolescence.

ANSWERS

Multiple Choice Questions

1. (*d*)	2. (*a*)	3. (*c*)	4. (*d*)	5. (*d*)

True/False Questions

1. True	2. False	3. False	4. True	5. False
6. True				

Case Study Based Question

Mr. Ramdass is a section engineer in a multinational company. He looks after assembly lines. Due to heavy absentism, the shift output is shrinking. This is true especially of night shift and before and after festivals. To counteract such uncertainties and fluctuating production output, he proposed a new plan to his boss Mr. Rehman. The plan involved transfer of surplus workmen from press shop to assembly lines. However, this plan was strongly opposed by the labour union. Mr. Ramdass again requested his boss to take up contractual workers. But this was again opposed by unions tooth and nail.

Left with no other choice, Mr. Ramdass requested management to provide trainee workmen to compensate for loss of manpower. Management refused to allow the trainees to do the job of an experienced worker. However, management is pressurising Mr. Ramdass to improve shift production.

1. What are the options available to Mr. Ramdass?
2. What will you do if you are asked to take charge from Mr. Ramdass?

❑❑❑

19

Human Factors in Maintenance

Learning Objectives

After reading this chapter, you will be able to:

- Analyse human errors in engineering systems.
- Understand classification of human errors.
- Comprehend methods of prevention of human errors.
- Understand importance of engineering ethics.

19.1 INTRODUCTION

Importance of human factor in maintenance got highlighted due to aircraft accident in 1988 (Aloha B-737). The accident took place during flight in a Boeing 737-200 aircraft. The upper cabin structure of aircraft blew-off due to structural failure. Post accident investigation revealed failure of two maintenance inspectors to detect 240 cracks in the body of aircraft.

Similar incident took place in 1991 in an embracer 120 aircraft. The aircraft carrying 13 passengers crashed mid-air due to structural failure. Post accident investigation revealed removal of attaching screw on horizontal stabilizer during a maintenance check up.

In another accident in 1990, British airways aircraft heading for Spain carrying 81 passengers and crew members had its windscreen blown-off. This happened at a height of 17,300 feet. This caused the pilot sucked through the windscreen aperture. Cabin crew and Co-pilot held the pilot by ankles till the aircraft landed safely. Investigation revealed that windscreen was replaced by using bolts of wrong thread. Alarmed by such accidents, survey was carried out by an American aviation industry which revealed some startling facts:

(1) 70–80% aircraft accidents are due to human factor.

(2) maintenance contributed to 15% of commercial jet accidents. This resulted in aviation industry "sitting up and taking notice" of human factor in maintenance. A lot of research was carried out and several systems were designed to minimize detrimental effect of human factor in maintenance.

As we will study later, all such human errors were due to a variety of reasons such as complacency, distraction, fatigue, job pressure, stress, lack of resources, lack of knowledge, lack of communication, sabotage etc. So a highly complex and sophisticated equipment may or may not perform as intended due to human factor. Hence, despite all technological progress, human being is still the single most important factor in successful operation of equipment and system.

In this chapter, we will explore reasons, causes and methods of prevention of human error. Also, effect of ethics will be discussed as to how these factors induce intended and unintended errors.

19.2 HUMAN ERRORS IN ENGINEERING SYSTEMS

Reliability of an equipment or a system among others is dependent upon human reliability. A large numbers of failures are due to human errors. Reasons for human errors are following:

(*i*) Poor motivation.

(*ii*) Complex task.

(*iii*) Poor verbal communication.

(*iv*) Distraction.

(*v*) Forgetfulness.

(*vi*) Wrong tools.

(*vii*) In-adequate training.

(*viii*) Poor equipment design.

(*ix*) Bad work environment (poor lighting, high temperature, high humidity, dust, noise vibration *etc.*).

(*x*) Poor work space layout.

(*xi*) Lack of standards or inability to follow standard operating procedures.

(*xii*) Sabotage.

(*xiii*) Fatigue.

(*xiv*) Job stress.

(*xv*) Poor physical and or mental fitness

All above reasons result in poor work quality, wrong assembly, in complete job *etc.* leading to failures and accidents.

19.3 CLASSIFICATION OF HUMAN ERRORS

All human errors can be classified into error of omission and commission *i.e.*–

(1) They should have done something but did not do.

(2) Failed to do something they should have done.

A study showed omission errors were

(*i*) Repairs and modification – 41%

(*ii*) Testing and calibration – 33%

(*iii*) Inventory control – 9%

(*iv*) Manual operation and control – 6%

Following are classification of human errors:

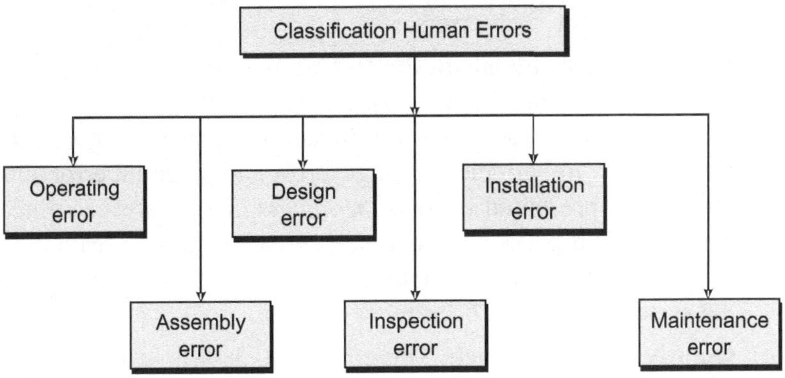

Fig. 19.1 Classification of Human Errors.

(*i*) **Operating error:** These errors are due to lack of proper operating procedures (SOP), task complexity, poor training, carelessness, poor environmental conditions *etc.*

(*ii*) **Assembly errors:** These errors are due to use of incorrect component, omitting a component, poor communication *etc.*

(*iii*) **Design errors:** These errors are due to wrong specification, wrong choice of material, different operating conditions *etc.*

(*iv*) **Inspection error:** These errors due to lack of knowledge about test procedures, failures to pay attention to minute details, omitting certain tests, incorrect fitment of components, lack of time *etc.*

(*v*) **Installation errors:** These errors are due to fitment of wrong part, error in drawing, error in manufacturing, electrical wiring discrepancies *etc.*

(*vi*) **Maintenance errors:** These errors are due to misfits, loose fasteners, adjustment made without correct gauges, wrong spare part, dirty lubricant, sabotage etc.

19.4 METHODS OF PREVENTION OF HUMAN ERRORS

There are several approaches to prevention of human errors. Some are listed below:

(*a*) **Man-machine system analysis method:** This technique analyses function and goals of the system taking into account situational characteristics such as ambient temperature, illumination level *etc.* Analysis of task is performed to estimate potential errors. A likelihood of estimate of errors being undetected is obtained. Accordingly consequences of each undetected errors is estimated and recommendations are made to change the system so as to prevent reoccurrence of such errors.

(*b*) **Error-cause removal method:** This method lays more stress on preventive measures rather than remedial measures. A team of 8-12 persons is constituted and co-ordinated by a person whose job is to reduce error. In periodic meetings, error and potential error situation reports are presented by team members. These are examined and discussed, and preventive measures are adopted. The team members act as specialists who evaluate each error-like situation and re-design the system to prevent occurrence of error.

(*c*) **Quality circle approach:** This concept took birth in 1963 in Japan. It consists of a team which investigates problem with the aid of pareto analysis and fish-bone (cause-effect) diagram. The team is given training on statistical quality control methods such as control charts, pareto diagram, normal distribution etc. the problem is identified and analyzed by the team. Finally, a solution is obtained to prevent reoccurrence of the problem.

(*d*) **Preventive measure approach:** This method consists of observing preventive measures to prevent error reoccurrence. These are listed below:

(*i*) Attention—Paying full attention to the job so as to avoid attention deficit errors.

(*ii*) Fatigue—Reducing fatigue so as to prevent occurrence of accidents.

(*iii*) Failure to note important indication—Paying attention to minute details.

(*iv*) Wrong control setting by operator—Proper training and sensitization of operator.

(*v*) Error in reading instruments—Paying attention to the job.

(*vi*) Noise, vibration, temperature etc. casing irritation—minimising environmental irritants.

(*vii*) Failure to follow standard operating procedures—Paying attention to follow procedures.

(*viii*) Proper task design—Proper task design helps to minimise errors due to understanding task content and their execution.

(*ix*) Proper work planning—Proper work planning helps to minimise wastage of time, wrong usage of spare parts, incorrect installation etc.

(*x*) Safety culture—Inculcating a safety culture in organisation through various awareness programs.

(*xi*) Physical fitness—Paying extreme importance to physical fitness to work.

19.5 ENGINEERING ETHICS

Engineering ethics deals with engineers' obligation to society, clients, employers and above all to profession. It is not merely remaining honest rather following professional code of conduct, legal rules and following good engineering practices. There are numerous such failures which cannot be classified as human error but ethical failures. Here, it is necessary to make distinction between human error' and 'ethical failure'. A human error' can be intentional or unintentional but ethical failures are always for some tangible or intangible gains and are deliberate acts.

Example of ethical failures are — Space Shute Columbia Disaster (2003), Space Shuttle Challenger Disaster (1986). Three Mile Island Accident (1979), Minamata Disease (1908-1973) etc.

Engineering ethics encompasses a wide area, some examples are following:

(*i*) Software piracy.

(*ii*) Copying of patented design.

(*iii*) Modification of patented design without permission.

(*iv*) Bribery.

(*v*) Fraud.

(*vi*) Endangering public safety.

(*vii*) Conflict of interest.

(*viii*) Legal non-compliance.

(*ix*) Leaking confidential information.

(*x*) Mis-representation or suppression of facts *etc*.

19.6 EXAMPLES OF ETHICAL FAILURE

This case study refers to an aircraft industry in U.S.A. which nearly caused disaster. On June 12, 1972, a DC-10 aircraft took-off with 67 passengers on- board from Detoit, U.S.A. After reading a height of 12000 feet, cargo door of the aircraft blew-off. This resulted in collapse of floor and disrupted all hydraulic controls to tail section. However, due to light load and pilot's effort a disaster was averted.

Upon detailed investigation, it was found that similar problem had occurred in a ground test in 1970. The problem was traced to faulty design of latch and floor. However, no remedial action was taken then.

After this near disaster incident, National Transportation Safety Board (NTSB) recommended design changes in door latch and floor design. This was agreed to by the company president. But again was not carried out. Such repeated failure of personnel to rectify major defects in past which were detected earlier is an extreme example of ethical failure. As the concerned personnel could fully understand result of non-compliance. The reason for non-compliance was the cost factor involved.

Similar case is with space shuttle challenger disaster which occurred on January 28, 1986. It broke apart in 73 seconds, killing all crew members.

The cause of disaster was failure of 'O' ring seal in fuel storage. The contractor which had designed 'O' ring seal had intimated concerned NASA official about probability of its failure due to sudden temperature change. However, modification of 'o' ring seal would have delayed the scheduled lift-off. Hence, the suggestion was ignored. Here, the concerned person was aware of result of 'O' ring failure but still ignored the suggestion from contractor for its modification resulting in disaster which could have been avoided.

19.7 PREVENTION OF ETHICAL FAILURES

Organizations and individuals follow path of gain, ignoring ethical calls. It is the duty of government to frame rules to prevent such occurrences. The professional societies can play an active role by framing such code of conduct and imposing them on members.

Apart from deterrance, the next step is raising awareness about ethical failures and educating the engineering fraternity about its pitfalls. One such example is the Whistle Blower's Protection Act in U.S.A which protects the person who informs regulatory authority about any such ethical wrong doings.

Summary

Successful operation of any equipment depends largely on the operating and maintenance personnel. Human intervention can control quality of product; minimize chances of accidents and enhance life of equipment through proper care. Even with proper training, standard operating procedures, chances of human error—both intentional and unintentional cannot be ruled out. Hence, reliability of equipment is dependent on human beings.

Human errors in engineering systems take place due to variety of reasons. Some of these reasons are complexity of task, lack of communication, bad working environment, fatigue, stress, lack of knowledge *etc.* These factors lead to poor quality of work, wrong assembly, incomplete job and may even lead to accident.

Contd...

All human errors can be broadly classified in two categories—error of omission and error of commission. In general, errors can be categorized into six distinct groups *viz* operating errors, assembly errors, design errors, inspection errors, installation errors and maintenance errors.

Four approaches have been discussed in this chapter for prevention of human errors. These are man—machine system analysis, error-cause removal method, quality circle approach and preventive measure approach. Man-machine system method analyses functions and goals of system, keeping in view situational conditions. This is used to estimate probability of errors going undetected and accordingly methods are devised to prevent such errors. While in error-cause removal method, stress is on preventive method to reduce errors. In quality circle approach, a team of operators discuss problem and find solution through brain storming. Preventive measure approach prevents error- recurrence through a combination of proper observation, environmental factor and standard operating procedures.

Engineering ethics deals with proper professional conduct of engineers. It means following applicable laws and good engineering practices. Ethical failures are deliberate act for some tangible and intangible gains. The temptations could be money, status or any other immoral, unethical or illegal gains. Act of ethical failures can result not only in financial loss but also can endanger public safety. Preventive measures include awareness and legal enforcement.

Review Questions

1. Explain why human factor is important for successful operation of plant and machinery?
2. Explain the factors responsible for human error—both intentional and un-intentional.
3. Discuss classification of human errors with example.
4. Discuss, how maintenance errors can be reduced?
5. Illustrate with example how to reduce errors at design stage.
6. Discuss various method of prevention of human error.
7. Explain in detail—quality circle method of error mitigation.

8. What is engineering ethics?
9. Differentiate between human errors and ethical failures.
10. Discuss method of prevention of ethical failures.

Discussion Questions

1. "It is difficult to design equipment that is "idiot proof". Do you agree?
2. "Majority of assembly, installation and maintenance errors can be taken care off at design stage" Justify.
3. Discuss – "Ethical failures are deliberate acts for some tangible or intangible gains".
4. Elaborate "To be ethical professionally, one has to be ethical in personal life".
5. Discuss, "Legal frame work is necessary to present ethical failures".

Multiple Choice Questions

1. Operational errors are due to:
   (*a*) Wrong choice of materials.
   (*b*) Task complexity.
   (*c*) Incorrect fitment.
   (*d*) None of these.
2. Maintenance errors are due to:
   (*a*) Wrong specification.
   (*b*) Missing data in blue print.
   (*c*) Wrong spare part.
   (*d*) Environmental conditions.
3. Human errors can be prevented by:
   (*a*) Man—machine system analysis.
   (*b*) Error—cause removal method.
   (*c*) Quality circle approach.
   (*d*) All of these.
4. Ethical failure is:
   (*a*) Bribery.
   (*b*) Conflict of interest.
   (*c*) Mis—representation of facts.
   (*d*) All of these.

True/False Questions

1. Majority of accidents and disaster can be averted by presence of mind.
2. Maintenance errors can be avoided by proper training and following standard operating procedure.
3. Man-machine system analysis method concentrates on preventive measures.
4. Quality circle approach uses statistical methods to reduce errors.

ANSWERS

Multiple Choice Questions

1. (*d*) **2.** (*c*) **3.** (*d*) **4.** (*d*)

True/False Questions

1. True **2.** True **3.** False **4.** True

Case Study Based Question

Mr. Subrahmaniam is a chief engineer in a large industrial enterprise. Being highly experienced, he also does consultancy during holidays. His employer is inviting consultancy bids for an expansion project. Mr. Subrahmaniam has also submitted consultancy bid for the project. His private practice is not known to his employers. His bid being lowest is awarded with the contract. During the course of work, employer come to know about Mr. Subrahmaniam's consultancy company doing the job. Job is 80% over to complete satisfaction of the management. Now the management is facing the dilemma whether to continue with the remaining part of contract or to terminate the contract.

You as the management representative on board, suggest measures to be taken. Supplement your answer with logic.

❑❑❑

PART V

Maintenance Engineering

Strategic Outline

Maintenance engineering is the "know-how" part. It deals with all technical aspects of maintenance.

Non-destructive and destructive testing deals with various techniques available and their area of applications.

Condition monitoring deals with objectives, advantages and characteristics. It explains steps in implementation process.

Lubrication practices deals with basics of lubricants including testing and reclamation.

Failure analysis and reliability engineering deals with types, causes and prevention techniques.

Maintenance of electrical machine explains types and cause of faults, various condition monitoring techniques including details of batteries and earthing.

Thermal insulation and refractory deals with function, classification and properties of thermal insulation including refractory material.

Energy conservation and energy audit deals with energy conservation opportunities in industries, energy efficient devices, green building, and energy audit.

Contd...

Chapters in this section

20

Non-Destructive and Destructive Testing

Learning Objectives

After reading this chapter, you will be able to:

- Understand various testing techniques.
- Develop a keen understanding of application of these techniques.
- Develop an insight into information conveyed by the test results.

20.1 INTRODUCTION

Non-Destructive testing (NDT) involves examination of materials (metallic and non-metallic), components and equipments without damaging their utility. It is one of the main tools to assess the condition of components, materials and equipments. These tests can be performed on-line as well as off-line. Defects can be detected in raw material stage or during manufacturing stage or during operational stage. While destructive testing involves examination of material resulting in damage or destruction of the test specimen.

In raw materials, defects like inclusion, lamination , cavity, thickness variation *etc.* can be detected. During manufacturing stage, defects like blow-holes, shrinkages, voids, slag, porosity, scans, pin holes, lack of penetration, lack of fusion, lack of bonding *etc.* can be easily detected. Other operational defects like corrosion, stress cracking, leaks, fatigue fractures, inter-granular cracking *etc*. can be detected. A timely detection of such defects helps to avoid an impending catastrophic failure. With new techniques being added, it has become a specialist's field.

NDT methods are not only used for impending failure detection but also for renovation, modernisation and life extension of critical equipments like boilers, turbines, alternators and structures (bridges, dams *etc*).

20.2 NDT METHODS

Various NDT methods used are following:

(*i*) Visual examination.

(*ii*) Dye penetration examination.

(*iii*) Magnetic particle examination.

(*iv*) Ultrasonic examination.

(*v*) Eddy current testing.

(*vi*) Radio-graphic examination.

(*vii*) Endoscopy (Fibroscopy/Boroscopy).

(*viii*) In-situ metallography testing.

(*ix*) Electron diffraction examination.

(*x*) X-ray diffraction examination.

(*xi*) Acoustic emission testing.

(*xii*) Wear particle analysis (spectroscopic wear particle analysis /magnetic chip detection).

(*xiii*) Vibration analysis (Frequency analysis/phase analysis).

(*xiv*) Infra-red imaging etc.

All NDT methods mentioned above are based on changes in structure of material. Accordingly, based on effect produced by the defects in the material. All NDT methods can be broadly classified into following groups:

1. Physical effects monitoring.
2. Particle effects monitoring.
3. Dynamic effects monitoring.
4. Chemical effects monitoring.
5. Temperature effects monitoring.
6. Electrical effects monitoring.

20.3 PHYSICAL EFFECTS MONITORING

Physical effect monitors changes in structure of material e.g. dimensional changes, development of cracks and fractures, visible effect of wear *etc*. Various methods are:

(*i*) Liquid dye penetration.

(*ii*) Electrostatic fluorescent dye penetration.

(*iii*) Magnetic particle test.

(*iv*) Ultrasound pulse echo technique.

(*v*) Ultrasound resonance technique.

(*vi*) Ultrasound frequency modulation.

(*vii*) Coupon testing.

(*viii*) Eddy current testing.

(*ix*) Radiography.

(*x*) Endoscopy.

(*xi*) In-situ metallography.

(*i*) *Liquid Dye Penetration:* This technique detects surface discontinuities due to fatigue, wear, surface shrinkage, heat treatment, corrosion stress, corrosion fatigue and hydrogen embrittlement.

In this method, surface is first thoroughly cleaned and a liquid dye is applied. After soaking for sufficient time, excess dye is wiped off. Thereafter, a developer is applied over the test surface which draws out the penetrant from cracks to the surface. The crack then easily shows up and is evaluated.

This technique can be used if the crack is large enough to allow the penetrant to seep through. However, this method cannot be used with porous or chemically sensitive material.

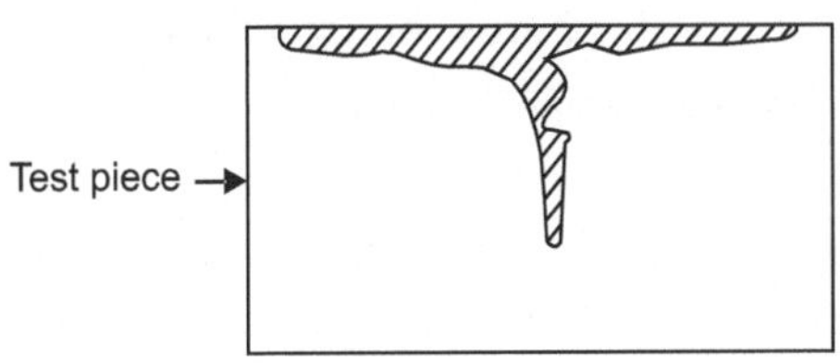

Fig. 20.1 Crack after Application of Penetrant.

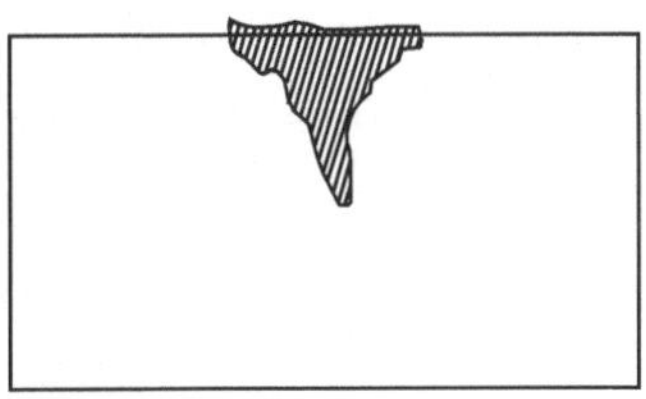

Fig. 20.2 After Removal of Excess Penetrant and Application of Developer.

(*ii*) *Electrostatic Flourescent Dye Penetrant:* This technique is very much similar to liquid dye penetrant except that in this method opposite polarities are induced in work piece and testing material. This ensures better visibility of crack due to even deposition of penetrant and developer.

(*iii*) *Magnetic Particle Inspection:* Surface and near surface cracks and discontinuities caused by fatigue, wear, lamination, surface shrinkage, grinding, heat treatment, corrosion stress are indicated by this method.

Only ferro-magnetic metals can be used in this test. The test can be conducted on welded surface, machined surface, boilers, steel structures *etc.*

In this method, magnetic field is induced in the part to be tested. A solution containing fine iron particles are sprayed on the part. Iron particles settle down on either side of the crack as magnetic flux gets concentrated due to presence of air gap. This makes crack visible.

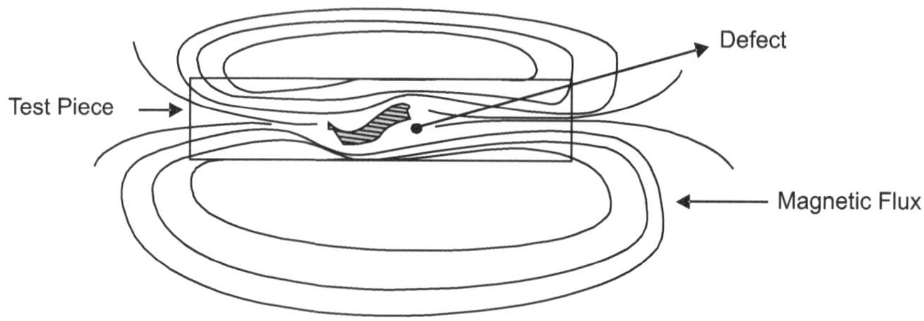

Fig. 20.3 Magnetic Particle Test.

(*iv*) *Ultrasound Pulse Echo Technique:* This technique is applicable to ferrous and non-ferrous weld, steel structures, boiler tubes, shafts *etc.* Surface and below surface cracks can be detected through this method.

An ultrasound (sound having more than 20 KHz frequency) is used to send ultrasonic pulses into the test piece. Pulse returned from the test piece is amplified and fed to an oscilloscope (C.R.O) The echo is a combination of return pulses from opposite side of test piece and from cracks present. Time elapsed between go and return signals indicates presence of cracks.

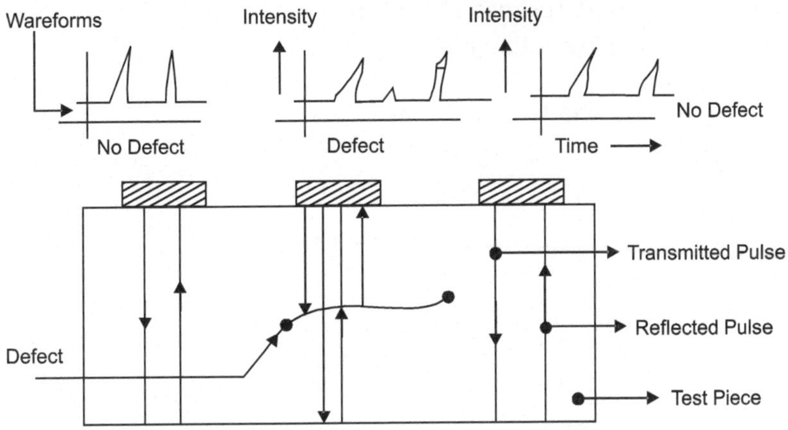

Fig. 20.4 Ultrasound Pulse Echo Technique.

(*v*) *Ultrasound Resonance Technique:* This technique is similar to pulse echo technique. It is based on the fact that presence of discontinuities cause the transmitted signal to fade or disappear. Absence of any discontinuity keeps the transmitted signal high.

(*vi*) *Ultrasound Frequency Modulation:* In this method, a transducer sends ultrasound waves continuously at changing frequencies and intercept the new changed frequency. By measuring the phase between frequencies, the location of defect can be detected.

(*vii*) *Coupon Testing:* Purpose of coupon testing is to detect localized erosion and corrosion of metal. It is a very widely used method of detecting corrosion in vessels, pipes *etc*. Coupons are made of mild steel or low carbon steel or from grade of material of vessel or pipes. These coupons are measured and weighed before being immersed in the process stream for a period ranging from few weeks to several months. Thereafter, coupons are removed and weighed for metal loss or pitting. From the observation, relative metal loss from pipes or vessels can be estimated.

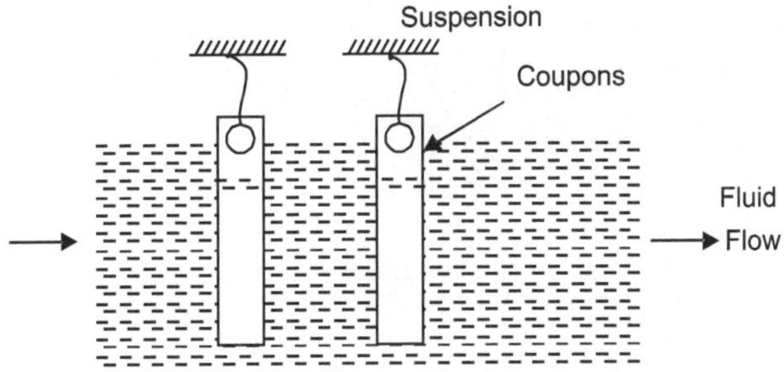

Fig. 20.5 Coupons Suspended in the Stream.

(*viii*) *Eddy Current Testing*: This technique is used to detect surface and sub-surface discontinuities

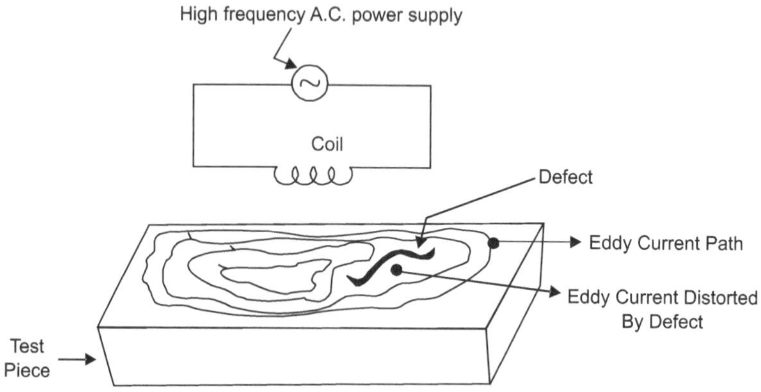

Fig. 20.6 Eddy Current Testing.

The method consists of a coil carrying alternating current at 0.1 to 4MHz frequency. This coil induces eddy currents in the test piece. Presence of discontinuities interrupts eddy current path. The eddy current detour around crack. The electrical reaction on the test piece is amplified and displayed in oscilloscope.

Ferrous metals used for boiler tubes, heat exchanger tubes, hydraulic tubing, hoist ropes, railway lines and overhead conductors are tested by this method.

(*ix*) *Radiography:* This technique is used to detect surface and sub-surface discontinuity in welds, steel structures, plastic material *etc.*

A radiograph is produced by passing X-ray or Gamma ray through the test object. A crack or a void appears as a dark patch

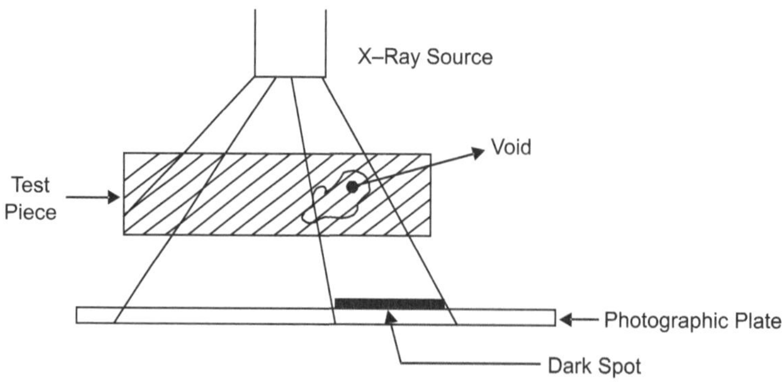

Fig. 20.7 Radiography

(*x*) *Endoscopy:* It is sophisticated instrument for visual examination of inaccessible internal parts through a periscope (used in submarines to detect ships) like instruments. It is used to detect surface cracks. Oxide films, weld defects, corrosion and wear fatigue. It is used for inspection of narrow tubes, bores of engines, pumps, turbines, compressors and boilers. These are used extensively in ship building, air-craft, power generation and chemical industries.

Endoscopes are of two types (*i*) flexible type called Fibroscope and rigid type called Boroscope.

It consists of thousands of glass strands which transmit light from light source to object. The light intensity is a high as 300 watt. The images are either taken into CCTV or photographic attachment for permanent record. The working length of instrument varies from 250 mm to 2050 mm with a diameter of 6 mm to 14 mm.

(*xi*) *In-situ Metallography:* It is an optical microscopy technique for the determination of micro-structural state of a component. This technique enables to determine the degree of degradation of a component.

Major modes of component degradation are micro-structural decomposition, propagation of fine cracks, micro-structural voids, corrosion attacks in the form of selective elemental leaching, oxidation attack at grain boundaries and pitting.

In this technique, surface to be examined is first prepared by grinding with a hand held grinder followed by a fine polishing using diamond and alumina paste of 1-5 micron particle size. Thereafter surface is washed and dried in acetone and then etched using a suitable reagent. The prepared surface is then examined using a portable hand held microscope with 100 to 500 times magnifying power.

20.4 PARTICLE EFFECTS MONITORING

This technique utilizes detection of particles of different shapes and sizes released into the system. Monitoring of these particles indicates potential failures.

Various methods are following:

(*i*) Ferrography.

(*ii*) Graded filteration.

(*iii*) Magnetic chip detection.

(*iv*) X-ray fluorescence.

(*v*) Blot test.

(*vi*) Light detection and ranging.

(*i*) *Ferrography:* This technique detects ferromagnetic wear particles upto 300 micro-metre in systems such as gas turbines, diesel engines, helicopter gearbox, *etc*. Ferrograph is an analytical instrument which determines the quantity and size of wear debris in an oil sample. The device can measure particle size distribution.

The process involves dilution of oil sample with a solvent to increase mobility of particles. It is then led into an inclined glass slide about 60 mm long. The oil flows down due to gravity through a magnetic field. The magnetic field is stronger at upper end of glass slide than at the lower end. With the result higher size particles are retained at the input to the glass slide and smaller particles are distributed down the slide. This slide is called "Ferrogram" which can be studied under bichromatic microscope or electron microscope to reveal details of shape, size and quantity of particles. If the number of particles and ratio of large to small particles increases it is an indication of severe wear. Generation of large wear particles signals imminent failure of wear surface. Increase in number and size of particles such as loops, spirals and bent wires indicates abrasive wear. An increase in particles such a spheres, chunks, laminar platelets indicates fatigue failure of rolling element bearings.

(*ii*) *Graded Filteration:* This technique is used to detect particles (iron, copper, lead, chromium, aluminium, silicon) in lubricating and hydraulic oils. These particles are generated in system due to wear, fatigue and corrosion. The method is used in closed loop oil systems such as lubricating and hydraulic oils systems in engine sumps, gear boxes and hydraulic system.

The process consists of passing diluted sample of oil through a series of calibrated membrane filters. Particles so collected are counted under a microscope for various size groups. A statistical analysis of particle distribution profile indicates level of wear (Fig. 20.9).

(*iii*) *Magnetic Chip Detection:* This technique is applicable to closed loop lubricating oil systems and is used in engine sumps, gear boxes, turbines, compressors *etc*.

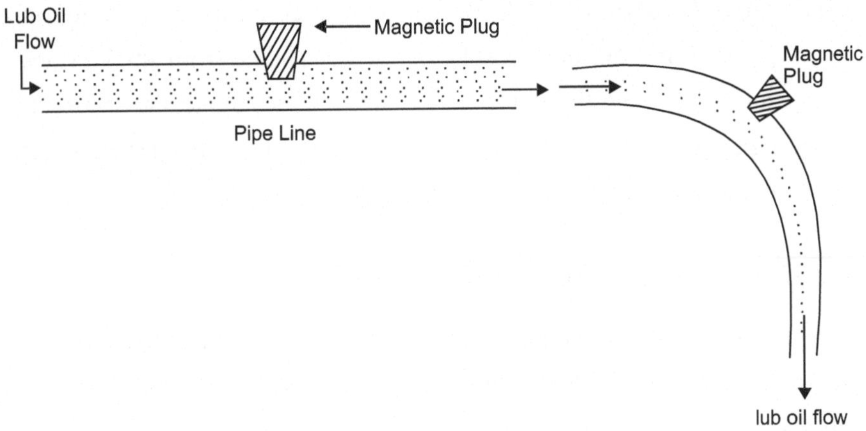

Fig. 20.8 Magnetic Chip Detection.

A magnetic plug is mounted at a suitable location in the lubricant oil circuit. Metal particles present in oil get attracted to the magnetic plug. These particles are later collected and analysed under a microscope.

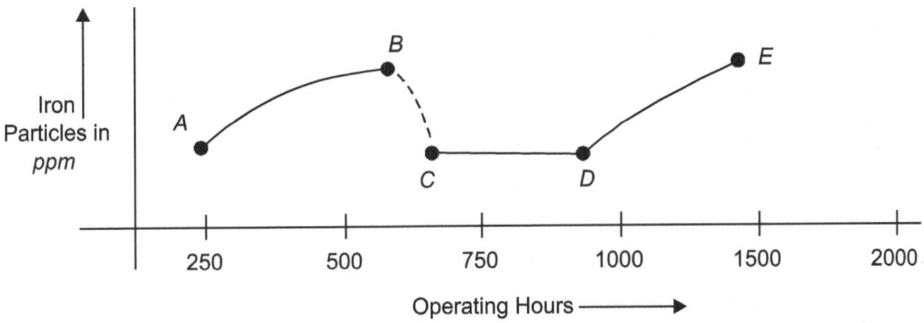

Fig. 20.9 Iron Contamination Curve.

(*iv*) *X-ray Fluorescence:* This technique uses exposure of oil sample to a source of radiation leading to emission of *X*-ray by the contamination. Type and quantity of contamination is identified by rate of X-ray emission and its spectrum.

The technique is used in closed loop lubricating and hydraulic oil systems such as engine oil sumps, gear boxes and hydraulic system.

(*v*) *Blot Testing:* In this technique, a drop of sample oil is deposited on a filter paper. This oil drop spreads out on the filter paper and contamination particles remain in the circle formed by the oil spread. After an interval of 24 hours, the filter paper is analysed photometrically.

This test generally detects corrosion particles, fatigue and wear.

(*vi*) *Light Detection and Ranging:* This technique uses detection of particles present in atmosphere. Such as smoke from smoke stack. A single wavelength light is passed through the smoke and scattering of light by the particles is measured. This gives clue to the particles present in the smoke.

20.5 DYNAMIC EFFECTS MONITORING

This technique detects potential failures which cause abnormal release of energy in the form of vibration, noise, shock pulse *etc*. Various such techniques are following:

(*i*) Broad band vibration analysis.

(*ii*) Octave band analysis.

(*iii*) Narrow band frequency analysis.

(*iv*) Proximity analysis.

(*v*) Shock pulse monitoring.

(*vi*) Kurtosis.

(*vii*) Acoustic emission monitoring.

(*viii*) Ultrasound leak detection

(*i*) **Broad Band Vibration Analysis:** All rotating machinery transmit vibrations. The normal vibration characteristics changes due to wear, mis-alignment, imbalance, fatigue, mechanical looseness *etc*. Hence, by monitoring vibration at bearings (journal and roller), gear boxes, shafts, engines, pumps, compressors *etc*., cause of change of vibration characteristic can be predicted.

A vibration meter (analogue or digital) is used with a probe. The meter can measure displacement (micron), velocity (mm/sec) and acceleration (mm/sec^2). Depending upon the speed of machinery either displacement or velocity or acceleration is measured. When speed is less than 600 rpm, displacement is measured, speeds between 600 to 60,000 rpm, velocity is measured and for speed more than 60,000 rpm, acceleration is measured.

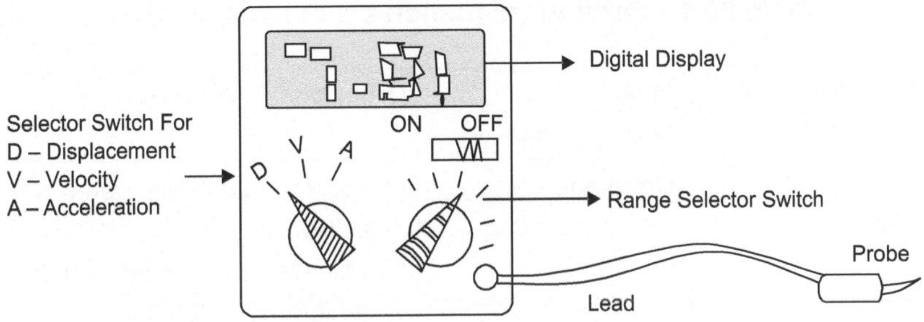

Fig. 20.10 Digital Vibration Meter.

The significant characteristics of vibration are:

(*a*) *Amplitude of Vibration:* It determines the severity of vibration. It is measured as displacement or as velocity or as acceleration.

(*b*) *Frequency of Vibration:* It identifies the problem with the component causing the problem.

(*c*) *Phase of Vibration:* It helps in identification of certain typical problems.

The vibration measured with vibration meter is analysed by trend monitoring. A graph is photted – frequency vs amplitude and frequency vs velocity. This helps to determine the root cause of the problem. This is known as vibration signature analysis.

Vibration limits are allowed as per acceptance criteria given by manufacturer or relevant BIS specifications.

(*ii*) **Octave Band Analysis:** This technique divides the frequency spectrum into series of bands of interest through fixed octave and fractional octave filters. Output from each filter is plotted successively on a logarithmic frequency scale. This plot is compared with the previous plot for fault detection.

(*iii*) **Narrow Band Frequency Analysis:** This technique uses an accelerometer which detects vibration and converts it into electrical signal. This electrical signal is amplified and fed into an analyzer. A constant bandwidth of 3.16 to 100 Hz and a frequency range of 2 to 200 Hz is chosen.

This technique helps to identify broad band vibration and multiple harmonics.

(*iv*) **Proximity Analysis:** This technique is used to detect misalignment, oil whirl, imbalance or bent shaft. Process consists of feeding signal from a transducer to the object under study. A plot is obtained between signal and time base. In some cases, two signals may be used to produce information on x-y diagram with a single impulse input from a transducer, a sinusoidal plot indicates presence of imbalance, bent shaft and oil whirl.

Table 20.1 Vibration frequencies and likely causes

Frequency in terms of RPM	**Most Likely cause**	**Other possible causes**
1 × RPM	Unbalance	(i) Eccentric Journals, gears and pulleys. (ii) Mis-alignment or bent shaft, if high axial vibrations. (iii) Bad belts. (iv) Resonance. (v) Electrical Problem.
2 × RPM Looseness	Mechanical	(i) Misalignment-if high axial vibrations. (ii) Reciprocating forces. (iii) Resonance. (iv) Bad belts.
3 × RPM	Misalignment	Combination of misalignment and excessive axial clearance (Looseness).
Less than 1 × RPM	Oil whirl (less than ½ RPM)	(i) Bad drive belts. (ii) Background vibrations. (iii) Sub-harmonic resonance. (iv) Beat vibrations.
Many times RPM (Harmonically related Frequency)	(i) Bad gears. (ii) Aero-dynamic forces. (iii) Mechanical Looseness. (iv) Reciprocating forces.	(i) Gear teeth times RPM of bad gear. (ii) No. of fan blades times RPM. (iii) No. of impeller vanes times RPM.
Synchronous (AC line frequency)	Electrical Problems.	(i) Broken rotor bar. (ii) Eccentric rotor. (iii) Unbalanced phase in poly-phase system. (iv) Unequal air gap
2 × Synchronous Frequency	Torque Pulses	Rare as a problem unless there is resonance.

(*v*) **Shock Pulse Monitoring:** This technique is used to detect surface deterioration and lack of lubrication in ball bearings, valves of internal combustion engine and pneumatic impact tools.

During operation of machinery, an accelerometer is used to detect shock waves. The signal thus obtained is passed through a band pass filter which selects frequencies higher than 10 KHz. This high frequency input is converted into square pulse. The peak value of these pulses indicate extent of damage.

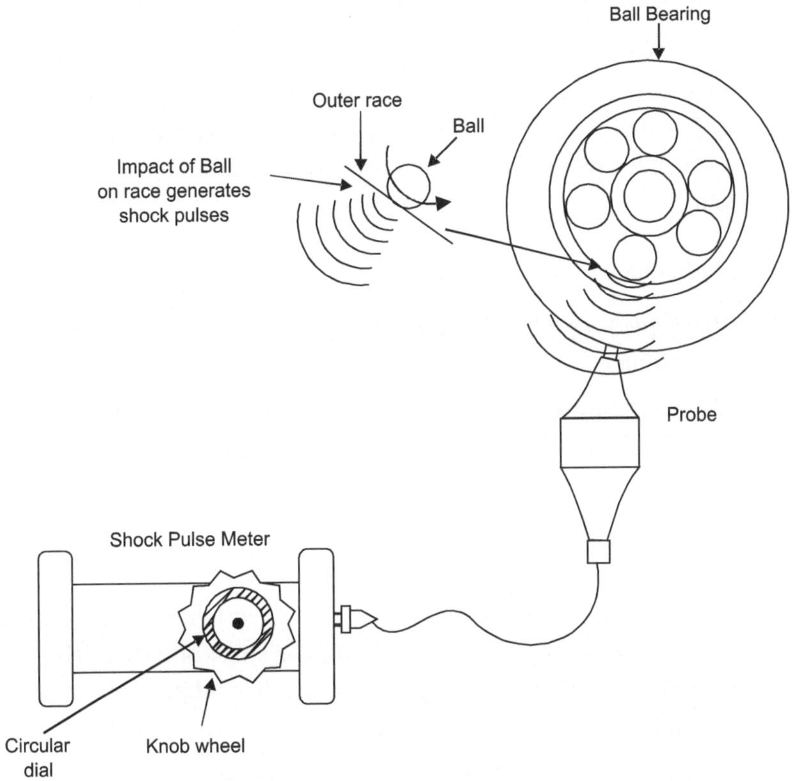

Fig. 20.11 Shock Pulse Measurement

The running surface of a bearing has certain degree of roughness even when it is new. This generates low level acoustic emission. But as the rolling element deteriorates the rolling surface suffers damage. Cracks and pitting appear and small particles of metal come off and get circulated in the bearing. As these debris pass through contact zone, a small "Knock" is generated. As the bearing further deteriorates, the "Knock" increases in amplitude. Shock pulse meter (SPM) measures these "Knocks".

Normalised = $dB_M = dB_{sv} - dB_I$

Where = dB_{sv} = Total shock pulse value.

dB_I = Initial shock pulse value.

(depends on speed and bore diameter of bearing)

dB_N = True measure of operational condition.

dB_N scale is divided into three zones as per following:

(*a*)	$dB_N < 20$	Good operational condition.	Green zone
(*b*)	$dB_N = 20\text{-}35$	Caution zone.	Yellow zone
(*c*)	$dB_N > 35$	Bad operational condition.	Red zone

(*vi*) **Kurtosis:** Kurtosis is a statistical method which examines "peakiness" of distribution of vibration amplitude. It is defined as fourth moment normalized by mean ($\bar{x}$) and standard deviation (σ) raised to power of four.

Mathematically, Kurtosis value $\beta = \int_{-\infty}^{\infty} \frac{(X - \bar{X})^4 p(\bar{X})}{\sigma^4} dx$

Where p(x) = probability density function is instantaneous amplitude (x) at time *t*.

When a rolling element strike damaged part of bearing, transient spikes are produced. These signals appear regularly at frequency equal to rotational frequencies of different parts of bearing. The kurtosis is a measure of these wave shapes.

A Kurtosis meter is used to measure kurtosis value.

(*vii*) **Accoustic Emission Monitoring:** This technique is used to detect deformation and cracks caused by fatigue, stress and wear. This method is used in metal structures, pressure vessels and pipelines.

For measurement of acoustic emission, elastic waves generated by sudden deformation in stressed material are sensed and converted into electrical signals. The acoustic emission instrumentation measures these signals and helps to evaluate behaviour of materials.

(*viii*) **Ultrasound Leak Detection:** This technique measures ultrasound (sound of more than 20 KHz frequency) generated by escaping high pressure fluids such as steam, gas, refrigerant, compressed air and high voltage "corona" discharge from electrical equipments.

Generation of ultrasound energy takes place due to molecular collision caused by escape of gas through a small orifice. This ultrasound energy is picked up by a high frequency ceramic microphone. The signal so obtained is amplified and fed to either a moving coil meter or audible signal headphone.

20.6 CHEMICAL EFFECT MONITORING

This technique is based on chemical effect monitoring of elements. Presence of certain elements indicate potential failure. Various materials released in the system and their origin are following:

Table 20.2 Various metals and their origin

S. No.	Metal	Origin
(i)	Aluminium	Piston rings.
(ii)	Antimony	Greases and bearings.
(iii)	Barium	Water additives and greases.
(iv)	Copper	Bearing and bushes.
(v)	Cadmium	Bearing.
(vi)	Chromium	Piston rings and liners.
(vii)	Cobalt	Bearings and turbines components.
(viii)	Iron	Gears, cylinder liners and crankshaft.
(ix)	Lead	Bearings, bushes, lubricants and paints.
(x)	Manganese	Shafts and Valves.
(xi)	Molybdenum	Piston rings and additives.
(xii)	Nickel	Shaft, gears and turbine components.
(xiii)	Tin	Bearing and coolers.
(xiv)	Zinc	Bearings and additives.

Various chemical effects monitoring techniques are following:

(*i*) Spectroscopic oil analysis.

(*ii*) Gas chromatograph.

(*iii*) Liquid chromatograph.

(*iv*) Infrared spectroscopy.

(*v*) Ultra violet and visible spectroscopy.

(*vi*) Electrochemical corrosion monitoring.

(*i*) *Spectroscopic Oil Analysis:* In this method, a small sample of oil is taken periodically from equipment's oil system. The oil sample is then analysed for the metal contents. The analysis is carried out either through emission spectroscopy or through absorption spectroscopy. Though absorption spectroscopy is more accurate but is time consuming.

In atomic absorption spectrometry, the oil sample is diluted and then sprayed into a flame and the atomic absorption spectrum is measured. In emission spectrometry, the sample oil is "ashed" then put into an arc to vapourise. The emitted radiation from the vapourised metal is measured. The essence of either technique is raising the molecules to an excited state by applying some form of energy (heat from a flame or very high voltage, 15000 V) to the sample and observing wavelength of resulting radiation. Hence, composition and quantity of metal present can be determined.

When the amount of certain metal rises above a pre-determined limit, this is interpreted as a signal to inspect a particular part which is composed of the metal. For example, increase in copper means bearing wear. These metals accumulate in oil system due to wear of associated metal components and leakage of metal particles into the oil system.

(*ii*) *Gas Chromatography:* In certain type of faults, gases are generated and Gas-chromatography detects presence of these gases such as Hydrogen, Sulphur Hexafluoride, Nitrogen, Air etc.

This technique is used in nuclear power plants and turbine generators having Sulphur Hexafluoride, and Hydrogen sealed systems.

In this process, gases are selectively absorbed by passing them through columns of finely divided absorbent material such as diatomaceous earth. The process is complicated and requires considerable skills.

(*iii*) *Liquid Chromatography:* This technique measures changes in lubricant property such as viscosity, acidity, alkalinity, ash, flash point and insoluble. This technique is useful in enclosed systems such as engine sumps, compressor sumps, hydraulic systems, transformer etc.

In this process, liquid is passed through absorbents such as diatomaceous earth. The instrument display or recorder shows concentration of various liquids.

(*iv*) *Infrared Spectroscopy:* This technique detects presence of gas such as sulphur hexafluoride, hydrogen, nitrogen, carbon-monoxide and ethylene. It is based on the principle that a sample in a beam of infrared light absorbs certain frequencies. These absorbed bands when plotted against wavelength indicate presence of these gases.

(*v*) *Fluorescent Spectroscopy*: This technique detects changes in properties of oil such as alkalinity, acidity *etc*.

In this process, the sample is irradiated and degradation products show fluorescence and phosphorescence spectra.

(*vi*) *Ultraviolet and Visible Spectroscopy:* This technique is similar to fluorescent spectroscopy except that the sample is subjected to intense ultra-violet light. The gases present absorb the radiation at specified wavelength.

(*vii*) *Electro-chemical Corrosion Monitoring:* This technique is used to detect corrosion of metals embedded in concrete e.g. structural elements.

It is based on the fact that presence of corrosion causes change of potential between ground and the structure.

In this method, small currents are passed between structure and a probe inserted in the ground near to the structure. The potential is measured by a cell in contact with the ground and close to structure. High currents indicates need for inspection.

20.7 TEMPERATURE EFFECTS MONITORING

Rise in temperature is a potential cause of failure in equipments and components. The various techniques of detection of rise in temperature are following:

(*i*) Thermography.

(*ii*) Fiber loop thermometry.

(*iii*) Temperature indicating paints.

(*i*) *Thermography*: It is based on the principle that all objects above absolute zero temperature (—237°C) emit infrared radiation. This infrared emission is detected by a infrared camera which produces real time thermal pictures. Variations in surface temperature are seen as light and dark areas.

It is used in power transmission lines, transformers, electrical switchgears, refractories, bearings *etc*.

The main advantage of this technique is that stationary or moving objects can be examined at a distance without touching or influencing the temperature of object.

(*ii*) *Fiber Loop Thermometry*: This technique is used to detect temperature variations caused by blocked cooling systems, insulation deterioration, leaks *etc*. Potential applications are in pipelines, engines, transformers, windings and power cables.

The method consists of passing light through fiber optic cable. A certain amount of light is reflected back towards light source and this reduces the strength of outgoing signal. There is a direct mathematical relationship between the time it takes the light to travel down the cable for a given distance, the amount of reflected light and temperature of cable. This relationship is used to calculate temperature at a given point in a cable.

(*iii*) *Temperature Indicating Paints:* This technique uses a silicon based paint which changes colour as temperature rises. The colour starts out green, changes to blue at 204°C and turns white at 310°C. Change of colour is permanent.

It is used to detect hot spots and insulation failure.

20.8 ELECTRICAL EFFECTS MONITORING

It uses changes in resistance, conductivity, dielectric strength, potential, high voltage *etc*.

The various techniques that monitor electrical effects are following:

(*i*) Insulation resistance/polarization Index measurement.

(*ii*) DC resistance measurement.

(*iii*) High voltage test.

(*iv*) Dissolved gas analysis.

(*v*) Frequency response analysis.

(*vi*) Acoustic partial discharge measurement.

(*vii*) Tan delta measurement.

(*viii*) Capacitance measurement.

(*ix*) DC leakage current test.

(*x*) ELCID Test.

(*i*) *Insulation Resistance/Polarisation Index Measurement*: Insulation resistance of all electrical equipments (motors, generators, transformers, switchgears *etc*.) is a very vital parameter for diagnosis of health of an electrical equipment. A low value insulation resistance may result in burn-out of the equipment when supply is switched on.

The range of meggar is from 0.25 to 10 kv. Insulation resistance is measured between phase and earth to phase. A low value of insulation resistance is a cause of concern.

For high voltage motors and generators, apart from insulation resistance, polarization index is measured. A polarization index of 1.5 or more indicates healthy state of insulation.

$$\text{Polarisation Index} = \frac{\text{Insulation resistance measured for 10 minutes}}{\text{Insulation resistance measured for 1 minute}}$$

(*ii*) *DC Resistance Measurement Test:* Cold resistance measurement of windings of electrical equipments provides a clue to condition of windings. Almost equal winding resistance of three phases of an electrical machine indicates healthy condition.

Resistance is measured with a sensitive ohmmeter or by ammeter-voltmeter method.

(*iii*) *High Voltage Test:* For high voltage equipments (transformers, alternators, motors *etc*), high voltage test at power frequency indicates condition of windings insulation. A successful high voltage test indicates good condition of windings insulation.

(*iv*) *Dissolved Gas Analysis:* This test is performed for power and distribution transformers of large rating. It is based on the fact that any electrical or thermal fault is transformer causes formation of gases. These gases (hydrogen, methane, ethane, acytylene, carbon dioxide, nitrogen, oxygen) are measured from oil samples. The rate of gas generation and history of previous faults are used for accurate interpretation. Following three methods are used:

- Ratio technique — In this method, 5 gases and 3 ratios are used, these ratios are CH_4/H_2, C_2H_4/C_2C_6 and C_2H_2/C_2H_4. The ratios indicate type of fault.
- "Key Gas" method — Few important gases are called "Key Gases". These gases are produced due to four types of faults:
  (*a*) Thermal (oil decomposition) (b) Thermal (cellulose decomposition) (c) Electrical (Corona) and (d) Electrical (arcing). Based on analysis of these gases, it is possible to predict the fault.

(*v*) *Frequency Response Analysis (FRA):* This technique is used for transformers. In this technique, a network/spectrum analyzer is used with three co-axial test leads. FRA scanning is carried out at 2 KHz,

20 KHz and 2 MHz. The transformer taps are kept at minimum, middle and maximum position. A sinusoidal signal input of 2V rms is given at one input and output is connected at other end. The voltage transfer function is measured for three standard frequencies. The amplitude and phase shift results are recorded for subsequent analysis.

Each winding design has a characteristic frequency response and any shift in winding or movement due to fault affects inductance and capacitance of windings. These changes show-up in frequency response curve.

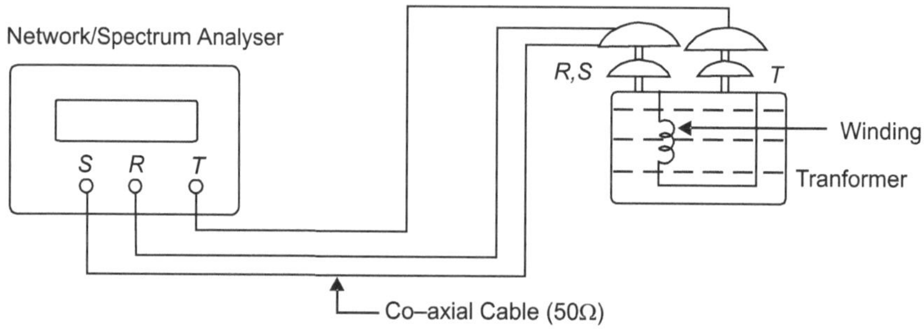

Fig. 20.12 Frequency Response Analysis.

(*vi*) *Acoustics Partial Discharge Measurement:* This test applicable to transformers and alternators. Accoustic partial discharge detector consists of an acoustic sensor which can pick up sound vibrations produced due to discharge in windings. These sound signals are converted into electrical signals. The assembly sensor has filter circuits to eliminate other environmental noise. The location of discharge can be located accurately.

(*vii*) *Tan Delta Measurement:* It is an important measure of quality of insulation. It directly indicates power dissipated by the insulation. A high dielectric loss results in thermal breakdown of the material. The variation of tan-delta with applied voltage provides useful information about the source of any imperfections in the insulation. Tan-delta value increases with insulation deterioration and serves as an early indicator of failure.

A low value of tan-delta is desirable. In a good insulation, tan-delta does not increase with increase in applied voltage. A higher value of tan-delta may indicate void, impurities, lamination *etc*.

(*viii*) *Capacitance Measurement*: Condition of insulation can be judged by measurement of capacitance. Capacitance value is dependent upon type of dielectric material and configuration of electrodes. Capacitance is measured with increase in applied voltage. The measured value of capacitance increases with increase in applied voltage then it indicates partial discharge due to voids.

(*ix*) *DC Leakage Current Test:* When a DC voltage is applied to insulation, three components of currents flow, the capacitive charging current, the absorption current which decay with time and a leakage current which is determined by the presence of moisture, dirt *etc*. The leakage current rises quickly and becomes stable. The test values of leakage current are taken after first two components of current become negligible.

The progressive increase in leakage current over a period of time indicates deterioration of insulation.

(*x*) *ELCID Test:* Defects in interlaminar insulation of stator (of alternator) cause fault currents to flow locally in the core. These currents can produce hot spots in the area and core may become progressively damaged.

In ELCID test (Electromagnetic Core Imperfection Detection), a small fraction of rated flux to generate fault current is picked up by the sensor.

This test can assess the condition of stator core and gives vital information for trend analysis.

20.9 DESTRUCTIVE TESTS

In destructive test, while determining various parameters of test piece, the sample losses its functional utility.

1. **Mechanical Parameter Monitoring:** Various Mechanical parameters monitored in destructive tests are:

(*i*) Hardness test

(*a*) Scratch hardness.

(*b*) Indentation hardness.

(*c*) Rebound or dynamic hardness.

(*ii*) Micro-hardness test.

(*iii*) Tensile test.

(*iv*) Impact test.

(*v*) Fatigue test.

(*vi*) Torsion test.

(*i*) **Hardness Test:** Hardness is the ability of a material to resist surface abrasion. It gives an idea regarding strength and heat treatment of metals.

There are three types of hardness measurement viz–

(*a*) Scratch hardness.
(*b*) Indentation hardness.
(*c*) Rebound or dynamic hardness.

(*a*) **Scratch hardness:** It is measured according to Mohr's scale consisting of ten standard minerals arranged in the order of their ability to be scratched. The softest mineral is talc having scratch harness '1' and hardest is diamond having scratch hardness '10'. It gives relative hardness but is not useful since the intervals are not evenly spaced.

(*b*) **Indentation:** In this method, a penetrator (steel ball or diamond) is pressed into the surface of a test piece using appropriate force. The surface area of indentation is measured and the hardness is derived by load (P) divided by surface area (A) Following two hardness scales are used to measure indentation hardness.

(*i*) Brinell hardness.
(*ii*) Vickers hardness.

(*i*) *Brinell Hardness:* This test consists in indenting the metal surface with 10 mm steel ball at a load of 3000 kg for 30 seconds. The ratio P/D^2 is measured where 'P' is the load and 'D' is the diameter of steel ball.

P/D^2 ratio for various material are given as an example:

(*a*) Steel and cast iron = 30
(*b*) Copper, copper and aluminium alloy = 10
(*c*) Aluminium = 5
(*d*) Tin, lead and their alloy = 1

(*ii*) *Vicker's Hardness:* In this test, metal is struck with square base diamond pyramid. The diamond pyramid has an included angle of 136° between opposite faces.

Vicker's hardness is calculated as per the following equation.

$$\text{VHN} = \frac{1.854P}{L^2}$$

where, P = Load
L = Average length of diagonal

(*c*) **Rebound or dynamic hardness:** In this test, a diamond tipped top is allowed to fall from a standard height. Hardness measured is expressed in terms of energy of impact.

(*ii*) **Micro Hardness Test:** This test is used to determine hardness over a very small area. A diamond indentor of specific geometry is used to produce indentations in a test piece.

Two methods are used:

(*a*) *Vickers Hardness Number (HV):* Vickers Hardness number (HV) is calculated as per following formula

$$HV = \frac{P}{A} = 1.8544\ P/d^2$$

Where, P = Load in kg

A = Surface area

d = Mean diameter (mm)

(*b*) *Knoop Hardness Number HK:* This method is same as described above but the geometry of indentor is different. It produces diamond shaped indentations with long and short diagonals in the ratio of 7:1. Indentor is rhombic based and included longitudinal edge angles of 172° and 130°.

Knoop Hardness Number (HK) = $14.229\ P/d^2$

(*iii*) **Tensile Test:** Tensile test provides basic information on strength of material that is ability to withstand static load.

In this test, the specimen is subjected to a continuously increasing uniaxial tensile force. It is continued till the material fails. The value so obtained indicates ultimate tensile strength such that is performed for compressive and shear stress.

(*iv*) **Impact Test:** Material may fail in a brittle or ductile manner. Brittle fracture involves little or no deformation. While ductile failure shows plastic flow of material.

Charpy impact test is used. In this test, a standard notch is cut in a standard test specimen which is struck by a pendulum. The striking energy which results in failure of specimen is calculated.

(*v*) **Fatigue Test:** When a metal is subjected to repetitive stress, it fails at a stress much lower than that required to cause fracture. Such failures which occur under conditions of dynamic loading are called fatigue failure. It occurs without any warning and it results in brittle appearing fracture with no gross deformation at fracture. Fatigue failure is enhanced by sharp corner, metallurgical stress concentration, corrosion, temperature, overload and residual stress.

A specialized rotating beam fatigue testing machine is used.

(*vi*) **Torsion test:** This test is carried out to determine torsional yield strength. Shaft axles and twist drills are subjected to torsional loading.

The test equipment consists of twisting head with other end fixed in the chuck and thus measures a twisting moment or torque. Twist is measure by measuring the angular displacement of a point near one end of the test section of specimen with respect to a point on the same longitudinal element at the opposite end.

20.10 ELECTRICAL PARAMETER MONITORING USING DESTRUCTIVE TESTING

Electrical equipments like transformers, alternators *etc*. are subjected to destructive test to measure fault withstanding capability.

Purpose of this test is to find out fault withstand capability of an electrical equipment. The test consists of subjecting the test equipment to actual fault condition and measuring and recording the fault MVA (Mega Volt Ampere) at which the equipment fails.

Such destructive tests can be performed at Central Power Research Institute, Bangalore and Bhopal.

Summary

All materials and components are examined for quality and various properties through non-destructive and destructive tests. These tests help to identify material defect, manufacturing defect and operational defects. Non-destructive test examines a specimen without impairing its functional utility while destructive test either completely or partially damages the specimen.

Non-destructive testing or NDT is a specialized field which is extensively used for impending failure detection in maintenance engineering. This knowledge is used for renovation, modernization and life extension of critical equipments. Wide variety of NDT methods are available based on six major effects *viz* - physical effect, particle effect, dynamic effect, chemical effect, temperatures effect and electrical effects. By monitoring the changes produced in the

Contd...

specimen, it is possible to anticipate various types of defects which can lead to failures.

Destructive tests include hardness testing, micro-hardness testing, tensile testing, impact testing, fatigue and torsion testing. All these are mechanical parameter monitoring. Example of electrical parameter monitoring is determination of fault-level withstand capacity of transformer, alternator *etc.*

All these tests provide a volumes of information which helps in various ways such as detection of impending failure, selection of appropriate raw material, testing of specimen after manufacturing and periodic monitoring of running equipments for operational fitness. Selection of appropriate test technique is important to achieve desired goal.

NDT method forms an important segment and these tests are performed either in laboratory or *in-situ* (at its position).

Review Questions

1. What is meant by non-destructive and destructive testing? Give an example.
2. Why is it necessary to perform non-destructive and destructive tests? What are the information revealed by these tests?
3. Enlist atleast ten NDT methods and explain any method in details.
4. What are various physical and chemical effects used to monitor condition of equipments ?
5. How will you monitor condition of insulation of a large electric motor?
6. What are the techniques used for monitoring corrosion of a vessel, pipeline and reactor inside lining?
7. Explain in detail Frequency Response Analysis method.
8. Describe various hardness tests.
9. Describe Spectroscopic Oil Analysis and Ferrography.
10. Describe various ultrasonic techniques used for surface/sub-surface flow detection.

Discussion Questions

1. "Vibration signature analysis is capable of detecting wide variety of problems in a machinery". Comment.
2. "Any physical deterioration sends out signals in various forms which if captured properly can prevent unexpected failures". Justify.

3. Discuss – "All types of NDT methods cannot describe complete physical status of a machinery".
4. Discuss – "Destructive tests provide volume of information which cannot be provided by non-destructive tests".

True/False Questions

1. Eddy current testing can be used only on insulating materials.
2. Dye-penetration test is not possible on non-magnetic materials.
3. Graded filtration method is useful in detecting particles in lubricating and hydraulic oils.
4. Magnetic chip detection method is useful is open loop system.
5. Shock pulse monitoring can be used on journal bearings.
6. Partial discharge measurement technique can be used for transformers and alternators.

ANSWERS

True/False Questions

1. False **2.** True **3.** True **4.** False **5.** False **6.** True

Case Study Based Questions

1. Mr. Sreeniwasa Rao is a senior engineer (maintenance) in M/S Ayangar Industries Ltd. Mr. Nilesh Satpaty, chief engineer, is interested in initiating predictive maintenance approach for the critical equipments. So Mr. Nilesh instructed Mr. Rao to select appropriate method for the following systems:
   (*i*) Testing of boiler tubes.
   (*ii*) Corrosion testing of chilled water pipelines.
   (*iii*) Detection of compressed air leakages.
   (*iv*) Testing of gas turbine blades.
   (*v*) Testing of hydraulic oil in a hydraulic press.
   (*vi*) Testing of high tension, 250 HP motor bearings.

You as a maintenance expert, suggest suitable testing methods for above situations.

❑❑❑

21

Condition Monitoring

Learning Objectives

After reading this chapter, you will be able to:

- Understand process of condition monitoring (CM) system.
- Gain an insight into application of various technique of CM system.
- Understand various steps involved in implemention of CMS.

21.1 INTRODUCTION

Condition monitoring is a maintenance strategy based on measurement of condition of an equipment to assess probability of future failure and taking necessary steps to prevent occurrence of failure. The technique is based on the fact that any running machinery undergoes wear and tear and hence sends out signals which indicates the extent of damage. By capturing these very signals through sophisticated instrumentation, it is possible to predict the current health of the machinery and in some cases impending failure. Hence, condition monitoring techniques enable to predict failure in a machinery before it actually takes place.

This can be illustrated with the help of an example of a human body. A normal healthy human being goes about his daily work enthusiastically. All signals given out by the body *viz* heart beat, pulse rate, body temperature, blood pressure, etc. indicates current health of a person which when measured gives a "Normal" reading. But, in the event of some abnormality in the body, all these readings change. By interpreting these signals, a doctor diagnoses the problem or a disease. Now, there are two ways in which a human being reacts—Reactive and Proactive. In reactive mode, a human being visits a doctor only when he is sick (all signals of body are in "Abnormal" range). In pro-active mode, a person undergoes periodic medical examination in order to gauge the current health status. Any abnormal reading can give an idea of an incipient disease.

For example, by checking "blood sugar", it is possible to find out if diabetes has set in or not. A person may be unaware that he has already crossed "Normal blood sugar" level and hence is in danger zone. Similarly, through colour-Doppler test it is possible to find out blockages in the artery which may lead to coronary disease in due course of time.

Hence, through periodic medical check-ups, it is possible to predict incipient diseases. In the same way, all critical machineries when periodically checked, indicates current health. There are numerous techniques of measuring, monitoring and recording parameters of machinery such as temperature, pressure, vibration, noise, leakage, thickness, odour *etc*. The various instruments used are Infrared gun (for temperature measurement), vibration meter (for measurement of vibration), Ferrograph and spectroscopic oil analysis (for measurement of contaminant in lubricant), visual test (for leakage, odour, noise *etc*), Ultrasonic test (for measurement of thickness of vessel) *etc*.

Hence, by prudently applying appropriate condition monitoring method, it is possible to state current as well as future health of machinery. This process of maintenance is "Proactive Maintenance" or "Predictive Maintenance". This process requires considerable amount of human effort and specialized expensive equipments. So, these techniques are used for "Identified Critical Equipments" in plant such as Turbines, Boilers, Reactor vessels, Alternators and other critical production machineries.

The cost and effort involved in implementation of condition monitoring should justify the benefits derived.

21.2 ADVANTAGES AND DISADVANTAGES OF CONDITION MONITORING

Following are the advantages of condition monitoring system:

(*i*) Reduction in downtime due to prior intimation about the impending failure.

(*ii*) Optimum spares cost as the spare requirements is indicated beforehand especially expensive insurance spares.

(*iii*) Efficient utilization of maintenance manpower.

(*iv*) Improved equipment reliability and availability.

(*v*) Improved equipment safety.

(*vi*) Extremely useful in areas where human intervention is difficult or hazardous *e.g.* in Nuclear power plants, aviation, space sciences *etc.*

(*vii*) Improved plant productivity.

(*viii*) Reduced maintenance cost.

Disadvantages of condition monitoring are following:

(*i*) Expensive system.

(*ii*) Non-availability of trained man-power.

(*iii*) Few select machines can be monitored due to cost factor.

(*iv*) Periodic up-keep and calibration of equipment is required.

(*v*) Data collection and analysis is time consuming process.

21.3 OBJECTIVES OF CONDITION MONITORING

Objectives of condition monitoring are following:

(*i*) Determination of current health status of machinery.

(*ii*) Basis for preventive maintenance interval.

(*iii*) Reduction of breakdowns due to incipient faults.

(*iv*) Trend analysis.

(*v*) Spare parts provisioning.

(*vi*) Basis for overhaul, reconditioning and retrofitting.

(*vii*) Basis for equipment replacement decisions.

(*viii*) Basis for performance improvement, modifications and design-out maintenance.

(*ix*) Reduction of failure cost.

(*x*) Improvement of equipment and system reliability and availability.

21.4 CHARACTERISTIC OF CONDITION MONITORING

Following are the characteristics of condition monitoring:

(*i*) Highly sophisticated process.

(*ii*) Required well trained manpower in use of monitoring, data collection and data interpretation.

(*iii*) Expensive test and analytical instruments.

(*iv*) May require service of external experts.

21.5 REQUIREMENTS FOR EFFECTIVE CONDITION MONITORING PROGRAM

Following are the requirements for effective condition monitoring program:

(*i*) Type of monitoring scheme and parameters to be selected are based on the ability to indicate "warning signals" due to the abnormality in the machinery.

(*ii*) Inspection interval to be based on criticality of equipment and breakdown record.

(*iii*) Number and location of monitoring points are to be selected based on past experience and original equipment manufacturer's (OEM) recommendations.

(*iv*) It is necessary to design suitable check list for all equipments under condition-monitoring plan.

(*v*) Instruments used must be calibrated at periodic intervals.

(*vi*) Interpretation of results should be done by well-trained and experienced person.

(*vii*) A proper documentation system need to be designed and implemented.

(*viii*) Review of effectiveness of condition monitoring program must be done atleast once in a year.

21.6 TYPES OF CONDITION MONITORING SYSTEM

Condition monitoring is carried out in two ways:

(*i*) On-line monitoring.

(*ii*) Off-line monitoring

(*i*) **On-line monitoring**: On-line monitoring is carried out when the machinery is in running condition and cannot be stopped for inspection purpose. Sensors are connected to the equipment which transmit relevant data to the receiving and recording unit. For example, vibration sensor is mounted on the turbine bearing housing and is wired to the receiving, display and recording unit. The data transmitted by the vibration probe is measured continuously. The data may be recorded for future analysis or may be simply displayed on a video display unit or LCD screen without recording.

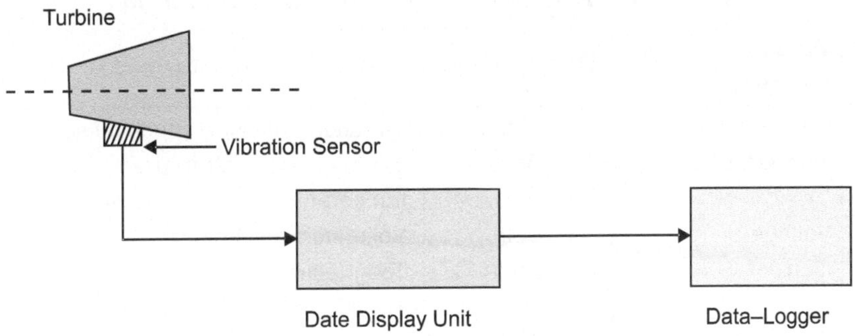

Fig. 21.1 On-line Vibration Monitoring.

(*ii*) **Off-line Monitoring:** Off-line monitoring is carried out on standstill machinery. This method is used during preventive maintenance or shutdown maintenance when machinery is stopped for general inspection and maintenance purpose.

For examples, Insulation resistance measurement of alternator windings can be done by off-line method only. Similarly, Boroscope inspection of gas-turbine blades can be carried out off-line.

Interval of off-line inspection is decided based on equipment failure history, manufacturer's recommendations, equipment availability for inspection *etc.*

21.7 PARAMETER SELECTION FOR CONDITION MONITORING SYSTEM

Choice of parameter for monitoring condition of a specific machinery is very crucial for success of a condition monitoring system. The parameter selected must indicate the desired health status to be monitored. A wrong selection of parameter will give non-relevant data about the machinery and will result in failure of program.

For example, for monitoring the health of insulation of electrical machinery, the important parameters are operating temperature, insulation resistance and winding resistance. By monitoring these parameters, the health of insulation can be gauged. However, a wrong selection of parameter will not give status of insulation and hence will be useless.

Various parameters used for monitoring are given in the following table:

Table 21.1 Parameters for Condition Monitoring

S. No.	Parameter measured	Online/off line	Instruments used
1.	Displacement	Both	Vibration analyser, shock pulse Meter.
2.	Temperature	Both	Thermometer, Infrared-temperature gun, Pyrometer.
3.	Voltage	Both	Voltmeter, multimeter.
4.	Current	Both	Tong tester, current Transformer with meter.
5.	Power	Both	Watt meter.
6.	Insulation resistance	off-line	Meggar.
7.	Speed	Both	Tachometer, stroboscope
8.	Wear	off-line	Thickness gauge
9.	Surface and sub surface flow detection	off-line	X-ray.
10.	Particle contaminant	Both	Spectroscopic oil analyzer, Ferrography.
11.	Corrosion	Both	Test Coupons, corrosion meter.

21.8 CONDITION MONITORING SYSTEMS

Condition monitoring systems are of two types *viz* preventive condition monitoring and predictive condition based on ultimate purpose.

(*i*) **Preventive condition monitoring:** Purpose of this type of monitoring is to prevent or resist the development of failure. The availability of information tells us whether the parameter being monitored is within limit or has reached alarm level.

For example, sealing water flow to pump seals need to be maintained at the designed pressure. Any drop-in the pressure level will lead to seal failure.

(*ii*) **Predictive condition monitoring:** Aim of this type of monitoring is to provide early warning so that impending failure can be avoided. Hence, a reliable prediction is required out of data gathered over a period of time so that a trend analysis could be carried out.

For example, bearing vibrations are measured at the turbine bearings. If data gathered over a period of time shows a rising trend, it is then indicative of a deteriorating condition which may lead to its failure.

21.9 Manual and Automatic Systems of Monitoring

Condition monitoring is carried out either manually or automatically.

(*i*) **Manual condition monitoring system:** It is cheaper way of monitoring health which does not require any inbuilt monitoring system in the machinery and can be used on wide variety of machines. Usually, instruments are used to measure various parameters by connecting the probe of the instrument to the machine at specified locations. The readings are taken and recorded for analysis. This method can be used on-line as well as off-line.

For example, vibration of a running machinery can be measured using a vibration meter. The probe of vibration meter is placed at various locations and readings are noted. Such measurements are carried out at periodic intervals.

Water sample of cooling tower is taken and analysed for total dissolved solids, PH, Alkalinity etc.

(*ii*) **Automatic system:** This system uses automatic monitoring of system parameter. It consists of transducer, transmitter, interfacing unit, computer with display, printer or data logger. Advantage of this system is that multiple parameters can be sensed and recorded by time-multiplicity.

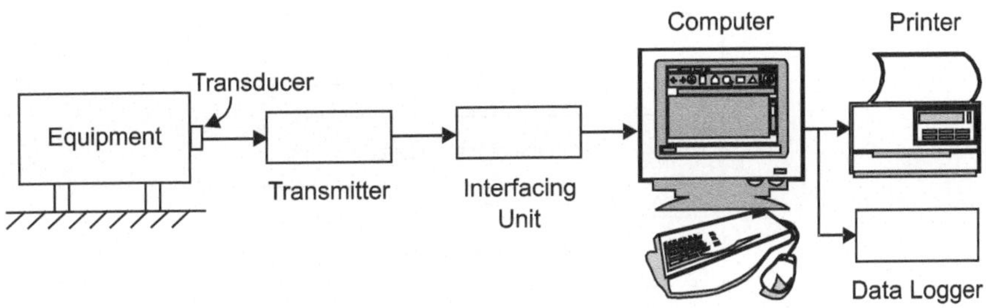

Fig. 21.2 Automatic Condition Monitoring System.

Functions of various elements are following:

(*i*) **Transducer:** It senses the parameter such as pressure, flow, temperature, vibration etc. and converts it into equivalent analogue electrical signal.

(*ii*) **Transmitter:** It receives signal from transducer and sends it through cable to an interfacing unit.

(*iii*) **Interfacing Unit:** It receives signal from transmitter and converts to the form required by the computer. Hence, it conditions the signal such as conversion from analogue to digital form or converting current into voltage signal or amplifying the signal *etc.*

(*iv*) **Computer:** It acts as display, data storage and control module.

(*v*) **Printing Device:** It can be continuous print out or print out as per demand. Data logger is used for additional storage of data.

The system can be without computer with data logger as data storage device.

21.10 PROCESS OF SETTING-UP CONDITION MONITORING SYSTEM

Establishment of condition monitoring system in a plant requires the following steps:

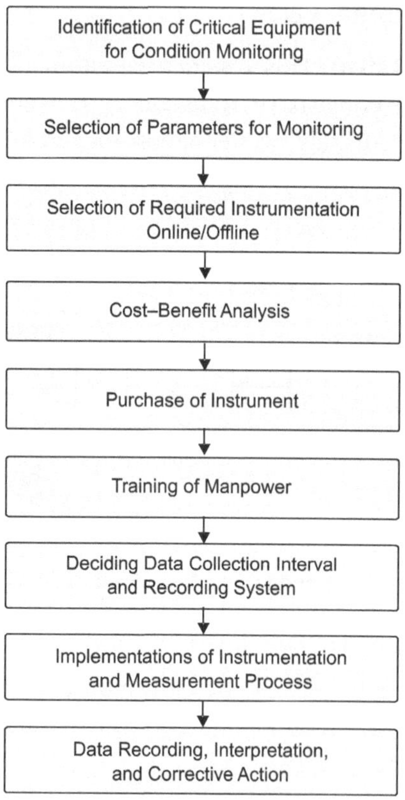

Fig. 21.3 Flow Chart

As explained in the above flow chart, establishment of condition monitoring system in plant requires identification of critical equipment on which the system is to be installed. Since this process is expensive hence only few critical machinery are chosen. So after selection of required instrumentation, a cost-benefit analysis is carried out as to determine the economic viability of this process. Thereafter, purchase of instruments, training of manpower and installation of instrumentation is carried out. Lastly, data obtained is analysed, recorded and subsequently necessary action, if required is taken.

21.11 IMPLEMENTATION OF CONDITION MONITORING SYSTEM

Implementation of condition monitoring system requires following key steps:

(*i*) Identification of critical machines that need to be monitored.

(*ii*) Selection of appropriate condition monitoring technique.

(*iii*) Setting of upper/lower limits.

(*iv*) Setting up of monitoring program.

(*v*) Creation of data recording system.

(*vi*) Training of maintenance personnel.

(*vii*) Establishment of monitoring schedule.

(*i*) **Identification of critical machines that need to be monitored:** Critical machines are these machines whose failure may cause production loss, unsafe condition, poor product quality *etc*. These machines have no standby or backup. These machines may handle dangerous or hazardous fluids or operate at very high temperature, pressure, voltage or speed. Even machines with imported critical spares with long lead time, comes under this category.

After identification of critical machines next step is selection of appropriate condition monitoring technique.

(*ii*) **Selection of appropriate condition monitoring technique**: Selection of appropriate condition monitoring technique depends on the part to be monitored. For example, for monitoring bearing — shock pulse meter can be used, for measurement of alternator windings — temperature measurement can be used. Similarly for monitoring lubricating oil — wear debris analysis can be used. But whatever the condition monitoring technique employed, it should be simple to use, robust and safe.

(*iii*) **Setting of upper/lower limit:** Condition monitoring technique monitors parameters such as temperature, pressure, vibration, contamination etc. So an upper and lower limit is to be set which will indicate severe operating condition needing immediate attention.

The limits can be based on following:

(*a*) Equipment manufacture's recommendation.

(*b*) Past experience of operating similar equipment.

(*c*) National standards such as BIS, BS, etc.

(*d*) Diagnostic equipment manufacturers recommendations.

(*iv*) **Setting-up of monitoring program:** Setting up of monitoring program requires component to be monitored, site condition, whether off-line or on-line, periodicity of measurement *etc.*

The periodicity of monitoring is to be decided based on operating condition, cost of monitoring, cost of downtime and personnel safey.

(*v*) **Creation of data recording system:** After initiation of condition monitoring, a large number of data are to collected, recorded and analysed. Based on the received data, analysis need to be carried out so as to determine the current condition or severity level. Based on the analysis, the result is to be conveyed to responsible person for further action. The recording system should preferably be computer based where it is possible to carry out analysis such as trending *etc.*

(*vi*) **Training maintenance manpower:** All condition monitoring systems are highly sophisticated and need good amount of prior training before handling them. Because, a mere collection of data has no meaning unless a person is trained to analyse and extract useful information. For this, concerned engineer and technicians need to be trained in proper handling of diagnostic equipment, appropriate method of data collection and analysis of collected data.

This training can be carried out either on-site or at equipment manufactures' premises.

(*vii*) **Establishment of monitoring schedule:** Like establishment of preventive maintenance schedule, a predictive maintenance schedule need to be chalked out. This will be based on equipment criticality, site condition, downtime cost, machinery manufacture's recommendations as well as diagnostic tool manufacture's advice.

21.12 VARIOUS CONDITION MONITORING TECHNIQUES

There are host of condition monitoring techniques available to detect initiation of degradation of a component before it fails completely.

The various monitoring techniques and its applications are tabulated below:

Table 21.2 Applications of CMS Techniques

S. No.	Condition monitoring Techniques	Area of application
1.	Visual inspection.	Surface cracks, winding overheating, colour of lubricant oil.
2.	Temperature measurement.	Temperature of winding of electrical machines, bearing, lubricants, oil coolant, furnace burner *etc.*
3.	Vibration monitoring.	High speed rotating machines for unbalance, oil whirl, mechanical looseness, damaged gears etc.
4.	Dye penetration test. Ultrasonic test.	For detecting surface and sub surface irregularities.
5.	Wear-debris analysis	Condition of lubricating oil.
6.	In-situ Metallography.	Detection of changes in metal micro-structure.
7.	Acoustic emission.	Detection of cracks and leakages in pressure vessels, pipelines etc.
8.	Tan-delta measurement.	Condition monitoring of insulation of electrical machines.

The list of condition monitoring technique is exhaustive and is dealt in more details in previous chapter of "Non-destructive and destructive tests".

So, various condition monitoring techniques can be used to detect changes or deterioration in equipment on physical, chemical electrical parameters. These changes enable to predict impending failures before they actually take place. Hence, a timely action helps to prevent an impending failure.

21.13 EXAMPLE OF SELECTION OF CONDITION MONITORING TECHNIQUES

Following section gives details of various condition monitoring techniques used for Turbine-Generator system.

Table 21.3 Condition Monitoring of Turbine-Generator System

S. No.	Main Par	Sub-part	Purpose	Technique Used
1.	Generator stator	(i) Stator coil (ii) Stator core	(i) Insulation condition. (ii) Looseness of winding wedge. (iii) Core varnish and looseness.	(i) Visual and Megger Test. (ii) Visual test. (iii) Visual test.
2.	Generator Rotor	(i) Shaft. (ii) Rotor wedge. (iii) Retaining ring.	(i) Development of crack.	(i) Visual, ultrasonic and dye penetration. (ii) Dye-penetration and hardness testing. (iii) Dye-penetration and ultrasonic.
3.	Hydrogen Cooler	(i) Tubes.	(i) Thinning of tubes.	(i) Visual test.
4.	Valves	(i) Seat.	(i) Surface wear.	(i) Dye-penetration.
5.	HP/LP/IP cylinders	—	(i) Crack.	(i) Visual and Magnetic particle test.
6.	Steam line weld.	(i) Joints.	(i) Cracks.	Visual, magnetic particle, ultrasonic, hardness test.
7.	Bolting Material	—	(i) Cracks.	Visual, ultrasonic, Magnetic particle, Dimensional test.

CASE STUDIES

In this section, real life case studies of various equipments have been discussed. It is important to understand that success of any condition monitoring system depends upon following:

(*i*) Use of appropriate condition monitoring system for the intended equipment.

(*ii*) Correct interpretation of readings or observations.

If above two factors are taken into account, rest is easy. Next job is to exactly pin-point the cause of defect and accordingly apply remedial measures.

Sometimes, different condition monitoring techniques are used in order to find right solution. It, then depends on ingenuity of the engineer to try appropriate alternative.

The following case studies help us to illustrate different possibilities in finding a solution.

Case Study-1

This is an interesting case and was carried out on 210 MW steam turbine unit in a state electricity board generating station.

The problem reported was sudden increase in vibration of steam turbine unit. The nature of increase in vibrations were frequent and random. So vibration signatures were taken under two conditions – (i) when machine was running with normal vibrations (ii) when machine was running with abnormal vibrations.

After detailed vibration analysis was carried out, it was found to contain one time the component during normal vibration and one time and 0.48 times the component during abnormal vibrations. It was inferred that oil whirl is the cause of vibrations.

Now, oil whirl can be caused by a good number of reasons such as bearing clearances, oil pressure, temperature of oil, viscosity of oil etc. Upon checking oil viscosity, water contamination was found. So it was established that oil whirl is due to water ingress in the oil. After removal of water from oil by oil purifier, the problem of abnormal vibration was solved.

So this case study illustrates, how exact cause of oil whirl was pin-pointed and problem was solved.

Case Study-2

This is another case study of a forced draught fan and its motor. Forced draught fans are used in thermal power plants to maintain draught in the boiler furnace. Problem reported by operational staff was high vibrations. This prompted to resort to vibrations signature analysis.

Vibration analysis contained 1 times the component. Vibration amplitude and phase were found to be fluctuating. These pulsations indicated an electrical problem. So to confirm this, power to fan motor were switched off and vibration reading were taken again. It was found that vibrations reduced immediately or waned gradually. So this confirmed electrical problem.

Electrical problem means either stator related or rotor related problem. To pin-point this, the vibration waveforms are observed in oscilloscope. A pulsating waveform on oscilloscope indicates stator related problem. Such problems can be:

(i) Open or shorted stator windings.

(ii) Unequal air gap.

(iii) Unbalanced phases.

However, if the waveform on oscilloscope is modulated, it is related to rotor. The rotor related problems can be:

(i) Broken rotor bar.

(ii) Open or shorted rotor winding.

(iii) Bowed rotor.

(iv) eccentric rotor.

After dismantling motor, following problems were found :

(i) Unequal air gap.

(ii) Unequal phases.

(iii) Rotor bar problem.

Since the motor was very old, it was decided to replace it with a new one.

Case Study-3

This case study pertains to 120 MW turbo-alternator operating in a thermal power plant. The turbo-alternator was overhauled (rotor was repaired and balanced) during annual maintenance shutdown. After overhauling, the turbo-alternator was started but it showed very high vibration on generator bearings.

So vibrations testing and analysis was carried out. The analysis showed 1 times component which was dominant. This indicated rotor unbalance.

So rotor dynamic balancing was carried out. After trial, desired result was obtained.

Case Study-4

This case study refers to condition monitoring of valves (controlling water flow in a power plant). Acoustic emission technique was used to monitor condition of eight such valves over a period of six months. During the observation period, it was found that some valves operated at steady acoustic emission level while in some others acoustic emission level rose progressively over several months.

So valves with rising acoustic emission level were taken up for testing. It was found that valves was passing 2.5 kg per second of water when fully closed at 200 bar pressure. So the valve was dismantled for maintenance, the valve seat was found to be visibly wornout. After replacement of valve seat, acoustic emission level reduced drastically. Following graph shows acoustic emission level of valves with good and progressively deteriorating seats.

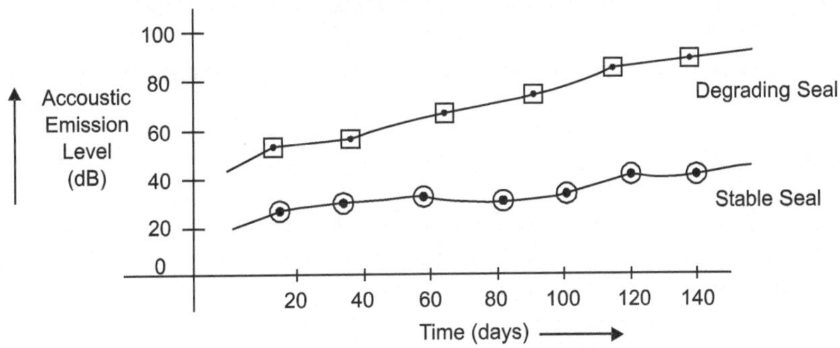

Fig. 21.4 Acoustic emission level from stable and degrading valves

Case Study-5

This case study refers to 210 MW steam turbine. Cracks were reported in HP turbines top casing. To check and confirm the extent of damage, dye-penetration and ultrasound testing were carried out. The tests indicated extent of damage. The measurement of cracks were carried out and are listed below.

Table 21.4 Locations and extent of crack

Location of Crack	Extent of crack
(a) Diaphragm linear Area.	Depth – 10 to 15 mm. Length – 400 mm.
(b) Another Linear Area.	Depth – 20 to 25 mm. Length – 30 mm.
(c) Parting plane.	Depth – 15 mm. Length – 400 mm.

After analysis, it was inferred that stress due to high temperature and cyclic thermal stresses are responsible for these cracks.

The case cracks were immediately repaired.

Case Study-6

This case study pertains to super heater header of a boiler in a thermal power plant. Super heater is used to superheat saturated steam before it is fed to steam turbine.

Routine condition monitoring of super heater header was carried out using fibroscopy. Inside the header, oxide scale formation was observed.

The cause was due to high temperature corrosion. Being a minor damage, it was decided to take up in next inspection.

Case Study-7

This case study also pertains to super heater header of a boiler in a thermal power plant. During annual inspection, header was found to be bent. Dimension measurements were carried out.
Total length = 5900 mm.
Maximum bent = 115 mm.
Upon inspection, it was found to be due to high temperature degradation. The bent header was replaced.

Case Study-8

This case study pertains to a 13.5 MW Turbo alternator. The unit tripped on vibration and repeated attempts to restart failed.

Vibration testing was carried out and unbalanced was suspected. After in-situ rebalancing of alternator rotor, the unit still had high vibrations. At this stage, it was noticed that vibrations reduced drastically once the excitation was switched off. This clearly indicated an electrical problem rather than a mechanical one (i.e. unbalance)

The DC resistance of rotor field windings (4 pole rotor) were measured but it did not yield any clue. So impedance of field windings were measured at 1KHZ frequency. It was found that impedance of one pole winding is "low" compared to other pole windings. So it was a case of inter-turn short of field winding (inter-turn short means shorting of two or more turns of same field winding).

The only option was rewinding of faulty field pole. The rotor was removed from alternator and was sent to original equipment manufacturer for rewinding. After rewinding, the unit was commissioned successfully.

Summary

Condition monitoring is a proactive approach which assesses probability of future failures in an equipment so that timely steps could prevent such occurrences. This types of maintenance is applicable to few identified critical equipments due to the cost and effort required.

Condition monitoring technique utilizes various signals such as temperature, vibration, noise *etc.* emanating from equipment. By monitoring such signals over a period of time gives sufficient information about the current health of equipment as well as anticipated failures. However, the most difficult part is data interpretation which only a trained person can do.

Principally, condition monitoring is carried out by on-line and off-line methods. In on-line method, data is recorded while the equipment is working, while is off-line method, data is recorded when equipment is stopped.

Manual and automatic both systems are used where data recording is done either manually or automatically. This is decided by cost factor.

Contd...

Implementation of condition monitoring system requires—identification of critical equipment, selection of appropriate parameter for monitoring, selection of appropriate instrumentation, installation of instruments, data collection and recording, interpretation and corrective action. Before data collection is started, concerned manpower need to be trained in data collection and interpretation.

The various condition monitoring techniques are visual inspection, temperature measurement, vibration monitoring, dye-penetration test, ultrasonic test, wear-debris analysis, acoustic emission *etc.*

Review Questions

1. What is meant by condition monitoring ? Illustrate with an example.
2. Discuss on-line and off-line methods of condition monitoring. How will you carry out condition monitoring of (i) motor insulation (ii) bearing condition (iii) corroding pipeline ?
3. Discuss various types of parameters that are measured for condition monitoring purpose.
4. How will you select a parameter for condition monitoring ?
5. Discuss objectives and characteristics of condition monitoring system.
6. Describe in detail process of setting-up of condition monitoring system in a cement plant.
7. Write briefly about data recording and analysis.
8. What types of instruments are needed to carry out condition monitoring of:
   (i) blades of a gas turbine (without opening casing).
   (ii) detecting hot-spot in a alternator winding.
   (iii) particle contamination in used lubricating oil sample.
   (iv) detecting sub-surface flaw in a welded joint of a pipe.
9. How condition monitoring can help to achieve equipment reliability and reduce ownership cost in the long run ?
10. Why condition-monitoring proves to be cheaper in the long-run?

Discussion Questions

1. Justify "condition monitoring helps to improve system availability".
2. Discuss, "condition monitoring is economically justified for high cost capital equipment".

3. Elaborate on – "Interpretation of data obtained from condition monitoring is the most difficult task".
4. Comment " Pressure on maintenance department makes condition monitoring activities unsustainable in the long run".

Multiple Choice Questions

1. Purpose of condition monitoring is
   (*a*) Replace preventive maintenance
   (*b*) Prevent failures
   (*c*) Reduce maintenance cost
   (*d*) None of these.
2. Corrosion measurement is
   (*a*) On-line process (*b*) Off line process
   (*c*) Both (*a*) and (*b*) (*d*) None of these.
3. Most crucial factor in CMS is
   (*a*) Relation of equipment (*b*) Selection of right parameter
   (*c*) Data collection (*d*) Training.
4. Surface irregularities can be measured by
   (*a*) Dye penetration (*b*) Ultrasound
   (*c*) Magnetic particle (*d*) All of these.
5. Condition monitoring increases
   (*a*) Reliability (*b*) Cost
   (*c*) Availability (*d*) All of these.

True/False Questions

1. Condition monitoring based system of maintenance can replace preventive maintenance completely.
2. Condition monitoring based system proves to be very expensive.
3. Application of condition monitoring techniques is not economically justified in all situations.
4. Selection of type of parameter to be monitored is most important factor in the successful operation of condition monitoring system.

5. Interpretation of data from condition monitoring equipment is most critical in whole process.

6. Condition monitoring helps to improve life of equipment.

ANSWERS

Multiple Choice Questions

1. (*b*) **2.** (*c*) **3.** (*b*) **4.** (*d*) **5.** (*d*)

True/False Questions

1. False **2.** False **3.** True **4.** True **5.** True
6. False

Case Study Based Questions

Mr. Harish Khola is a chief engineer of a medium size industry manufacturing colour picture tubes. He is a traditional engineer and is skeptical of modern instruments used for condition monitoring. He relies on his experience and fined tuned senses. Preventive maintenance was carried out on a 3 MW gas-turbine set about a month ago. Upon routine measurement of vibration of a air blast oil cooler, the shock pulse meter readings of blower were found to be on a higher side. The concerned engineer advised Mr. Khola to go for bearing replacement as is may fail in near future leading to turbine tripping.

Since preventive maintenance was carried out only a month ago, hence Mr. Khola termed this as "wild imagination" and rejected the idea of bearing replacement.

1. What options are open to the engineer to avoid an impending failure ?
2. How can the engineer be sure that it is the bearing problem and nothing else?

❑❑❑

22

Lubrication Practices

Learning Objectives

After reading this chapter, you will be able to:

- Understand types and properties of lubricants.
- Explore application areas of these lubricants.
- Understand testing and reclamation methods.

22.1 OVERVIEW OF FRICTION AND WEAR

Every moving machinery is subjected to friction due to movement of parts sliding or rolling over one another. The movement of parts gives rise to friction due to surface irregularities. Due to gradual use, the mating parts wear out.

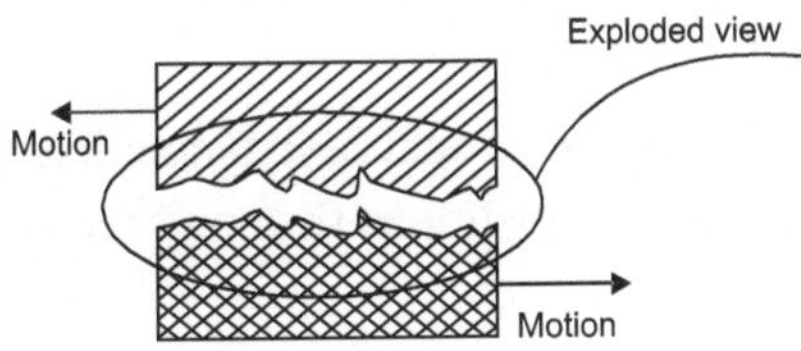

Fig.22.1 Mating of two Moving Part.

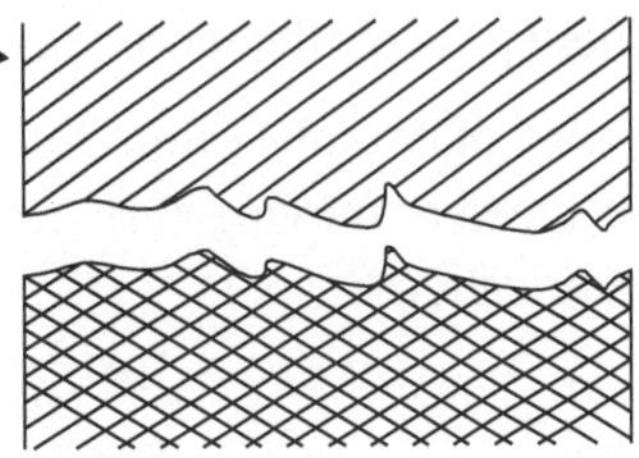

Fig. 22.2 Exaggerated View of Mating Surface.

As shown in above figures — the two parts are moving against each other. An exaggerated view of mating surfaces is shown in Fig. 22.2. All surfaces have ridges and valleys and are not smooth. So when these surfaces slide against each other, these ridges and valleys oppose the motion thereby producing friction. When the motion is continued despite friction, wear and tear of mating parts takes place. Wear refers to gradual loss of material from the part.

In order to reduce friction and hence wear, a lubricant is used in between two sliding surfaces. The lubricant forms a layer which fills up ridges and valleys of the two moving parts. So the part can slide against other with very low friction.

Figure 22.3 shows lubricating film over the moving surface formed by a lubricant. Formation of film lowers the sliding friction and hence the wear.

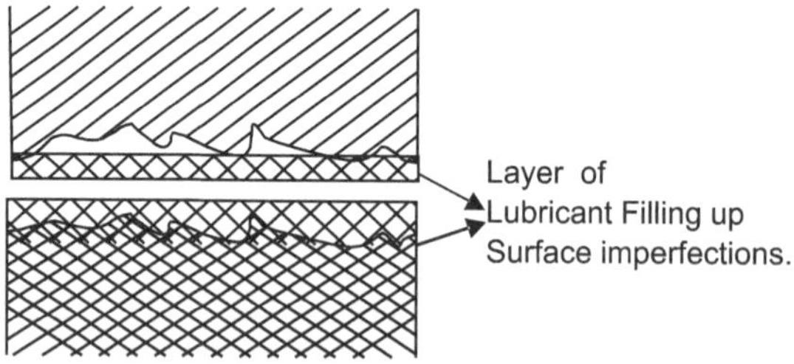

Fig. 22.3 Mating Surface with Lubricating Film.

So depending upon friction, lubrication can be:

(*i*) Dry friction – no lubrication.

(*ii*) Mixed friction – boundary lubrication.

(*iii*) Liquid friction – complete liquid lubrication.

According to Coulomb's law, frictional resistance 'F' is approximately proportional to force P with which the moving surfaces are pressed against each other. This is the case with dry friction and mixed friction.

So, $F \propto P$

or, $F = \mu P$

Where μ is coefficient of friction

In case of complete liquid lubrication, frictional resistance depends upon relative speed of moving surfaces, thickness of liquid film and area of contact under lubrication.

$$\therefore \quad F = \eta A \frac{dV}{dy}$$

Where A = Surface area

$\frac{dV}{dy}$ = Surface gradiant of liquid film.

η = Coefficient of internal friction of liquid or viscosity of liquid.

22.2 PURPOSE OF LUBRICATION

A lubricant is a substance that reduces frictional resistance between two moving parts.

Purpose of lubricants is to form a film between moving surfaces so that friction, wear and tear *etc*. of moving parts is reduced. A lubricant is required to do the following jobs:

(*i*) Lubricate *i.e*. form a film between highly loaded moving parts.

(*ii*) Act as coolant to remove frictional heat generated by motion.

(*iii*) To receive and carry away contaminants generated from internal and external sources.

(*iv*) To transmit hydraulic pressure.

(*v*) To protect against wear of heavily loaded parts.

(*vi*) To protect against rust and corrosion.

(*vii*) To resist aeration and foaming.

(*viii*) To separate water coming from external sources.

(*ix*) To prevent accumulation of deposit such as sludge and varnish.

(*x*) To act as seal.

22.3 CLASSIFICATION OF LUBRICANTS

Lubricants are classified according to following:

(*i*) According to their physical state:

- (*a*) Liquid lubricants.
- (*b*) Semi solid lubricants.
- (*c*) Solid lubricants.

(*ii*) According to specific service:

- (*a*) Engine oil.
- (*b*) Hydraulic oil.
- (*c*) Circulating oil.
- (*d*) Refrigerant oil.
- (*e*) Gear oil.
- (*f*) Heat transfer oil.
- (*g*) Machinery oil.
- (*h*) Special machinery lubricants.

(*i*) Cutting and metal working fluids.

(*j*) Turbine oils.

(*k*) Spindle oils.

(*l*) Quenching oil.

(*m*) Marine oil.

(*iii*) According to viscosity of oil— based on SAE (Society of Automotive Engineers) or API (American Petroleum Institute) grading.

(*iv*) According to origin of oil – Such as mineral oil or synthetic oil.

22.3.1 Liquid Lubricants

A liquid lubricants is manufactured from –

(*i*) Mineral oils.

(*ii*) Animal and vegetable oils.

(*iii*) Synthetic oil.

(*i*) **Mineral Oil:** Mineral oils are derived from crude petroleum of suitable stock. These are extensively used due to low cost and stability. These are composed of two important components — base oil and additives. Base oils are derived from refining of crude oil while additives are chemicals that are added to provide certain characteristics such as rust inhibitor, oxidation inhibitor, viscosity index improver, anti wear *etc*.

Generally three categories of base oil are used – Napthenic base oil, paraffin base oils and aromatic base oils.

Paraffin base oils have high viscosity index but contains wax that settles out at low temperature. Napthenic base oils contain asphalt which at high temperature leaves sticky substance. Aromatic base oils have poor viscosity index and leaves behind sludge and sticky substance.

(*ii*) **Animal and Vegetable Oils**: Vegetable oils used are olive oil, castor oil, Rapeseed oil, cotton seed oil *etc*. Animal oils are tallow oil, Lard oil *etc*. These oils have tendency to get oxidized when exposed to air and get hydrolysed in contact with moist air. Hence, mineral oils are preferred over animal and vegetable oils.

(*iii*) **Synthetic Oil**: These are polyalkylene glycol and their derivatives. Silicones are also used which are derived from coal and sand. Synthetic oils exhibit very good properties compared to mineral oils but are very expensive.

22.3.2 Semi-solid Lubricants

These are lubricants having viscosity higher than oil, wax etc. Various grades of greases are example of semi-solid lubricants. These lubricants are used where machine parts are subjected to high load and slow speed and where the entry to dust and moisture is harmful.

Greases are produced by mixing petroleum oil with soap. Thickening agents are also added such as sodium, calcium, lithium, barium *etc.*

22.3.3 Solid Lubricants

These lubricants are used where high temperature and pressure does not allow lubricating film to be maintained between moving parts. Examples of solid lubricants are graphite, Molybdenum disulphide *etc.*

Solid lubricants are either used directly or as a paste mixed with oils and greases. These can be used upto 375°C.

22.3.4 Engine Oil

Majority of engine oils are fed from a sump in crank case where rotary motion of crankshaft is converted into reciprocating motion. The oil is forced into crankshaft main bearing, connecting rod bearing, cylinder walls *etc.* Oil is also pumped into crankshaft bearing and valve gear. Engine oils deal with extreme temperatures, broad speed changes and contamination.

These oil have high viscosity index and high oxidation stability. Certain chemical additives are also added to enhance properties such as — dispersants, anti-wear agents, pour point suppressants, oxidation inhibitors, rust inhibitors, foam inhibitors *etc.*

22.3.5 Hydraulic Oil

These are basically circulating oils in a closed system with storage tank. Main features of hydraulic oils are:

(*i*) Excellent resistance to oxidation.
(*ii*) Good wear protection.
(*iii*) High film strength to minimize friction.
(*iv*) Good anti corrosion properties.
(*v*) Quick water separation properties.
(*vi*) Reduced tendency to foam.
(*vii*) Good air release properties.

Hydraulic oils have long service life. Some hydraulic systems use non-petroleum, fire resistant fluids. These are synthetic materials that have high oxidation resistance and can operate upto 150°C. The auto-ignition temperature (temperature of spontaneous ignition in air) is more than 540°C.

22.3.6 Circulating Oils

Such systems are needed where large amount of oil is required for lubrication and cooling. For example, in paper mill bearings on steam heated drier rolls, run hot and there can be chances of oxidation of oil. Oil may get contaminated with dirt, processing liquid, fibrous paper *etc.*

22.3.7 Refrigerant Oils

These oils are used in compressors of refrigeration system. High temperature and contamination are root causes of oil deterioration. For example, ammonia and Freon are widely used refrigerant gases. Air can infiltrate the ammonia system and can oxidize lubricant. Oxidation products react with ammonia refrigerant to form heavy, oil insoluble sludge.

Freon though thermally stable reacts with both water and air producing corrosive byproducts.

22.3.8 Gear Oil

Main function of gears is to transmit power. There is friction at the mating gear teeth that generates heat. Gears operating under continuous heavy load requires oil coolers. Lubricating oils in this service can oxidize and this will increase viscosity. Higher viscosity means more friction loss and even higher temperature. This forms a chain reaction. Viscosity monitoring is an important check point for gear oils.

22.3.9 Heat Transfer Oils

These are used as medium for transferring heat in a variety of industrial application. These oils are formulated with oxidation inhibitors and detergents. These oils operate at high temperature upto 325°C for long period of time. Inspite of closed system, oxidation of oil takes place with formation of deposits. These deposits prevent heat transfer. This results in development of hot spots and carbonization of oil.

Hence, monitoring of flash point, viscosity and oxidation are essential in service life.

22.3.10 Food Machinery Oils

Food machinery is designed to minimize leakage of lubricants into the food being processed. Seals are so designed that processed fluid will leak into lubricants.

22.3.11 Special Machinery Lubricants

These lubricants are used for special machinery such as precision machine tools, rock drill oils, textile machinery lubricants, wire rope lubricants, metal cutting fluids *etc*.

22.3.12 Classification Based on Viscosity

Viscosity is the most important property of oil. A viscosity numbering system is used by Society of automotive engineers, USA. For classification of oils, oils are given numbers based on their viscosity. So viscosity number helps to identify an oil for a particular application.

22.3.13 Classification of Oil Based on Origin

Mineral oils are derived from crude petroleum. Majority of lubricating oils used are mineral based. Animal and vegetable oil finds very limited use.

For special purpose applications, synthetic oils are also used.

22.4 PROPERTIES OF LUBRICANTS

Properties of lubricants play an important role in selection of lubricant for a particular application. Monitoring of these parameters give an indication of residual life of oils.

The various properties of lubricants are:

(*i*) **Viscosity:** It is a measure of resistance of oil to flow. A thicker oil has higher viscosity such as mustard oil while a thinner oil has lower viscosity such as kerosene oil.

Viscosity of oil in a system changes due to oxidation temperature and contamination of metals such as copper, iron, lead, zinc *etc*. Low viscosity lubricants are used for low pressure and high speed applications while high viscosity lubricants are used for high pressure and low speed applications.

Viscosity increase is often accompanied by increase in acidity and viscosity decrease is generally accompanied by flash point decrease.

(*ii*) **Viscosity index:** Viscosity Index is defined as rate of change of viscosity of a liquid with the variation in temperature. A good lubricant is one whose viscosity does not change much with variation in temperature.

Viscosity Index is a numerical figure showing viscosity temperature relationship of an oil. A high viscosity index indicates a relatively small change of viscosity with temperature and vice-versa.

An arbitrary scale from 0-100 is chosen to represent viscosity Index. However, with latest manufacturing technique and introduction of synthetic lubricants, oils with viscosity index higher than 100 are available.

(*iii*) **Flash point**: It is the lowest temperature at which an oil catches fire momentarily. A good lubricant should have high flash point.

Pensky Mortens closed cup and cleaveland open cup tests are used to determine flash point.

(*iv*) **Pour point:** It is the lowest temperature at which the oil will flow when cooled under standard condition. It is related to ability of oil to start lubrication when a cold machine is put into operation. A good lubricant should have as low pour point as possible.

(*v*) **Cloud point:** It is the temperature at which the wax or other solid matter begins to separate from solution when the oil is cooled under standard condition.

(*vi*) **Carbon residue or ash content:** It is related to quantity of incombustible material present in the sample. Ash contents give an indication of quantity of metallic additives, non-combustible solid contaminants *etc.*

(*vii*) **Neutralisation value:** It is a measure of acidic impurities present in the oil. It is expressed in mg of potassium Hydroxide necessary to neutralize 1gm of oil. Mineral oils have low acidic matter compared to vegetable oil. With the passage of time acidity increases which may lead to corrosion.

(*viii*) **De-emulsification number:** It is defined as number of seconds required for an oil to separate when it is emulsified and separated under specified conditions. Hence, it is a measure of how fast water will separate from oil. It is commonly applied to turbine oils.

(*ix*) **Saponification number:** It is a measure of potassium hydroxide required to completely saponify oil. It is expressed as mg KOH pre gm of oil. Presence of acids, fats and oxidized products are indicated by this test.

(*x*) **Rust preventing characteristics:** Many oils such as turbine oils, hydraulic oils are doped to prevent rusting of steel in presence of water. The test indicates if rust prevention ability is good or impaired.

(*xi*) **Oxidation stability:** These tests help to determine efficacy of antioxidants used in oil.

(*xii*) **Air release value:** It is a measure of ability of oil to separate entrained air. This is applicable to hydraulic and turbine oils. It is defined as number of minutes for air in oil to reduce in volume to 0.2% under test condition.

(*xiii*) **Foaming:** Foaming is rising of quick bubbles to the surface of oil. Presence of additives in mineral oil produce foaming tendency.

(*xiv*) **Total Base Number (TBN)**: It indicates acid neutralizing power of an oil. It is determined by titration of oil with HCL or perchloric acid.

22.5 ADDITIVES

Additives are chemicals that are added to oils to improve certain properties. Following are various additives used in the oil:

(*i*) **Antiwear and extreme pressure additives:** These provide load carrying capacity where needed and prevent scuffing of moving parts under conditions of boundary lubrication.

(*ii*) **Inhibitors:** These are:

(*a*) *Anti-rust inhibitor*: These control rusting of mating surfaces.

(*b*) *Oxidation inhibitor*: These prevent oxidation of lubricating oil.

(*c*) *Corrosion inhibitor*: These prevent corrosion of parts where different metal surfaces (such as copper, iron, steel etc) mate with each other.

(*d*) *Viscosity improvers*: These increase the viscosity of lubricating oil.

(*e*) *Pour point depressants*: These decrease the pour point of lubricant oil.

(*f*) *Depressants and detergents*: These control the deposit formation tendency throughout the system.

(*g*) *Foam inhibitors*: These reduce foaming tendency of lubricant oil. Silicon inhibitors are normally used.

22.6 LUBRICATING GREASE

Grease is basically soap dispersed in lubricating oil. It is prepared by saponification of fat with alkali and adding hot oil to it.

Due to higher viscosity, grease is used for heavy load at lower speeds.

Different types of grease are:

(*i*) **Calcium base grease:** These are emulsion of petroleum oil with calcium soap. These are cheap and insoluble in water, but can be used for low temperature application.

(*ii*) **Sodium base grease:** These are emulsion of petroleum oil with sodium soap. These are water soluble and can be used upto 175°C.

(*iii*) **Lithium base grease:** These are emulsion of petroleum oil with lithium soap. These are water resistant and can be used for low temperature (upto 15°c). These are used to produce multi-purpose grease.

(*iv*) **Alluminium base grease:** These have same properties as calcium base grease but with higher melting point.

(*v*) **Mixed base grease:** These are obtained by mixing solid lubricant such as graphite or molybdenum disulphide to improve anti-scuffing qualities.

Grease are inferior to oil as lubricants and are not suitable for high speed and high temperature applications.

22.7 PROPERTIES OF GREASE

Following are the properties of grease:

(*i*) **Consistency or yield value**: It is the ability of grease to flow. A penetrometer is used to measure the consistency of grease. It is measured as distance in tenth of mm that a standard cone penetrates vertically into the sample under standard condition of load (150 gm), temperature (25°c) and time (5 seconds)

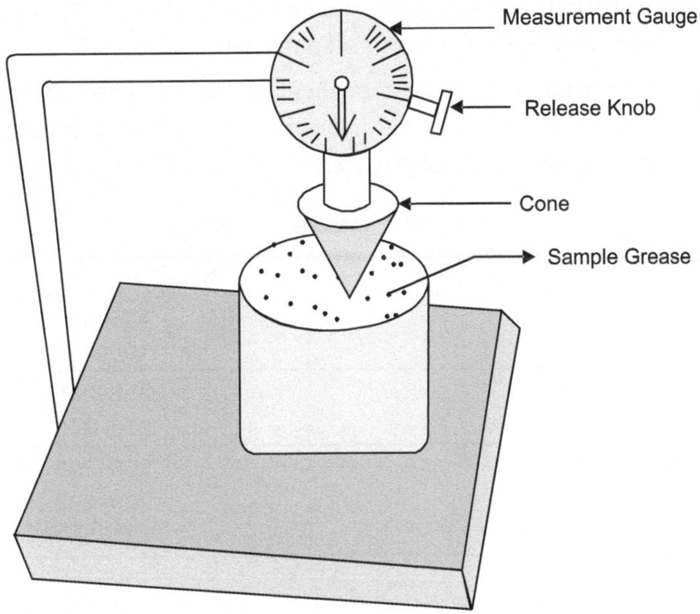

Fig. 22.4 Penetrometer

(*ii*) **Drop point**: It is defined as the temperature at which it passes from semi-solid state to liquid state. It helps to determine upper temperature limit of applicability.

Drop test is carried out in a apparatus with a hole in the bottom. The cup holds grease and thermometer. This arrangement is then kept inside air tight chamber and immersed in a water heating bath.

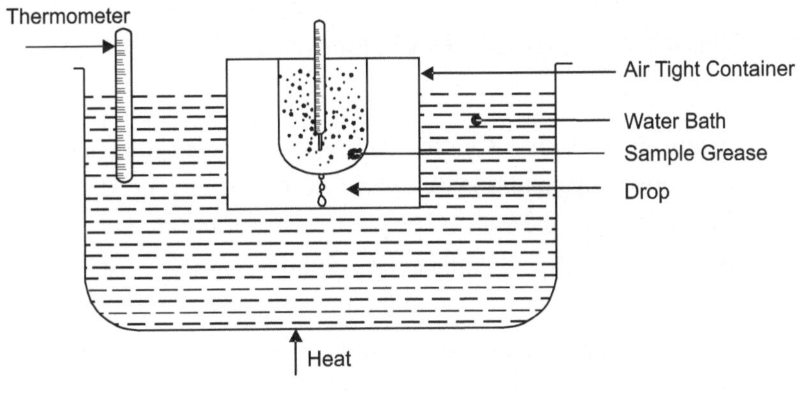

Fig. 22.5 Drop Test Apparatus.

(*iii*) **Bleed rate**: It is resistance of oil to migrate out of thickner. Hence, it is a measure of separation of oil and soap.

(*iv*) **speed rating**: This test establishes viscosity of base oil. Grease are classified based on softness or hardness. ISO (International Standards Organisation) classifications are:

Table 22.1 Grease Rating

	ISO-2137 Penetration
Very soft greases	445–470 400–430
Soft greases	355–385 310–340
Medium greases	265–295 220–250 175–250
Hard greases	130–160 85–115

22.8 SOLID LUBRICANTS

These are used where lubricating films cannot be maintened. Also, there are chances of contamination, grit, high operating temperature and load. Examples of solid lubricants are — Graphite and Molybdenum disulphide. Both these materials have laminar structure and operate by interposing small laminar films between two mating surfaces. These are used for high pressure and low speed applications. Other solid lubricants are Boron Nitride, PTFE (Poly-tetra Flouro Ethylene), Calsium Fluoride and Tungsten disulphide.

22.8.1 Characteristic of Solid Lubricants

Following are the charactereistics of solid lubricants:

(*i*) **Oxidation stability:** It is the ability of lubricant to resist oxidation under operating condition. A good lubricant must not undergo oxidation changes when used within applicable temperature range.

(*ii*) **Thermal stability**: It is the ability of lubricant to withstand high and low temperature changes without undergoing structural degradation.

(*iii*) **Crystal structure:** Lubricant should possess laminar crystal structure because it has low shear strength.

(*iv*) **Volatility:** A good solid lubricant should have low vapour pressure.

(*v*) **Chemical reactivity:** Lubricant should not react with the mating surface material.

(*vi*) **Electrical conductivity:** Certain applications require low electrical conductivity.

22.9 CAUSES OF LUBRICATION FAILURE

Following are fundamental reasons for failure of lubrication:

(*i*) Additives in lubricants during service are depleted and as a result, the lubricants loses some of its functional properties imparted by these additives.

(*ii*) Oil becomes contaminated from internal and external sources.

(*i*) **Loss of additives**: Many additives added in oil are consumed during their service life. So they need to be replenished by make-up oil or by oil replacement. Some additives are temperature sensitive and they evaporated when operated beyond the design limit. In addition, heat and loss of additive can damage the base oil.

For example, depletion of oxidation inhibitor causes oxidation of hot oil when exposed to air. Oxidation results in formation of acids which attack or corrode the metals. The metal salts thus formed and products of oil oxidation act as catalyst and speed up further oil oxidation leading to a chain reaction. Continued oxidation produces insoluble such as varnish and sludge in the oil and oil darkens.

(*ii*) **Contamination of oil**: Oil is contaminated by dirt, water, corrosive fluids, wear metals, rust, soot etc. Contamination of oil with solid, liquid and gaseous elements, changes properties of oil due to oxidation, emulsification, foaming, *etc*. This leads to breakdown of lubrication film resulting in lubrications failure.

Table. 22.2 Causes of Lubrication Failure.

	Problem	Reason	Result
I	**Loss of:**		
	(i) Oxidation control.	Oxidation. (Inhibitor used up)	Oil viscosity increases, acids are formed and metal surface is corroded.
	(ii) Rust control.	Rust inhibitor used up.	Further rusting takes place.
	(iii) Load carrying capacity.	Additives consumed.	No protection against scruffing.
	(iv) Dispersant	Dispersant is contaminated.	Varnish and sludge formation.

Contd...

II Contamination with solids:		
(i) Dirt	Entry of dirt.	Promotes wear.
(ii) Wear metals.	Wear of internal parts of system.	Shortens machine life.
(iii) Rust.	Reaction with iron, oxygen and water.	Promotes wear.
(iv) Carbon.	Oil overheated.	Deposits block passages.
III Liquid Contamination:		
(i) Water and process fluids.	Condensation and leaking seal.	Deterioration of oil.
(ii) Other lubricating oils.	From internal or external system.	Changes properties of lubricating.

Table. 22.3 Problems Associated with Oils in Different Systems

Service	Problem
Turbine.	Oxidation and contamination.
Hydraulic.	Contamination.
Circulating.	Oxidation and contamination.
Refrigeration.	Oxidation.
Gears.	Oxidation.
Heat transfer.	Oxidation.
Insulating.	Oxidation and contamination.

22.10 SELECTION OF LUBRICANTS

Selection of lubricants is application specific *i.e.* whether required for steam turbine or refrigeration system or gear boxes *etc*.

It depends on the following factors:

(*i*) Area of application.

(*ii*) Service condition.

(*iii*) Ambient conditions with respect to temperature, moisture, dust, gases and other possible sources of contaminations.

(*iv*) Compatibility with other liquid or gases present in the system (such as in refrigeration)

(*a*) Features required for steam turbine oils

(*i*) Oil should provide excellent long-term protection against rust and corrosion.

(*ii*) Oil should readily separate from water.

(*iii*) Should have good oxidation stability.

(*iv*) Should be non-foaming

(*v*) Should release entrapped air

(*b*) Features required for hydraulic oils:

(*i*) Excellent resistance to oxidation

(*ii*) Good wear protection

(*iii*) High film strength to minimize friction

(*iv*) Good anti-corrosion properties

(*v*) Should readily separate from water.

(*vi*) Reduced foaming tendency.

(*c*) Features required for refrigerating systems:

(*i*) Excellent resistance to corrosion.

(*ii*) Minimize carbon deposits on value plates

(*iii*) Should provide starting at low temperature.

(*iv*) Provide protection against rust and corrosion.

(*v*) Compatible with refrigerant.

22.11 TESTING OF LUBRICANTS

Lubricating oils deteriorate during service due to internal breakdown or contamination (internal and external both). Hence, it is necessary to monitor quality of oil on periodic basis so as to keep the machinery in healthy state.

Need for testing of lubricating oil are following:

(*i*) to ward off major breakdowns.

(*ii*) to detect any impending failure.

(*iii*) to check metal wear rate.

(*iv*) to get long service from oil and machinery.

(*v*) to check if right oil is being used.

22.11.1 Various Methods of Lubricating Oil Testing

Lubricating oil during service gets deteriorated due to entry of dust, grit, wear metals, corrosive gases, water and consumption of additives while in service. Hence, lubricating oil loses its properties and the contaminants in lubricating oil cause damage to the system. Hence, it is essential to test the lubricating oil in order to check its fitness for further service. Periodicity of testing is specified by equipment manufacturer, oil manufacturer and also from the operating conditions.

After fresh charge of oil into the system, the various parameters of oil are tested and it serves as a base for future tests.

Various tests can be broadly categorized into three:

(*i*) Sensory tests.

(*ii*) Physico-chemical tests.

(*iii*) Spectrometric tests.

(*i*) **Sensory Tests:** Sensory tests refer to appearance, colour and odour of lubricant oil. These tests serve as first indicators before detailed physico-chemical tests are carried out. These tests can be performed by experienced plant personnel on weekly or monthly basis.

Various sensory tests are:

(*a*) *Appearance:* It is simplest of all tests. Lubricant oil sample is taken in a clean bottle and checked under bright light. A good oil is clear and bright. A hazy or cloudly appearance indicates water contamination. Foaming in oil indicates either too much churning or loss of foaming additives.

(*b*) *Colour Test:* This test can be used with light coloured oil. A dark colour indicates oxidation or contamination. However, diesel crank case oils darken during service.

(*c*) *Odour Test:* Lubricants oils have typical oily odour. A pungent odour indicates oxidation and overheating of oil.

(*d*) *Crackle Test:* This test is used to check presence of water in oil. If appearance is hazy or cloudly then this test can be performed.

A few drops of sample oil is placed on a hot plate metal surface (100°C). If these is little crackle sound, the sample oil contains more than 100 ppm water.

(*e*) *Blotter Spot Test:* This test is used to monitor engine crank-case oil. An ordinary blotting paper is used to determine the presence of sludge in oil.

One or two drops of sample of oil are placed on a blotting paper. It is then allowed to dry and then examined. It may be interpreted as under:

Oil spot fades out within oil boundary → there is no sludge in the oil.
Sharply marked oil boundary → presence of sludge is indicated.

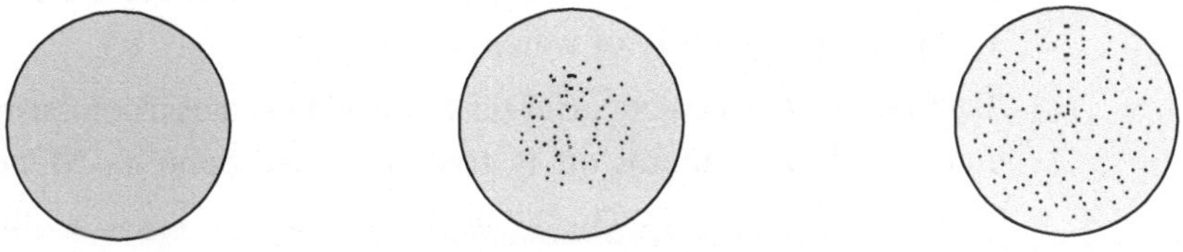

No Sludge Presence of Sludge Dispersed Sludge

Fig. 22.6 Blotter Test.

If presence of sludge is indicated then further tests such as Ferrography, UV, IR etc. are necessary.

(*ii*) **Physico-chemical Tests:** These tests are following:

(*a*) *Specific Gravity*: This test is performed if adultration of oil with lighter or heavier oil is suspected. Specific gravity of oil under test can be compared with the grade of oil being used.

(*b*) *Viscosity:* It is most significant property of lubricating oil. Kinematic viscosity of oil is determined at 40°C and 100°C. It can be determined as per IS 1448 or ASTM D445/D2161.

The viscosity of oil changes with use. An increase in viscosity indicates oxidation of oil. Similarly, lowering of viscosity indicates contamination with lower viscosity oil. For example, in internal combustion engines, crank case oil gets diluted due to mixing of fuel oil. Increase in viscosity leads to higher acidity of oil causing corrosion of metal surfaces. A decrease in viscosity decrease the flash point of oil. For proper lubrication, viscosity need to be maintained within specified range.

(*c*) *Neutralisation Value:* It is a measure of ability of lubricating oil to react with a alkaline reagent. During service, oil undergoes oxidation which increases its acidity. This test measures the acidity of lubricating oil. The sample oil is diluted with a neutralized mixture of Benzene and Alchohol containing phenolphthalein as indicator. This mixture is titrated with decinormal solution of KOH. The amount of KOH required in milligram to neutralize one gram of oil is the neutralization value.

(*d*) *Water Content:* Lubricants oil gets contaminated with water during service. It can be determined quantitatively by two methods: (i) Karl-Fischer method and (ii) Dean and Stark method.

Dean and Stark method is easier one. It requires oil to be heated with a water immiscible solvent (xylene or toluene) upto 100 to 120°C. After distillation, solvent and water are separated. Test provides quantity of water present in oil.

(*e*) *Ash Content*: Ash content in lubricanting oil indicates incombustible material. Pure mineral oil is free from ash content. Metallic additives added to the oil to enhance its properties. These additives generate ash content. Apart from additives, contaminants such as dust and dirt also generate ash.

This test is used to determine the level of depletion of additives in engine oils. The test can be carried out as per ASTM – D482

Other tests that are carried out on lubricating oil are pour point, flash point, foaming, Total Base No., Pentane and Benzene insoluble etc. These are performed as per IS1448 or ASTM standards.

(*iii*) **Spectrometric Tests:** These are sophisticated instrumental techniques used to analyse oil minutely. The various techniques are:

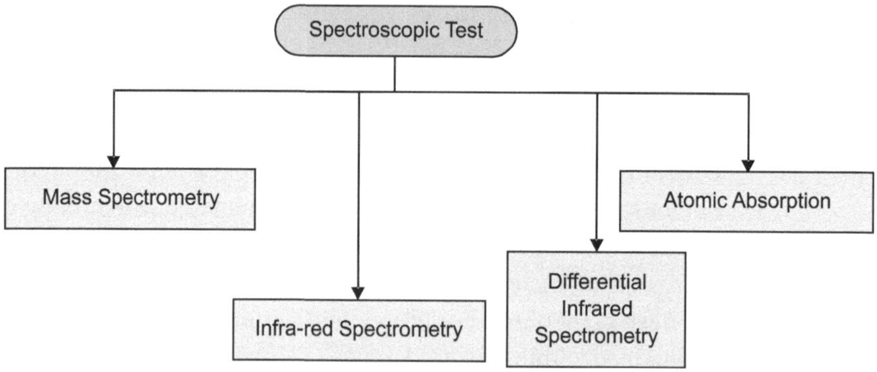

(*a*) *Mass Spectrometry:* This test is used to determine various components of substance present in oil sample due to typical behaviour exhibited by molecules of different substance under electric and magnetic field.

In the test, material under test is ionized and this ionized beam is directed in a curved path under the influence of electric and magnetic field. Heavier ions tend to to deflect out of curved path due to higher inertia. This produces a spectrum of mass of various components present in the oil. This is recorded on a photographic plate.

(*b*) *Infrared Spectrometry:* This test shows amount and pattern of infrared absorption by various components of substance present in oil sample. The spectrum so obtained shows characteristic peaks and its area which are used to identify substance and its concentration.

The result indicates changes in molecular structure of oil sample due to use.

(*c*) *Differential Infrared Spectrometry:* In differential Infrared spectrometry, new d oil samples are compared with respect to absorption of infrared rays. The absorption band indicates change in used oil sample with respect to fresh oil sample. A downward movement of absorption band indicates presence of new substance.

This method is needed to determine contamination of oil with coolant, fuel oil *etc*. Hence, it is possible to indicate expected future life of oil.

(*d*) *Atomic Absorption:* The method is based on absorption of radiation in narrow bands of various elements. So presence of various metals can be quantified in the oil sample. This indicates wear of various components of equipment.

In this method, arc is produced in an oil sample by an electric arc and radiant energy is measured by the photo electric spectrometer.

It is possible to estimate quantity of various metals such as copper, aluminium, lead, chromium *etc*. This gives an idea of wear and depletion of oil additives.

22.12 LUBRICANTS OIL RECYCLING AND DISPOSAL

Mineral lubricating oils are produced from refining of crude oil. All crude oils are not suitable for manufacture of lubricating oils especially crude oil produced in India are not suitable for manufacture of lubricating oil due to high aromaticity and wax content. Crude oil produced by Saudi Arabia, Iraq and Iran, Venezuela etc. are suitable for lubricating oil production. Hence, due to scarcity and fast depletion of fossil oil, it is imperative that lubricating oils are conserved and recycled.

However, lubricating oil, undergoes deterioration during its normal use. A part of the oil is consumed due to equipment design, operating conditions, maintenance practices and lubricants oil quality, Hence, oil need to be changed after a certain fixed interval apart from periodic topping-up with fresh oil.

Let us first understand causes of oil deterioration. Two major factors that lead to deterioration of lubricants oil are:

(*i*) Contamination.

(*ii*) Chemical breakdown.

(*i*) **Contamination:** Contamination is the unintended mixing of foreign material into lubricating oil. Source of contamination can be internal, external or both. Various contaminants are:

(*a*) *Metal Particles:* Metal particles are generated from the wear of the system components *e.g.* from piston rings, bearing races, shafts, turbine components, valves *etc*.

(*b*) *Corrosion Products*: Corrosion products are generated when acidic lubricating oil attacks metal parts. These corrosion products then enter lubricating oil system.

(*c*) *Combustion Products:* In an internal combustion engine, products of combustion enter into oil through blow bys contaminants are carbon, soot, acid, water *etc*.

(*d*) *Unburnt Fuel:* In internal combustion engine, unburnt fuel enters crankcase from engine cylinders. This causes contamination of lubricating oil.

(*e*) *Contamination from Surroundings*: Atmospheric dust, dirt, moisture, gases *etc* contaminate lubricating oil after gaining entry into lubricating oil system.

(*ii*) **Chemical Breakdown:** Lubricating oil also deteriorates due to chemical breakdown. It takes place due to reaction of lubricating oil with oxygen or other chemicals. Oil also undergoes oxidation due to temperature. Presence of certain metals such as copper, nickel, lead, zinc *etc* act as catalyst and accelerate oxidation process. This oxidation produces acids which attack metal surfaces.

However, there is little change in composition of mineral lubricating oil during service. Barring little deterioration, major part of mineral hydrocarbon oil can be regenerated.

22.12.1 Various Methods of Used Oil Reclamation

Reclamation techniques can be broadly classified into following categories:

(*a*) Mechanical filtration process.

(*b*) Chemical treatment process.

(*a*) **Mechanical Filtration Process:** This method employs various physical methods such as filtration, centrifugation, dehydration, decantation, settling, distillation *etc*. Generally, a combination of above mentioned methods are used to make oil suitable for further use.

Mechanically separation of contaminants in oil uses the following techniques:

(*i*) Use of strainers.

(*ii*) Use of absorbent or adsorbent filtering media.

(*iii*) Centrifugal separation.

(*iv*) Magnetic filtration.

(*v*) Chemical treatment with acid, alkali and solvents.

(*i*) *Use of Strainers:* Strainers are perforated filtration device through which oil is forced. The solid contaminants of size more than the size of perforation are held back. Filter elements used are porous metallic or non-metallic material such as metal, ceramic, glass-powder *etc.*

(*ii*) *Use of absorbent or adsorbent filtering media*: Absorption process removes moisture, acids, coarser particles. Filter media used are nylon, felt, cotton waste, paper, jute, mineral, woollen *etc.*

While adsorption process uses media such as charcoal, silica-gel, Bentonite, fullers earth, activated alumina *etc.*

(*iii*) *Centrifugal separation*: Centrifugal process is used to remove solids and water from oil. The oil is taken in a centrifuge and rotated or whirled at high speed causing contaminants such as water, solid *etc.* to separate out.

(*iv*) *Magnetic filteration:* This process is used to remove ferrous metal contaminants from low viscosity oils and water soluble oil coolants. Presence of magnetic material causes separation of ferrous metal impurities.

(*v*) *Chemical treatment with acid, alkali and solvents*: Oils such as hydraulic oils, cutting oils, turbine oils, quenching oils, circulating oils *etc.* can be reclaimed by methods described earlier. But engine oils require both chemical and physical treatment methods of separation.

(*b*) **Acid and clay treatment process**: This method is oldest and was devised by Indian Institute of Petroleum, Dehradun and Defence Research Laboratory, Kanpur. This process involves the following steps:

(*i*) Settling and Decantation.

(*ii*) Dehydration.

(*iii*) Acid treatment and neutralization

(*iv*) Clay treatment.

(*v*) Filtration.

(*i*) *Settling and decantation*: Used oil is allowed to stand in a oil tank with conical bottom. Insoluble matter, water and sludge settles down at the bottom of the tank and is removed.

(*ii*) *Dehydration:* Oil obtained after settling and decantation is dehydrated under vacuum to remove dissolved water and some lighter volatile impurities.

(*iii*) *Acid treatment and neutralization:* Dehydrated oil is treated with sulphuric acid to remove asphaltic matter, oxidized products, spent and unspent additives. The impurities are removed as precipitate.

(*iv*) *Clay Treatment:* Acid treated oil is then passed through activated clay at high temperature under vacuum. Unburnt fuel contaminants, dissolved impurities and acid sludge from previous step are removed. It also removes colour and odour.

(*v*) *Filteration*: After clay treatment, oil is passed through filter cloth to give clear lubricating oil base. This oil is then further processed to get finished product.

There are various other processes used to reclaim oil. Some are Phillips process, Recyclon process, Berc process, KTI process, Snam Progetti process, Revac process *etc.*

22.13 MECHANISM OF LUBRICATION

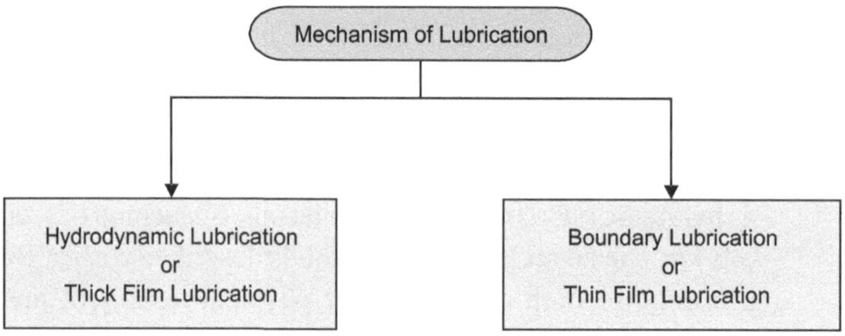

(*i*) **Hydrodynamic lubrication:** In this type, moving surfaces are separated by a thick film of lubricant. This prevents direct contact between moving surfaces. This method of lubrication is used is journal bearings.

(*ii*) **Boundary lubrication:** This method is used where it is not possible to maintain a continuous film of lubricant due to high load, low speed, low oil viscosity *etc*. The sliding surface is lubricated by adsorption of lubricant by the surface.

22.14 METHODS OF LUBRICATION

Depending upon the application, method of lubrication changes. Following are various methods of lubrications:

(*i*) **Splash Type Lubrication:** This method is used for lubrication of connecting rod, cylinder *etc.* in an IC engine. A cup is used to splash the oil onto surface by the motion of the crank shaft.

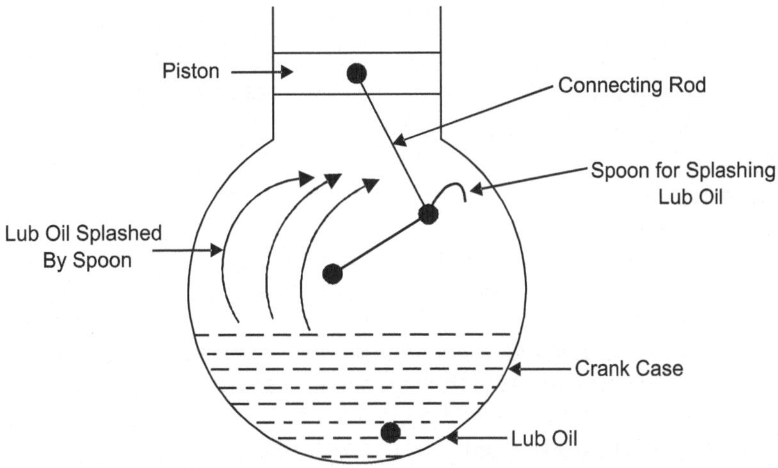

Fig. 22.7 Splash Type Lubrication.

(*ii*) **Wick Lubrication:** It is based on siphon principle. A wick is placed in a oil reservoir and due to siphon action oil is carried by the wick to the place of lubrication.

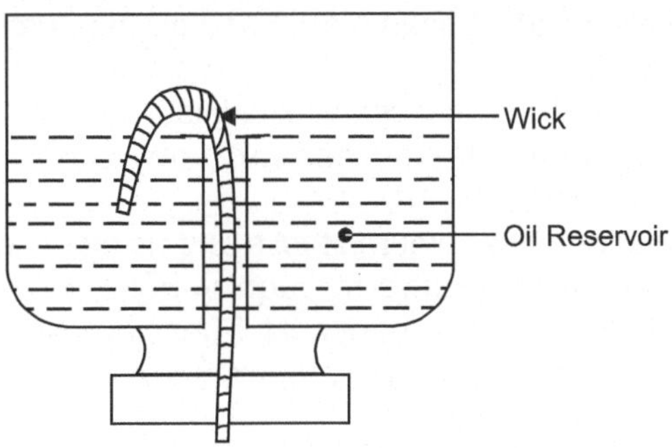

Fig. 22.8 Wick Lubrication.

(*iii*) **Forced Lubrication:** In this method, lubricating oil pump is used to pressurize the oil so that it can reach desired place. All complex machinery use this method.

(*iv*) **Mixed Lubrication:** This method is used in low power two stroke IC engines. Oil and petrol are mixed and then sent to engine cylinder. An oil scrapper ring removes excess oil which is exhausted to atmosphere.

(*v*) **Dry Sump Lubrication:** A separate oil sump is used containing very little oil. A pump is used to lubricate the parts and excess oil flows back to the sump. This method can be used in radial or inverted engines *e.g.* air-craft engines.

(*vi*) **Wet Sump Lubrication:** An oil sump contains oil and a pump is used to lubricate parts. The sump is full of oil.

Summary

A lubricant is used primarily to reduce friction between moving surfaces. Apart from reduction of friction, it acts as coolant, caries away contamination, prevents corrosion *etc.* A wide variety of lubricants are available in liquid, solid and semi solid phase. Petroleum oil is used as a source but synthetic lubricants are also available.

A lubricant is characterised by various properties such as viscosity, flash point, pour point, ash content, neutralization value, de-emulsification number *etc.* These properties act as a guide for specific application. Properties of lubricants are improved through the use of additives. Additives are chemicals which either enhances or suppresses certain properties. For example, additives are used to prevent frothing tendency when lubricating oil comes in contact with water.

Apart from liquid lubricants, there are semi-solid (greases) and solid lubricants. Semi-solid lubricants or greases are essentially soap dispersed in oil. These are used for heavy loads operating at lower speeds. While solid lubricants are used for high pressure and low speed applications.

Testing of lubricants is necessary to maintain its quality and is carried out by three major tests—sensory, physico—chemical and spectrometric tests.

Due to depletion of petroleum stocks and rising prices of oil, spent lubricants are regenerated through various processes. These processes help to reclaim lubricants. These processes are broadly are of two types—mechanical filteration process and chemical treatment process.

There are two methods of lubrication hydrodynamic and boundary lubrication. Hence, lubrication forms a very vital part of maintenance engineering.

Review Questions

1. Define friction. What are the laws of friction ?
2. What are the various classification of lubricants ?
3. What are the properties of hydraulic oils?
4. What is an oil additive ? What are various additives used in engine oils ?
5. Discuss various properties of lubricants.
6. What is a grease ? What are the properties of grease?
7. Discuss solid lubricants and their applications.
8. What are the various causes of failure of lubricants?
9. Write short note on selection of lubricants?
10. Describe various methods of testing of lubricants.
11. Why recycling of spent lubricating oil is done?
12. What are the various methods of regeneration of spent lubricating oil ?

Discussion Questions

1. "Lubricating oils are like blood in a human body". Justify
2. "Identification of metal particles in lubricating oil can indicate an impending failure". Comment.
3. Discuss – "Life of machinery depends on quality of lubricating oil".
4. Discuss "Quality of base oil decides the quality of lubricants".
5. Discuss "Recycling of lubricant is a way to conserve fast depletion of oil resources".
6. Comment "All lubricating oil are not same but are application specific".

Multiple Choice Questions

1. A lubricant does not:

   (*a*) Reduce corrosion. (*b*) Reduce wear.

   (*c*) Form deposit. (*d*) Act as coolant.

2. Engine oil should possess a property of

   (*a*) Low viscosity index. (*b*) High oxidation stability.

   (*c*) High pour point. (*d*) None of these.

3. A hydraulic oil should not have:
   (*a*) Air release property.
   (*b*) High foaming tendency.
   (*c*) Resistance to oxidation.
   (*d*) Anti corrosion property.

4. A good gear oil should not have:
   (*a*) High viscosity index.
   (*b*) Good heat transfer capability.
   (*c*) Corrosion resistance.
   (*d*) Thermal stability.

5. Neutralisation value of oil indicates:
   (*a*) Presence of alkaline impurities.
   (*b*) Presence of acidic impurities.
   (*c*) Presence of solid contaminates.
   (*d*) None of the above.

6. Air release value describes property for
   (*a*) Engine oils.
   (*b*) Heating oils.
   (*c*) Electrical oils.
   (*d*) Hydraulic oils.

True/False Questions

1. Synthetic lubricating oils are better than mineral oils.
2. Thermic oils should have high operating temperature.
3. Viscosity of oil decreases with increase in temperature.
4. Low viscosity lubricants are suitable for low pressure and high speed applications.
5. Dispersant and detergent additives control foaming tendency of oil.
6. Spectrometric testing of oil provides accurate result.

ANSWERS

Multiple Choice Questions

1. (*c*) **2.** (*a*) **3.** (*b*) **4.** (*a*) **5.** (*b*) **6.** (*d*)

True/False Questions

1. True **2.** True **3.** True **4.** True **5.** False **6.** True

❑❑❑

23

Failure Analysis and Reliability Engineering

Learning Objectives

After reading this chapter, you will be able to:

- Understand causes, processes and modes of failure.
- Develop keen understanding of failure analysis techniques.
- Comprehend concept of reliability and ways of improving it.
- Understand various techniques of reliability enhancement.

23.1 INTRODUCTION

In the recent past, failure analysis has grown to be an independent field. Failure analysis being a daunting task needs multidisciplinary expertise and experience.

Failure analysis determines and describes the factor responsible for the failure of components. The proper application of failure analysis technique can provide a valuable feedback to overcome design flaws and material limitations. It can play a valuable part in design since most sophisticated simulation testing can never duplicate the varied conditions found in actual service.

The theme of failure analysis is important to aero-space, automobile, defense, engineering, chemical, petrochemical and power generation industries among others.

23.2 CASE HISTORY OF FAILURES

One of the oldest cases of reported failure is falling of a mill of Oldham, England, in 1844. The accident took place due to fracture of a cast iron beam and led to the death of 20 factory workers. This was a case of material failure. Another case was of bridge collapse in Scotland. The Montrose suspension bridge over river Esk in Scotland collapsed in 1830 with great loss of life.

Most spectacular of these failures were the fractures of two tanker ships "Schenectady" and "Esso" in Manhattan. Both of these broke in calm weather from the middle.

All the failures led to active and extensive research for understanding the causes of failures.

Other Cases of Failures

(1) This pertains to 10,300 foot long single track rail road bridge built in 1878 in Scotland. A portion of the bridge consisted of 13 wrought iron spans, each 240 feet in length and 88 feet above water supported by cast iron piers.

On 28 December 1879, a gale developed with a wind speed of 75 miles per hour. A passenger train while crossing the bridge plunged into river together with 13 centre spans. Seventy-five passengers and crew members lost their lives. The cause of failure of bridge structure was lateral force produced by the wind on passenger car and bridge structure. Such wind loading had not been taken into account in the design stage. *Hence, this was a case of design failure.*

(2) In January 1919, a steel tank containing 2,300,000 gallons of molosses suddenly collapsed. Twelve persons lost their lives. The reason was brittle fracture in steel tank. *This was a case of poor material selection.*

(3) At 5 PM, 15th December 1967, a suspension bridge over Ohio river suddenly collapsed, 46 people lost their lives. The cause of failure was traced to two adjacent small semi-circular flaws of the order of few mm in radius. These eye bars were arranged in pairs so that four bars met at a pin connector. The structure was so designed that if one member failed the resulting load on its companion members would be sufficient to bring failure. One of the flaws had developed due to combination of corrosion, fatigue and stress corrosion cracking. This led to the bridge collapse. *Hence, this was a case of design failure.*

(4) In April 1912, the RMS Titanic struck an ice-berg while steaming at 20+ knots speed and sank-off New foundland in less than three hours with loss of 1500 people. The wreckage was located in 1985. A piece of hull material was recovered and charpy impact test was carried out. It was found that steel fractured in a 100% brittle fashion at ice-berg temperatures. *Hence, this was a case of poor material selection.*

From the past accident analysis, the various causes were attributable to the failures *e.g.* poor material selection, design shortcomings, unexpected stresses, extreme ambient conditions, poor maintenance, wear and tear, corrosion, chemical attack *etc*. Hence, the design, fabrication, errection, commissioning and maintenance aspects are devised such that the system remains functional for the duration of its expected life without any failures. It is not only the design and material selection which contribute to failure, other factors like maintenance, fabrication procedure (for example improper stress relieving in mechanical components working under stress may lead to premature failure), method of errection and commissioning and testing equally contribute to the life of engineering systems (structures/equipments/plant *etc.*)

23.3 DEFINITION OF FAILURE

Failure is the inability of a component to perform its intended function for given length of time.

Failure is determined on the basis of past history (*e.g.* frequent repairs *etc.*) or on the basis of performance (*e.g.* loss of efficiency indicating component decay, increase in noise and vibration, initiation of leakage *etc*) or on the basis of inspection (*e.g.* changes in clearances, metallurgical changes *etc.*).

23.4 REASONS FOR FAILURE

Following are the various reasons for failure:

(*i*) Fatigue/fracture of material.

(*ii*) Structural overload.

(*iii*) Electrical overload.

(*iv*) Wear.

(*v*) Chemical attack.

(*vi*) Oxidation.

(*vii*) Material removal.

(*viii*) Radiation.

(*ix*) Environment causes.

(*i*) **Fatigue/Fractures of Material:** A failure that is the result of repetitive or cyclic loading is known as fatigue failure. So, alternate cycle of sudden load followed by no load repeated number of times on a material can cause failures at a stress much lower than designed stress level. A fatigue failure occurs without any obvious warning. It results in a brittle appearing fracture with no gross deformation at the fracture.

Fatigue cracks can be initiated at various places such as scratches, abrupt changes in cross-section, corrosion pits, weld configuration defect *etc*. In some cases, micro-cracks may be present such as quench cracks, hot/cold cracks from welding. This results in development of bigger cracks leading to failure. Other failures due to metallographical causes are following:

(*a*) **De-cohesive fractures:** A De-cohesive fracture is one which shows no bulk plastic deformation and does not occur by dimple rupture, cleavage or fatigue. This type of fractures results due to reactive environment or a unique micro-structure such as weakening of atomic bonds, rupture of proactive films *etc*.

(*b*) **Brittle fracture:** Brittle fractures are promoted by low service temperatures, high strain rates, presence of stress concentrators and certain environmental conditions.

The transition of metal from ductile to brittle range is influenced by strain rate, stress state, composition, micro-structure, grain size and specimen size.

(*c*) **Embrittlement due to intergranular corrosion:** Generally, metallic corrosion occurs uniformly all over the metal however under certain conditions attack is localized at the grain boundaries leading to intergrannular corrosion. Aluminium-copper and aluminium-magnesium alloys are susceptible to intergranular corrosion in certain environment. For example, die-cast Zinc-Aluminium alloys can fail in salt water.

(*d*) **Stress-corrosion cracking:** It occurs from combined effects of tensile stress and corrosive environment. These type of failures tend to be catastrophic. The cracks may propagate transgranularly or intergranularly. In most cases, there is little evidence of influence of corrosion. Certain types of embrittlement are season cracking of brass, caustic embrittlement of reveted carbon steel structures *etc*. Alloys are susceptible to such cracks especially if minor amounts of impurity elements are present.

(*ii*) **Structural Overload:** Any load bearing member may fail due to excess load than its designed capacity.

For examples, excess load on a column of a building may lead to its failure (the load might have been increased over a period of time due to expansion/renovation).

(*iii*) **Electrical Overload:** An electrical equipment may fail due to overload leading to burn-out. For example, an electrical motor if overloaded may cause failure due to excessive temperature rise.

(*iv*) **Wear:** Wear is progressive loss of material from the surface due to relative motion with each other. It occurs as a natural consequence when two surface rub against each other. Wear depends upon load, crystal structure and mechanical properties of material. Different types of wear are—adhesive, abrasive, corrosive, surface fatigue and minor types of wear are erosion, cavitation and impact chipping.

Various factor influencing wear are effect of surface film (oxide films, boundary lubrication, soil lubricants, other surface layer). Examples of wear are—normal wear of bearings and bushes during their expected life.

(*v*) **Chemical Attack:** Material may fail due to reaction with corrosive chemicals *e.g.* acid, alkali *etc*. These may be solid, liquid or gaseous chemicals either present in the atmosphere or in other surrounding media such as lubricating oil, coolant *etc*. It could be introduced by maintenance operation such as descaling.

(*vi*) **Oxidation:** Oxidation of material causes brittleness leading to failure. The oxidation can take place due to elevated temperature, chemical reaction *etc*.

(*vii*) **Material Removal:** Material removal (*e.g.* chipping) may cause failure due to reduction in load carrying capacity.

(*viii*) **Radiation:** Radiation exposure leads to failure of material. This aspect is extremely important in nuclear power plants.

(*ix*) **Environmental Causes:** These could be high/low temperature, humidity, dust, corrosive gases, vibration *etc*. present in the surrounding atmosphere.

23.5 TYPES OF FAILURES

Various types of failures are following:

(*i*) **Catastrophic Failures:** These are sudden and unexpected failures.

(*ii*) **Degradation Failures:** These failures result over a period of time *e.g.* due to normal wear and tear.

(*iii*) **Intermittent Failures:** These are not permanent failures. The device may sometimes functions normally but at other time may fail to operate.

23.6 EFFECTS OF FAILURE

Inception of failure leads to following changes:

(*i*) **Noise:** There may be sudden increases or decrease in noise level from the normal *e.g.* a bearing failure in motor leads to higher noise level.

(*ii*) **Erratic Operation:** The device may start/operating erratically.

(*iii*) **Inoperability:** The device may not operate at all.

(*iv*) **Instability:** The device operation may become unstable.

(*v*) **Intermittent Operation:** The device may operate and stop suddenly.

(*vi*) **Impaired Control:** It may not be possible to control/regulate the equipment/device/machinery.

(*vii*) **Unpleasant Odour:** Failure may give rise to unpleasant odour *e.g.* burning of choke of tube lights gives out a typical smell.

23.7 CAUSES OF FAILURES

Generally, causes of failure is combination of several factors. Following are causes of failures:

(*a*) **Poor Design:** Poorly designed components result in failure. The design should take care of actual operational conditions. For example, a metallurgical object may fail due to presence of a notch.

(*b*) **Poor Material Selection:** Material selection depends upon many factors such as cost, availability, expected life, fabricability *etc*. A wrong selection of material may result in failure *e.g.* due to wear, corrosion, excess load, hygroscopic nature (electrical insulation) *etc.* Hence, it is most important that appropriate material is selected for the particular application anticipating normal and abnormal operation possibility.

(*c*) **Material Imperfection:** Presence of material imperfections such as voids, porosity, lamination *etc.* promote failure due to propagation of cracks.

(*d*) **Fabrication and Processing:** Failure may result due to unsuitable processing methods, incorrect specifications, inability to follow correct procedures *etc.* Some mechanical operations such as bending, expanding, drawing, shearing *etc.* result in development of stresses. These stresses if not relieved properly may result into reduced resistance to fatigue, corrosion *etc.* This may lead to failure ultimately. For example, acid pickling and electroplating operations cause hydrogen embrittlement in steel. This is a potential source of failure.

(*e*) **Assembly:** Failures may result due to assembly errors. Assembly errors are due to operator's negligence, poor skill, incomplete specification of procedures *etc.*

(*f*) **Inspection and Testing:** Assembly errors as discussed above can happen if inspection fails to pinpoint them. So poor inspection may result in failure. Testing is an important part after assembly. Testing helps to detect problems in assembly, material, *etc.* If the testing is not comprehensive and rigorous, it may lead to failure.

(*g*) **Service Condition:** The component designed for a particular environment may fail if used in entirely different environment. For example, a component designed for a normal ambient condition may fail if used in corrosive environment.

(*h*) **Preventive Maintenance:** Poor preventive maintenance or lack of it enhances chances of failure.

(*i*) **Unexpected Exposure to Overload:** Components exposed to stresses more than the designed level will lead to failure.

(*j*) **Maloperation:** A component may fail if it is not designed for the intended operation.

23.8 FAILURE MODEL

One of the oldest models of failure is the "Bath Tub Curve" (named due to its shape).

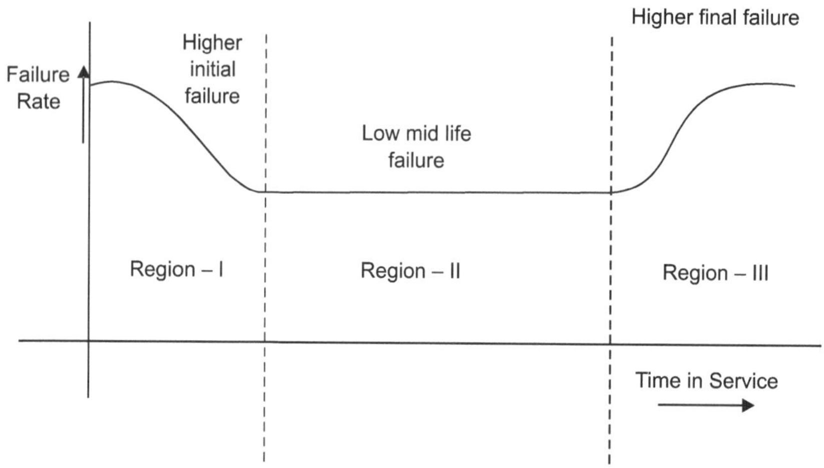

Fig. 23.1 Bath Tub Curve

The curve can be divided into three distinct regions:

Region-I: Initial failure or infant mortality. These failures take place during the beginning of operation. These are due to poor workmanship, poor handling, poor process control, wrong type of material, poor quality of material, contamination and improper installation.

Region-II: It is the period when very few failures take place and hence rate of failure is constant. These chance failures take place due to insufficient safety factor in design, higher than expected random load, human error and mis-application.

Region-III: It is the wear out phase when the failures increase due to end of useful life of the parts. These failures take place due to aging, degradation in strength, material fatigue, creep corrosion and poor maintenance.

23.9 FAILURE MODES

It refers to the route through which the failure took place. The various failure modes are following:

(*i*) Cracking.

(*ii*) Deformation.

(*iii*) Wear.

(*iv*) Corrosion.

(*v*) Loosening.

(*vi*) Leaking.

(*vii*) Sticking.

(*viii*) Electrical shorts.

(*ix*) Electrical open circuit.

(*x*) Oxidation.

(*xi*) Vibration.

(*xii*) Fracturing.

23.10 FAILURE ANALYSIS

The purpose of failure analysis is to prevent occurrence of failure through careful design, by introduction of redundancy (stand-by equipments), reconfiguration, modification of preventive maintenance plan, use of alternative material, change in operating procedure, introduction of tests, etc.

Procedure for failure analysis are following:

(*i*) **Problem Definition:** Problem is defined in broad aspect which includes all possible causes of failure.

(*ii*) **Background Information:** All relevant information such as specifications, drawings, available standards, operation and maintenance procedure, logbook readings, maintenance history, statement of witness are taken into account.

(*iii*) **Laboratory Examination:** Laboratory examination of samples is carried out through non-destructive testing, X-ray analysis, microscope, metallography, measurement of mechanical properties such as hardness, tensile strength *etc*.

(*iv*) **Reconstruction of Failure Sequence and Causes:** A hypothesis is developed regarding failure sequence and causes taking into account all available information, field observation and lab examination result.

(*v*) **Testing of Hypothesis by Actual Simulation:** The material is subjected to possible failure sequence and its result is compared with step (iv).

(*vi*) **Report Writing:** It is the final step involving testing, analysis and final conclusion.

23.11 FAILURE MODES AND EFFECTS ANALYSIS (FMEA)

FMEA takes into account every possible failure mode and its effect on the product, the criticality of this effect and the actions which must be taken to compensate this effect.

Depending upon the result of FMEA, it may be beneficial to:

- Design, introduce redundancy (standby equipments), improve preventive maintenance.
- Select alternative material.
- Change operating procedure.
- Introduce diagnostic tests.

Steps of FMEA are:

(*i*) Breakdown of system into components.

(*ii*) Determination of all possible failure modes for each subsystems.

(*iii*) Determination of cause and effect of failures for each mode through fish bone diagram or fault tree analysis.

(*iv*) Determination of frequency factor for each causes and severity factor for each effect.

(*v*) Identification of corrective measures.

In FMEA, all failures are analysed independent of each other.

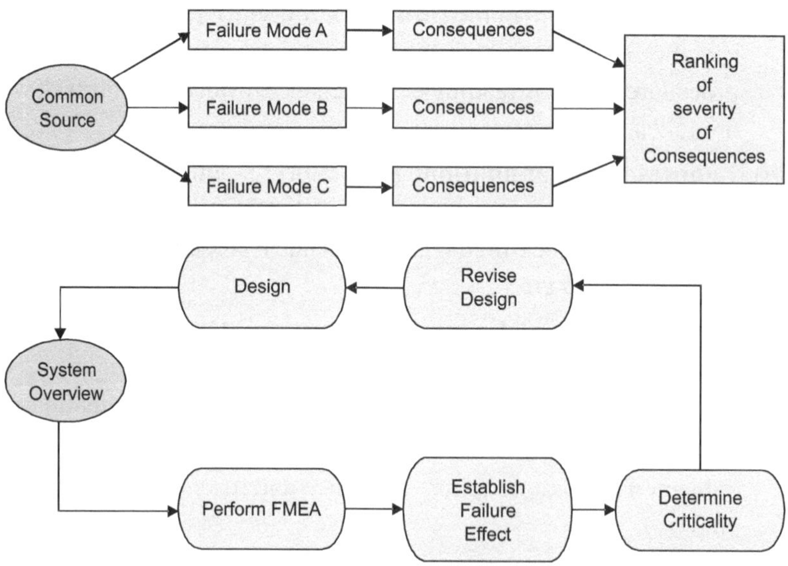

Fig. 23.2 FMEA Cycle.

23.12 FAILURE RATE AND DISTRIBUTION FUNCTIONS

(*a*) **Failure rate is defined as number of failures per unit of time:**

$$\text{Failure rate } (\lambda) = \frac{\text{No. of Failures}}{\text{Total Test Time}} = \frac{f}{T}$$

For a properly designed system, failure rate is constant during its useful life. Smaller the value of failure rate, higher the system reliability.

(*b*) **Mean Time between Failure (MTBF):** It is average time between failures of a repairable item.

$$\text{MTBF} = \frac{1}{\text{Failure rate}} = \frac{1}{\lambda}$$

MTBF is the time of satisfactory operation of the system. Higher the MTBF, higher the reliability of the system. Failure rate and MTBF are used where the item is repairable.

(*c*) **Mean Time to Failure (MTTF):** MTTF is defined as:

$$\text{MTTF} = \frac{1}{n}\sum_{i=1}^{n} t_i$$

Where, n = no. of items

$t_i = t_{1,} t_2, t_3 \ldots\ldots\ldots t_n$ *failure times*

So, MTTF of n items with failure times $t_{1,} t_2 \ldots\ldots\ldots t_n$ is given by

$$\text{MTTF} = \frac{t_1 + t_2 + \ldots\ldots + t_n}{n}$$

MTTF is used for non-repairable items such as fuse, incandescent bulb, missile *etc.*

(*d*) **Maintainability:** It is the probability of performing a successful repair action within a given time. It is a measure of ease with which a system can be repaired.

$$\text{Maintainability } M(t) = 1 - e^{-\mu t}$$

Where, μ = repair rate

$$\text{Mean time to repair (MTTR)} = \frac{1}{\mu}$$

(*e*) **Availability:** It is the probability that a component or a system is operational at a given time. There are three types of availability:

(*i*) Inherent availability $- A_i = \dfrac{MTTF}{MTTF + MTTR}$ – for single component

$$A_i = \frac{MTBF}{MTBF + MTTR} \text{ – for a complicated system}$$

(*ii*) Achieved availability $- A_a = \dfrac{MTBM}{MTBM + M}$

Where, MTBM = Mean time beween maintenance.

M = Maintence time.

(*iii*) Operational availability $- A_0 = \dfrac{\text{Uptime}}{\text{Operating cycle}}$

Overall system reliability is a function of reliability of individual components of system. With increase in number of components, reliability of system decreases. This is the reason why a complicated system fails more often.

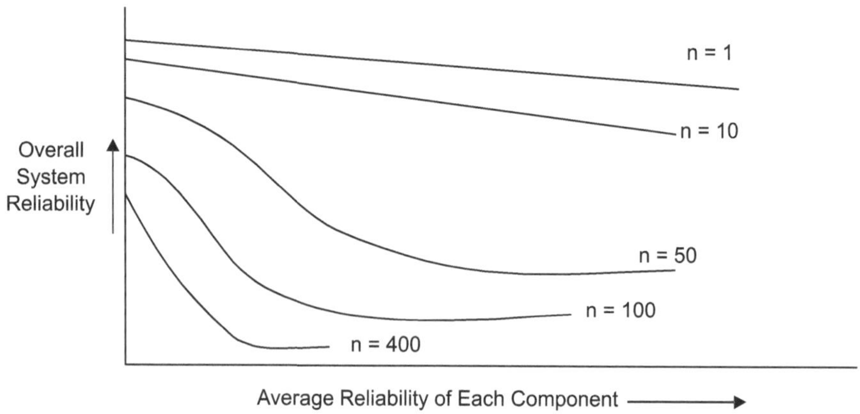

Fig. 23.3 Decrease of system reliability with increase in components.

Statistical distribution curves are used for predicting reliability of systems and sub-systems. Different items follow different statistical distribution. Hence, it is necessary to establish the type of distribution which describes the failure phenomenon.

Hence, to describe reliability of a component, first general failure pattern is established and thereafter parameters are estimated which completely define the distribution.

The various typical failure patterns are shown below:

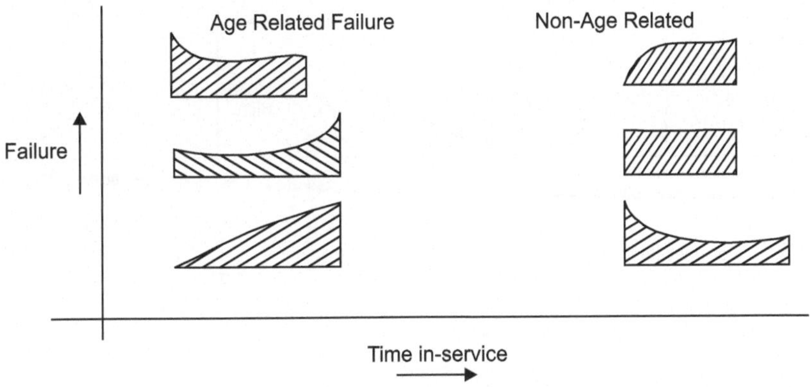

Fig. 23.4 Failure Patterns

When early failures have been eliminated in the "Bath Tub Curve", the constant failure rate is represented by exponential distribution. Similarly, the wear-out failures follow normal distribution. For certain items, statistical distribution has been established based on experience.

Table 23.1 Statistical distribution of select items

Item	Statistical Distribution Applicable
Valves	Exponential
Rubber components	Raleigh
Mechanical components	Weibull
Ball bearing	Normal

Graphical form of typical distributions are following:

(*i*) Normal Distribution

$$f(t) = \frac{1}{\sigma\sqrt{2\pi}} e^{-(t-m)^2/2\sigma^2}$$

Where, t = Critical period

m = Mean life

σ = Standard deviation

(*ii*) Exponential Distribution

$$f(t) = \frac{1}{m} e^{-t/m}$$

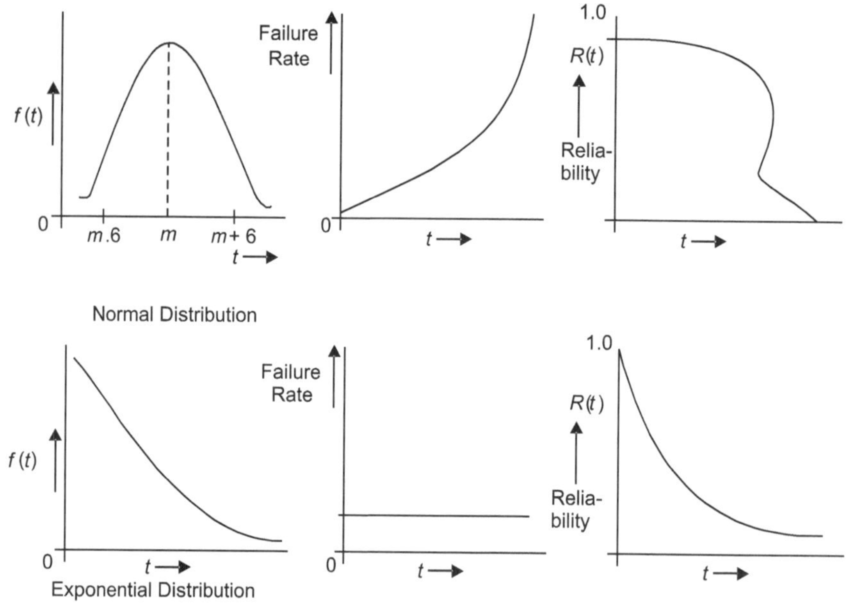

Fig. 23.5 Normal and Exponential Distribution Curves.

(*iii*) Weibull Distribution

$$f(t) = abt^{b-1}e^{-at^b}$$
$$a = \text{Scaling parameter}$$
$$b = \text{Shaping parameter}$$

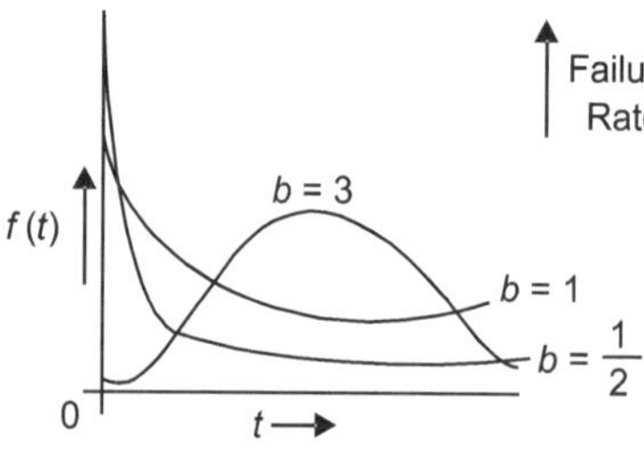

Fig. 23.6 Weibull Distribution

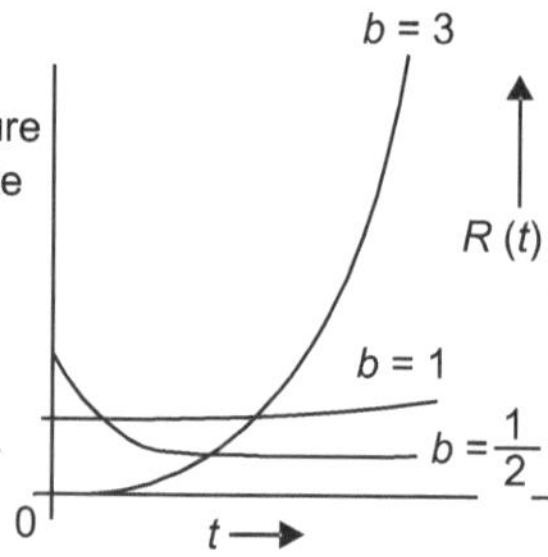

Fig. 23.7 Failure Rate

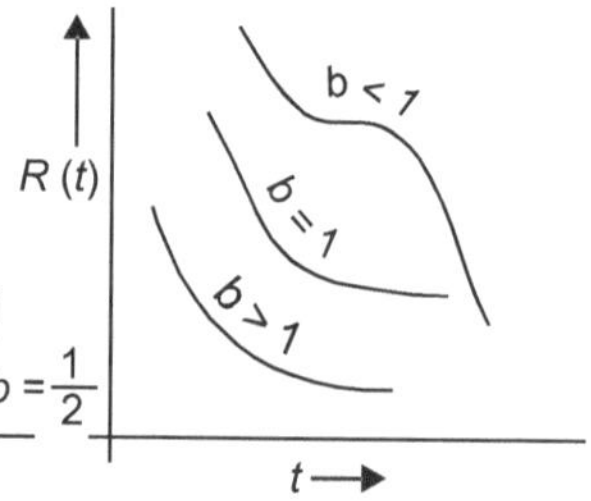

Fig. 23.8 Reliability Function

Other probability distribution functions are following:

(*iv*) Binomial distribution.

(*v*) Poisson distribution.

(*vi*) Geometric distribution.

(*vii*) Lognormal distribution *etc.*

(*viii*) Rayleigh distribution *etc.*

23.13 FAILURE PREVENTION TECHNIQUES

Majority of degradation failures can be detected by following methods before they fail completely.

(*i*) **Non-destructive Tests (NDT):** Various non-destructive tests are available which enable to detect a failure at its initial stage. These are Dye Penetration test, Ultrasonies, Radiography, Visual and optical, Magnetic particle test, Eddy current test, Acoustic Emission, Infrared, Absolute pressure testing *etc.*

These tests allow to detect failures without causing any damage to the material and help to take corrective action in time.

(*ii*) **Metallographic Tests:** These tests help to determine micro-structural state of a component and helps to detect degree of degradation of a component. These tests help to detect micro-structural decomposition, propagation of fine cracks, corrosion attacks, pitting, oxidation attack at grains boundaries *etc.*

Above techniques helps to detect and apply corrective action before complete failure takes place.

23.14 CONCEPT OF RELIABILITY

In 1952, department of defence, USA, set-up an advisory group on reliability of electronic equipment. Purpose was to predict reliability of various military equipments. Later on, this concept found widespread use in other fields.

Reliability is defined as a probability that a device will operate without failure for a given period of time under given operating condition.

Reliability is always a function of time and also depends upon environmental condition. Since it is a probability, its numerical value is always between zero and one.

Reliability Function: This can be explained by taking a suitable example. Consider batch testing of 'N' incandescent bulbs (ordinary lighting bulbs) coming out of factory assembly line.

Assume, 'n' number of bulbs pass the test and "m" number of bulbs fail the test.

Then, reliability $$R = \frac{n}{N}$$

and failure $$F = \frac{m}{N} \qquad \text{Where } N = n + m$$

Also $\quad R + F = 1$ or $R = 1 - F$

$\therefore \quad R = \left(1 - \frac{m}{N}\right)$

Differentiating above equation w.r.t. t

$$\frac{dR}{dt} = \frac{d}{dt}\left(1 - \frac{m}{N}\right)$$

$$\frac{dR}{dt} = -\frac{1}{N}\frac{dm}{dt}$$

$$\frac{dm}{dt} = N\frac{dR}{dt}$$

dividing both sides by n

$$\frac{1}{n}\frac{dm}{dt} = \frac{-N}{n}\frac{dR}{dt}$$

In the above equation $\frac{1}{n}\frac{dm}{dt}$ is known as failure rate (λ)

$\therefore \quad \lambda = \frac{1}{n}\frac{dm}{dt}$

$\therefore \quad \lambda = \frac{-N}{n}\frac{dm}{dt}$

$$\lambda = \frac{-1}{R}\frac{dR}{dt} \qquad \left(\because R = \frac{n}{N}\right)$$

or $\quad \lambda dt = \frac{-dR}{R} \Rightarrow \int_o^t \lambda\, dt = -\int_o^R \frac{dR}{R}$

$$\int_o^t \lambda dt = -\log_e R$$

or $\quad \log_e R = -\int_o^t \lambda dt$

or $\quad \boxed{R = e^{-\int_o^t \lambda dt}}$

If failure rate (λ) is constant over a time t, then reliability $R = e^{-\lambda t}$

23.15 RELIABILITY OF SYSTEMS

Any system say a Boiler or a Turbine or Pump is connected or interconnected in an industrial set up. For example, in a thermal power plant, fuel is fed to the Boiler; output of Boiler *i.e.* steam goes to Turbine; Turbine output *i.e.* mechanical energy is used to drive alternator to produce electricity.

Hence, all above systems are connected in series as output of one element goes as input to the next element. This can be explained by following diagram.

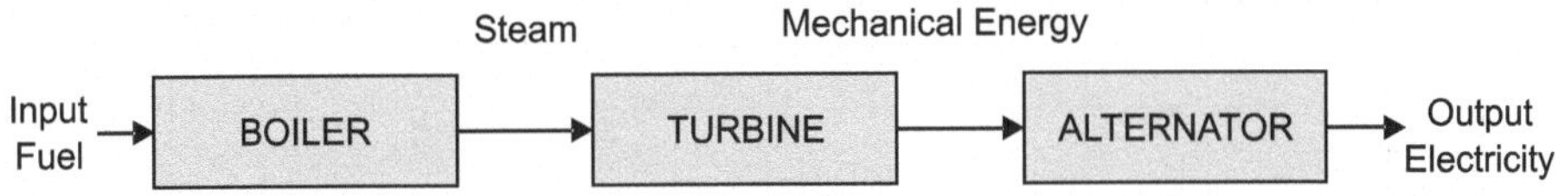

Fig. 23.9 Boiler-Turbine-Alternator Series System.

In the above series set-up, if any element fails, the whole system fails.

Let us consider a following system connected in series having individual reliabilities as R_1, R_2, R_3 .... R_n

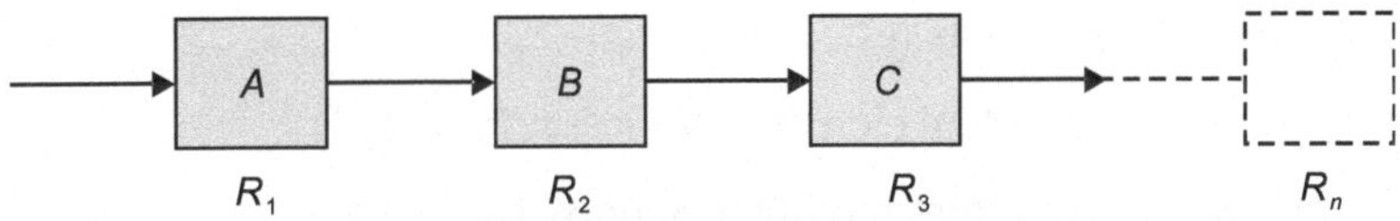

Fig. 23.10 Reliability of Series Element.

The reliability of this series system = $R_1 . R_2 . R_3 . ...R_n$

Let us consider a system having individual elements connected in parallel.

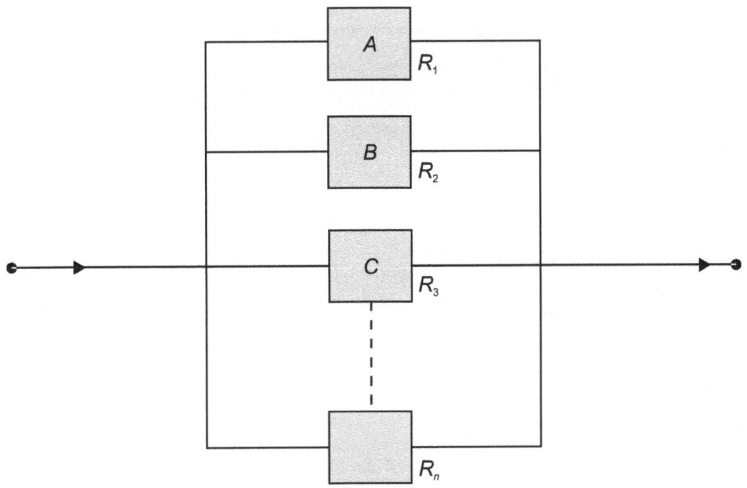

Fig. 23.11 Reliability of Parallel Connected System.

Let the reliabilities of individual elements be $R_1, R_2, R_3, \ldots\ldots\ldots\ldots R_n$

The overall reliability of the above parallel system

$$= 1 - (1 - R_1) \times (1 - R_2) \times (1 - R_3) \ldots\ldots.(1 - R_n)$$
$$= 1 - [F_1 \times F_2 \times F_3 \ldots\ldots\ldots F_n]$$

Where $F_1, F_2, F_3 \ldots\ldots\ldots F_n$ are unreliability or failures of individual systems.

Now consider a system composed of series and parallel elements

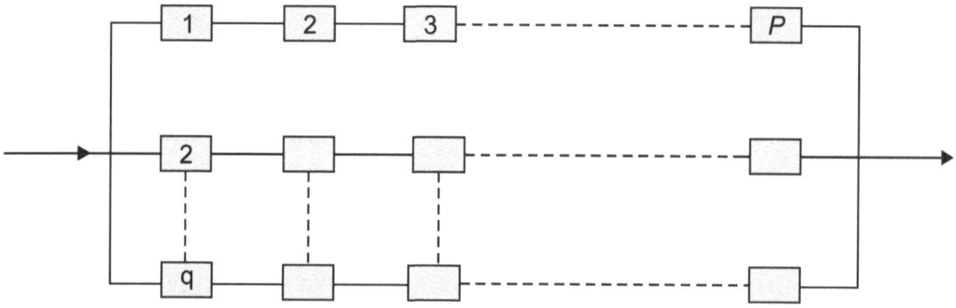

Fig. 23.12 Reliability of Series Parallel Connected System

Reliability of the above system = $\left[1-(1-R)^q\right]^p$

(Assuming all individual elements possess reliability R)

Let us consider a system having a standby (redundancy).

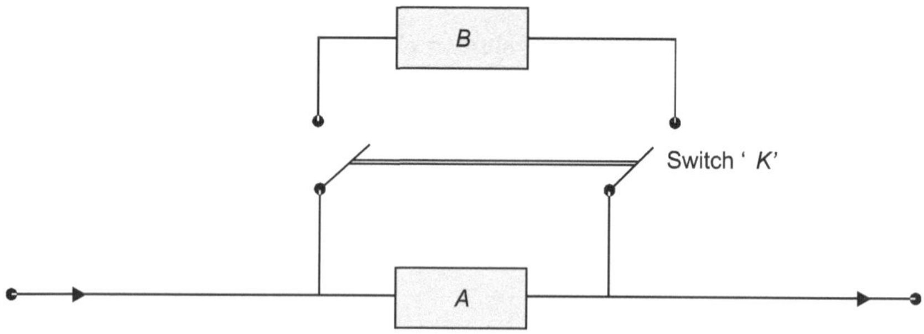

Fig. 23.13 Standby System

Here, component '*A*' is in normal operation. In case of failure of '*A*', component '*B*' is brought into service by automatic closure of switch '*K*'. A perfect sensing device closes the switch '*K*' in the event of failure of '*A*'.

Component *A* is operating upto time *t*. Component A fails at time *t*, and component *B* operates from t_1 to *t*.

$$\therefore \qquad R(t) = \int_t^{\infty} f_a(t)\,dt + \int_o^t \left(f_a(t)\,dt \int_{t-t_1}^{\infty} f_b(t)\,dt_1 \right)$$

The first term represents that component 'A' will operate until time t. The second term (excluding outside integral) is the density function of component 'A' failing exactly at t, and B operating for the remaining time $(t - t_1)$

For a exponential case, $f_a(t) = \lambda_q e^{-\lambda at}$ and $f_b(t) = \lambda_b e^{-\lambda bt}$ where λ_a and λ_b are failure rates of components A and B respectively.

$$R(t) = \frac{\lambda_b}{\lambda_b - \lambda_a} e^{-\lambda_a t} - \frac{\lambda_a}{\lambda_b - \lambda_a} e^{-\lambda_b t}$$

If $\lambda_b = \lambda_b = \lambda$

Then, $R(t) = (1 + \lambda t) e^{-\lambda t}$

For 'n' components of equal failure rate, the reliability is given by

$$R(t) = e^{-\lambda t} \sum_{r=0}^{n-1} \frac{(\lambda t)_r}{\gamma!}$$

This probability distribution function is of poission distribution of 'n' terms.

$$\text{MTBF} = \int_o^\infty R(t)\,dt$$

$$= \int_o^\infty e^{-\lambda t} \times (1 + \lambda t)\,dt$$

$$= \frac{2}{\lambda}$$

It is seen that, mean time between failure of redundant system consisting of two components is twice that of single component.

23.16 IMPROVEMENT OF RELIABILITY OF SYSTEMS

To improve system reliability, following techniques are used:

(*i*) **By having built-in redundancy:** Systems are built with standby equipments. For example, boiler feed pump always has a standby pump. So in case of failure of running pump, standby pump is switched on.

In nuclear power plant, the primary as well as secondary reactor coolings pumps have number of standby pumps.

(*ii*) **Having fewer components:** More the number of components in a given system, higher is the probability of failures. Hence to improve reliability, the number of components need to be reduced to bare minimum.

(*iii*) **Better material selection:** To improve reliability, a better material is chosen for the application.

(*iv*) **Module-to-system test arrangement:** Modern complex control systems are made in modular form *i.e.* the entire system is divided into functional sub systems which can be separately tested. The system is then tested together. In case of failure of any particular sub system, the affected system is replaced quickly. This helps in quick and correct fault diagnosis and repair.

(*v*) **Over design:** The systems are designed with extra safety margins.

(*vi*) **Proper protection and interlocks:** Systems are provided with protection against wrong operation, extreme over or under parameter variation, environmental causes such as corrosion, erosion, chemical attacks *etc.*

(*vii*) **Quality control:** Product reliability can be improved by better quality control during manufacturing stage.

(*viii*) **By better maintenance**: Quality and periodicity of maintenance affects the reliability directly. It can be improved by analysis of failures, condition monitoring, technical upgradation, health assessment of critical components, adherence to operating parameters *etc.*

23.17 FAULT TREE ANALYSIS (FTA)

Fault Tree Analysis was developed at Bell Laboratories, USA in 1960s. It was first deployed in Minuteman Missile Program in USA to determine various events that could lead to an accidental launch of a missile. This technique is extensively used in atomic reactor design, space mission, design of weaponry system *etc.*

This method focuses on functional defects and product components and follows a 'top-down' approach. The fault tree consists of logic gates such as AND, OR, NOR etc. (used in Boolean algebra) and are interconnected. Each event follows a sequence such as:

(*i*) Listing of events on priority basis.

(*ii*) Connections of events with causes.

(*iii*) Developing ways to avoid these causes

The main objective is to identify causes of events that lead to failures. Thereafter, causes are eliminated so as to prevent occurrence of failure. Basic components of a fault tree are shown in table 23.2.

23.17.1 Benefits of Fault Tree Analysis

(*i*) Improved reliability.

(*ii*) Identification of safety issues at design stage.

(*iii*) Identification of root causes of failures.

Table 23.2 Symbols Used in FTA

Symbol	Meaning
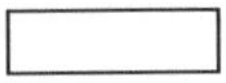	Event.
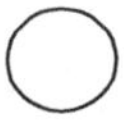	Failure or undesirable event.
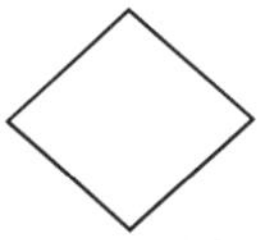	Event not expanded further due to lack of information.
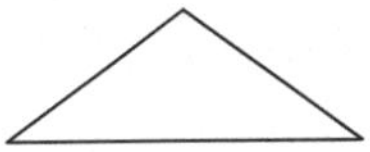	Connection of one branch of fault tree with another.
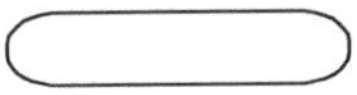	A conditional event.
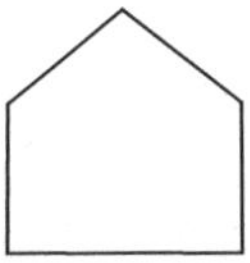	An event expected to occur in normal operation.
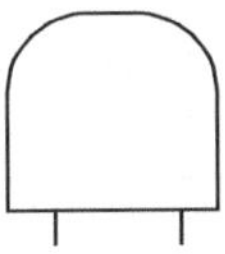	AND gate (output fault event occurs only when all input fault events occur)
	OR gate (output fault event occurs when one or more of input events occur)

Fault Tree Algebraic Events

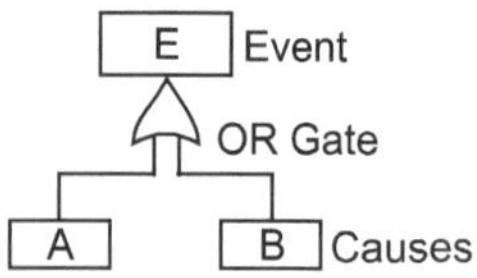

Fig. 23.14 OR Logic

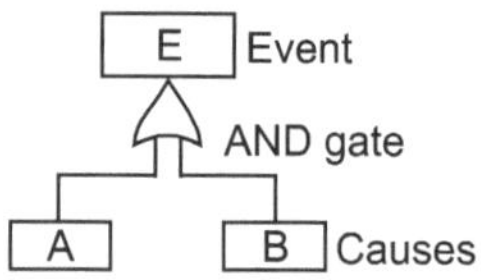

Fig. 23.15 AND Logic

Probability of occurrence of event E is P(E)

$$P(E) = P(A) + P(B)$$
$$= P(A \text{ or } B)$$

Probability of occurrence of event E is P(E) = P(A). P(B)

$$= P(A \text{ AND } B)$$

23.17.2 Steps Involved in Fault Tree Analysis

(*i*) Define system.

(*ii*) Identify events leading to failure.

(*iii*) Construct fault tree.

(*iv*) Collect data such as failure rate of components, probability of occurrences *etc.*

(*v*) Evaluate fault tree.

(*vi*) Suggest corrective measures to avoid failures.

23.17.3 Example of Fault Tree Analysis

Let us consider a simple case where a lamp does not work on pressing a switch.

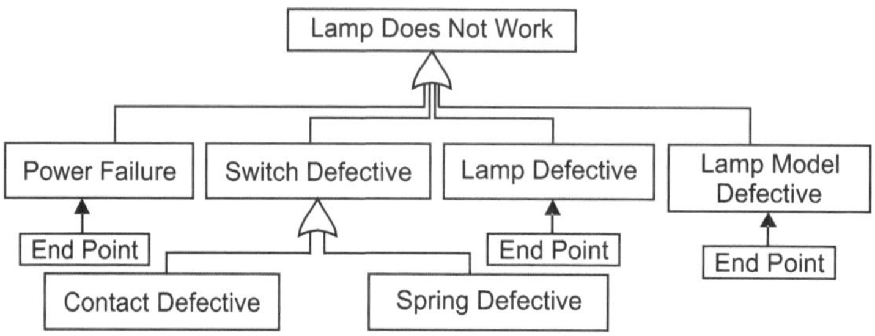

Fig. 23.16 Example of FTA

If lamp does not work, the probable causes could be either power failure or defective lamp or defective switch or defective lamp holder. Now, power failure, defective lamp and defective lamp holder events cannot be expanded further as the lamp will work once these causes are removed. However, defective switch could be due to either defective contact or defective spring.

Let us consider another complicated problem. Consider a system in which hot coolant is cooled through a heat exchanger and after cooling, the coolant flows to a sump from where it is again pumped into control valve. The system is provided with a bypass line which allows excess pumped coolant back to the sump. This is drawn below.

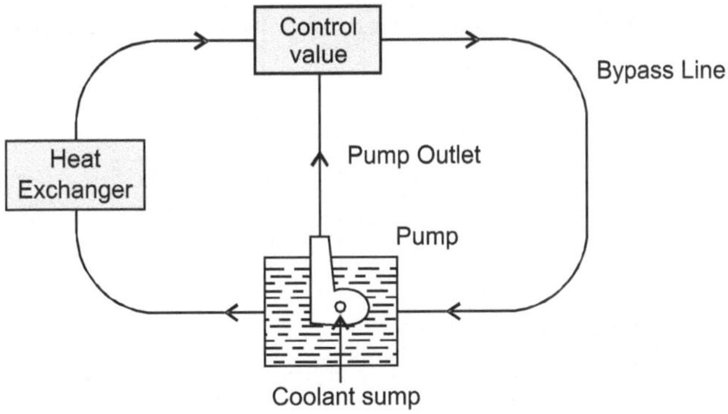

Fig. 23.17 Closed Loop System.

Let us find out possible causes of loss of coolant flow in heat exchanger using Fault Tree Analysis method.

Fault tree analysis shows that loss of coolant flow to heat exchanger could be due to either rupture of coolant line or due to no flow from control valve. These two possibilities could be due to two independent events and hence represented by OR gate. The second possibility *i.e.* No flow from control valve is expanded further. This could be due to rupture of control valve or flow stoppage within control valve and no flow in control valve inlet. All these events occurring independently can stop flow from control valve.

Flow stoppage within control valve could be due to foreign objects in control valve or failure of valve stop of control valve in maximum or minimum flow position. Extending same logic, other probabilities are expanded. The probable causes of failure are picked-up from the fault tree and their chances of occurrence are reduced by improving upon the system. So this way, the reliability of system can be improved by eliminating or reducing the probability of occurrence of such events.

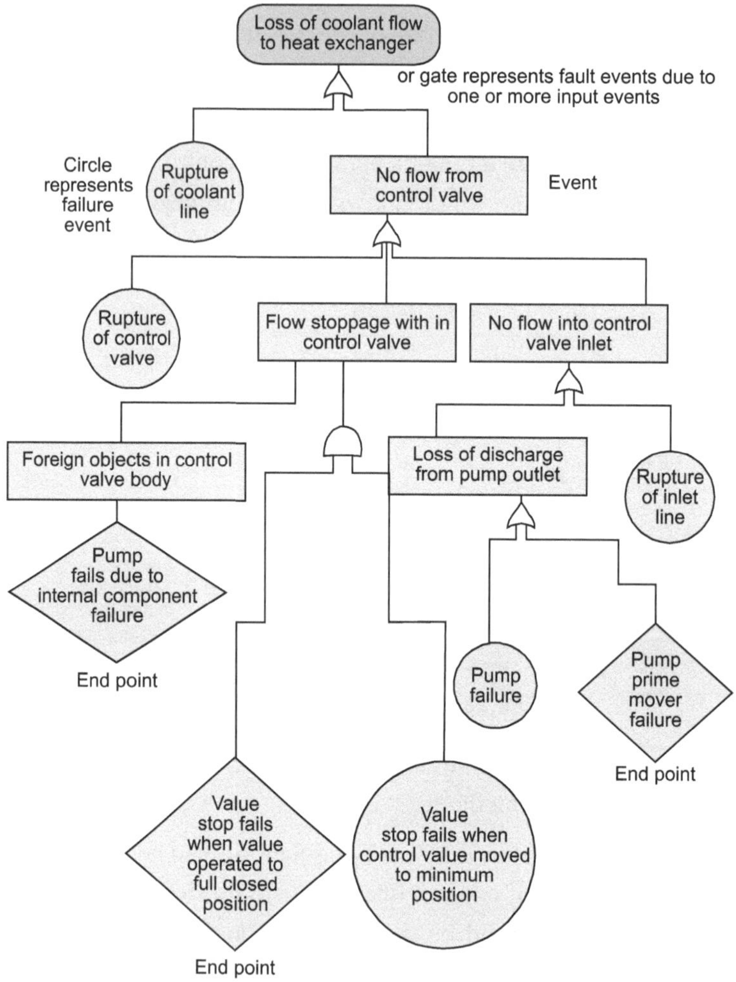

Fig. 23.18 FTA of Close Loop System.

CASE STUDIES ON FAILURES

Case Study-1

1. **Steam Turbine Failure:** This case study, refers to failure of steam turbine *i.e.* tripping of turbine on "vibration" signal. After, stoppage of turbine. Following checks were carried out.

   The turbine and alternator were checked by running turbine without load. The vibration levels were normal. Thereafter, alternator was

Contd...

excited (d.c. field supply was switched on), the vibrations suddenly increased leading to tripping of turbine. Hence, it was obvious that cause of tripping is of electrical origin.

The impedance of field windings were measured at 1 khz. It was found that one of four field poles had low impedance level. Hence, an inter-turn short in the field winding was suspected. This defective field pole was causing unbalanced magnetic field leading to vibrations. So, rotor of alternator was sent to OEM for repair.

After rewinding of field pole, tests were carried out and it was re-synchronised sucessfully.

Hence, it was failure of few turns of field winding of an alternator.

Case Study-2

Cooling Water Pump Shaft Failure: Cooling water pump of vacuum forming machine broke down with a loud noise. Upon dismantling, the pump shaft was found broken. The shaft was sent to a lab for testing. The result indicated shaft material failure.

Case Study-3

Failure of 1000 KVA Transformer: A 1000 KVA, 11000/433 volt transformer got burnt. Upon investigation, it was found that the portion of winding near off-load tap changer was burnt while rest of the winding was normal.

Later on, with further investigation, it was found that an operator tried to change tap of off-load tap changer while the transformer was without load but energized from 11000 V supply. This led to burn-out of winding near tap changer.

This failure was due to wrong operation as off-load tap changers are to be operated when transformer is completely switched-off.

Case Study-4

Failure of DG Set: A low speed marine engine DG set had seized piston (8 cylinder engine). Upon investigation, it was found that cooling water pump supplying water to engine jacket tripped in the early morning and engine continued to run at full load. This caused excessive heat generation leading to piston seizure.

This was a case of human negligence.

"Failure is the opportunity to begin again more intelligently"

—Henry Ford (1863-1947)

Summary

Failure is inability of a component to perform its intended function for a given length of time. Purpose of failure analysis is to determine causes of failures so as to devise ways and means to prevent failures. It is a multi-disciplinary approach involving knowledge of different disciplines.

There are different types of failures that can occur such as catastrophic (sudden), degradation and intermittent. These failures can lead to inoperability, instability, intermittent operation, erratic operation *etc.* Reasons of such failures could be fatigue, overload, wear, oxidation, chemical attack, radiation *etc.*

To understand failure process and predict reliability, different failure models are used. These models are based on certain probability distribution functions such as normal, weibull, exponential *etc.*

Failure analysis is necessary to improve component or a system reliability. Some of the methods widely used are fault tree analysis technique, failure nodes and effects analysis technique *etc.* among others.

Non-destructive and metallographic tests are used as failure prevention techniques. Other methods of improving system reliability are built-in-redundancy, better material selection, modular design, higher factor of safety, quality control *etc.*

In short, purpose of failure analysis is to improve component and system reliability so that the devices continue to perform their intended function during their economic useful life.

Review Questions

1. Define failure. Why is failure analysis critical?
2. Elaborate on various causes of failures.
3. Classify various types of failures.
4. Discuss failure model in detail. What is the effect of preventive maintenance on the life of equipment?
5. Discuss various modes of failures.
6. Write briefly on failure analysis.
7. Describe in detail failure modes and effects analysis.
8. Define terms (i) failure rate (ii) MTTF (iii) MTBF (iv) Availability (v) Maintainability.
9. Discuss various failure prevention techniques.
10. Discuss fish bone diagram and fault-tree-analysis techniques.

Discussion Questions

1. Elaborate "Failure analysis is the basis of a robust design."
2. "Choice of appropriate material and robust design drastically reduces chances of failure". Comment.
3. "Human factor contributes towards large cases of failures". Discuss.
4. "Design for improved reliability is a function of cost". Justify.

Multiple Choice Questions

1. Fatigue failures are not caused by
   (*a*) Cyclic temperature variation.
   (*b*) Cyclic load variation.
   (*c*) High stress concentration.
   (*d*) Environmental reasons.
2. Initiation of failure causes following changes
   (*a*) Erratic operation.
   (*b*) Increase in noise.
   (*c*) Instability.
   (*d*) All of the above.

3. Infant mortality rate or initial failure rate is higher during initial stages of operation due to
   (a) Insufficient safety factor. (b) Improper installation.
   (c) Degradation. (d) Corrosion.
4. Mechanical components observe the following statistical distribution
   (a) Exponential. (b) Releigh.
   (c) Weibull. (d) Normal.

True/False Questions

1. With increase in reliability, cost of system design decreases.
2. Metallurgical changes can lead to failures.
3. Fatigue failures result due to cyclic loads.
4. Overload can cause electrical failures.
5. Catastrophic failures can be prevented by good design.
6. Systems connected in series are more reliable than connected in parallel.

ANSWERS

Multiple Choice Questions

1. (d) **2.** (d) **3.** (b) **4.** (c)

True/False Questions

1. False **2.** True **3.** True **4.** True **5.** False **6.** False

Caselet

1. A, 1100 KVA, low speed diesel generating set had been operating for the past three years. One fine morning, the piston of one of the engine seized. Upon investigation it was found that out of two cooling water pumps circulating cooling water in engine jacket, one had failed. Even though one cooling water would not have caused piston seizure. The possibility could be severe scaling in cooling water circuit.

   As a maintenance engineer, what steps will you take to not only to start the operation but also remove the cause of failure?

❑❑❑

24

Maintenance of Electrical Machines

Learning Objectives

After reading this chapter, you will be able to:

- Understand basic types of ventilation and various faults in electrical machines.
- Develop keen understanding of condition monitoring techniques of electrical machines.
- Understand batteries and earthing system.
- Understand the basic of electrical safety.

24.1 INTRODUCTION

Operation and maintenance of electrical machines present a different class of problems due to different characteristic as compared to mechanical machines such as pump, compressor, turbine, boiler, evaporator *etc.*

It is comparatively easier to detect and monitor signs of deterioration in a mechanical machine than in an electrical machine. Due to recent advances in condition monitoring, latest techniques are available for monitoring health of electrical machine. However, each type of electrical machine requires specific techniques of maintenance and condition monitoring.

Focus of this chapter will be on basics of electrical machines, faults and their causes, condition monitoring and maintenance procedures for major electrical machines.

24.2 CLASSIFICATION OF ELECTRICAL MACHINES

All electrical machines can be broadly classified into the following categories:

(*i*) Based on type of supply.

(*ii*) Based on level of voltage.

(*iii*) Based on construction, whether static or rotating.

(*i*) **Classification based on type of supply:** Electrical machines can be classified based on type of supply *i.e.* AC or DC. Accordingly, these are known as AC machine or DC machines. Transformers, Induction motors, synchronous motors, synchronous generators *etc.* are examples of AC machines while DC generators, DC motors (series, Shunt and compound) are examples of DC machines. AC machines can be further classified as single phase and three phase.

(*ii*) **Classification based on level of voltage:** All electrical machines can be broadly classified under three heads — low voltage machines (voltage upto 660 V), medium voltage machines (voltage 660 V to 1000 V) and higher voltage machines (voltage more than 1000 V).

(*iii*) **Classification based on construction:** All electrical machines can be classified as static and rotating machine. Transformer, switch gears *etc.* are static devices while motors, generators *etc.* are rotating devices.

24.3 HEATING AND VENTILATION OF ELECTRICAL MACHINES

Ventilation of electrical machine is critical aspect of design, operation and maintenance. Heat is generated in an electrical machine due to various losses taking place in it. These are copper losses taking place in conducting material, iron losses taking place in magnetic material, dielectric losses taking place in insulating material, friction and vintage losses due to moving parts and a small amount of stray losses. These losses raise the temperature of inside parts of machine. The dielectric material used is sensitive to temperature rise. The life of dielectric material depends on temperature. Hence, ventilation and cooling methods play an important role in life of electrical machines.

Ventilation methods employed in electrical machines, can be broadly divided into three categories:

(*i*) Radial ventilation.

(*ii*) Axial ventilation.

(*iii*) Combined radial and axial ventilation.

(*i*) **Radial Ventilation:** In this system, air flows along the radial direction.

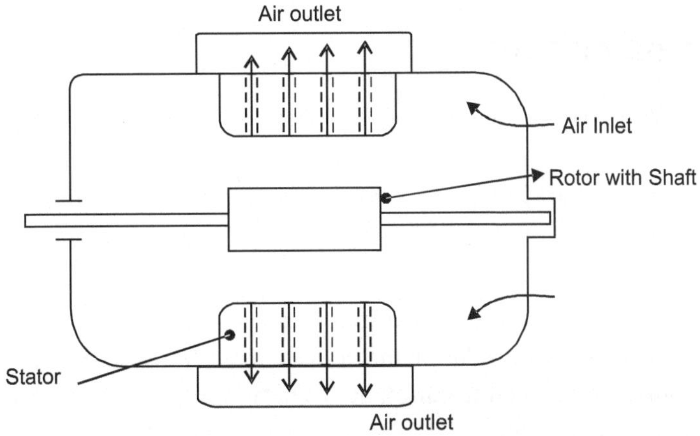

Fig. 24.1 Radial Ventilation System.

Air enters from one end of machine and passes through radial ventilating ducts through stator (stationary part of machine) and rotor (rotating part of machine). The air is then either discharged to atmosphere or is passed through a heat exchanger.

This method is most widely used due to uniform cooling.

(*ii*) **Axial Ventilation:** This method of ventilation is used for medium output and high speed machines. The air flows along the direction of shaft axis.

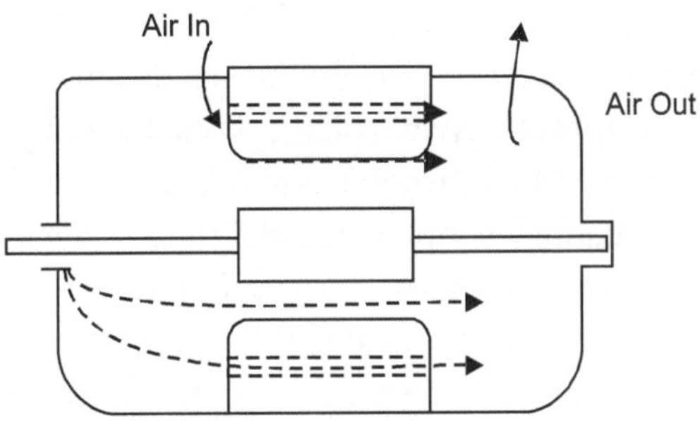

Fig. 24.2 Axial Ventilation System.

(*iii*) **Combined Radial and Axial Ventilation System:** This method is used for large machines. The machine has radial as well as axial ducts for passage of air.

Apart from ventilation methods discussed above, forced and induced drought ventilation is also used employing a fan at one end of machine.

All large machines employ forced method of ventilation so as to increase the heat transfer rate.

24.4 TYPES OF FAULT

Generally, three types of fault take place in electrical machines. These are following:

(*i*) **Open Circuit Fault:** This type of fault involves opening of a joint of a broken conductor. This results in stoppage of current flow in the circuit.

(*ii*) **Short Circuit Fault:** All current carrying parts of electrical machines are insulated by insulation material such as PVC, XLPS, paper, Varnish, enamel *etc*. If the insulation between two current carrying parts fail, a short circuit fault takes. This results in heavy flow of current in the affected part. This type of fault causes extreme damage to the machine/equipment/system.

(*iii*) **Earth Fault:** All current carrying parts of electrical machines are insulated from earth or body of the machine. If insulation between a current carrying part and body of the machine fails, an earth fault takes place. It involves heavy rush of current from conductor to earth. This type of fault also causes damage to the machine/equipment/system.

24.5 CAUSES OF FAULTS

Various causes of faults in electrical machines are given below:

(*i*) **Overload:** All electrical machines are designed to carry a certain amount of current at a given voltage. If the load carried by the machine is more than the rated then the extra current flowing in the machines causes over heating and may lead to failure of the machine. Although, all machine are designed to carry a fixed percentage (*e.g.* 110% or 115%) of rated load as overload for a short duration without any harmful effects. But sustained overload or overload above designed value may lead to deterioration of insulation of machines. This may result in failure of machine.

(*ii*) **Insulation failure:** Insulation of electrical machine may fail due to various reasons such as overheating, surges, lightning, fault *etc*.

(*iii*) **Moisture in insulation:** Any idle machine may absorb moisture from surrounding. This results in lowering of insulation resistance of machine. This increases chances of failure when the machine is switched on.

(*iv*) **High ambient temperature:** A high ambient temperature aids failure of electrical machine. This is due to the fact that higher operating temperature of machine causes insulation to deteriorate ultimately leading to its failure.

(*v*) **Single phasing:** All machines used in industry are three machines (except very small rating machines). Loss of one phase leads to flow of higher current which subsequently causes overheating. If the temperature rise due to overheating is sufficiently high then failure of machine may take place.

(*vi*) **High or low voltage**: All electrical machines are designed for certain voltage range *e.g.* 415 V ±10%. Any marked deviation from the specified value may damage the machine.

(*vii*) **Failure of cooling system:** Any blockage or obstruction or failure of cooling system due to any reason may lead to failure of machine.

(*viii*) **Stalled rotor:** In a motor, if the rotor is unable to rotate due to any reason then the machine draws large current and may result in failure.

(*ix*) **Increased duty factor:** All electrical machines are designed for a particular duty application such as continuous duty, intermittent duty, intermittent periodic duty *etc*. Any change in duty factor may lead to machine failure.

(*x*) **Repeated switching:** Rotating machines such as motors are designed for a certain number of switching operations per hour. Because every switching on causes heavy rush of starting current and hence a certain time is required by the motor to cool down. A frequent switching causes sudden rise in temperature which may ultimately lead to failure.

Although in a system, there may several other reasons for failure of electrical machines but above mentioned reasons are the major causes of machine burn out. A good care goes a long way in achieving a long life of machine.

24.6 CONDITION MONITORING OF ELECTRICAL MACHINES

First, we shall discuss general techniques for condition monitoring. This will be followed by techniques used for specific electrical machines.

(*i*) **Insulation Resistance Measurement:** Health of insulation of electrical machine determines the operational readiness and life of machine in the long term. Hence, condition monitoring starts with monitoring the status of insulation in an electrical machine.

For measurement of Insulation Resistance (IR), an instrument called Meggar is used. Meggar is available for different voltages and its use is determined by the voltage level of the winding under test.

Table. 24.1 Selection of Meggar Voltage

Rotated Voltage of Winding	Meggar Voltage (volts)
Less than 1000 V	500
1001 – 2500 V	500-1000
250 – 5000 V	1000-2500
500 – 12000 V	2500-5000
More than 12000 V	5000-10000

The recommended minimum insulation resistance after one minute at 40°C can be determined from following:

$$KV + 1 \quad \text{Mega – ohm.}$$

where, KV = Rated machine terminal voltge in kilovolt

(*ii*) **Polarisation Index:** It is defined as ratio of insulation resistance measured at the end of 10 minutes to the insulation resistance measured at the end of one minute.

$$\text{Polarisation Index (PI)} = \frac{\text{Insulation resistance at the end of 10 minutes}}{\text{Insulation resistance at the end of one minute}}$$

PI is a pure number and can be considered independent of temperature Following table interprets values of PI.

Table 24.2 Insulation Condition Vs. Polarisation Index

Polarisation Index	Insulation Condition
Less than 1	Poor
1 to 2	Doubtful
2–4	Good
2–5	Good
More than 5	Very good

Polarisation test also depends on the class of insulation used.

Table 24.3 Minimum PI Value for different class of Insulation.

Insulation Class	Minimum PI Valve
Class A	1.5
Class B	2.0
Class F	2.0
Class H	2.0

(*iii*) **Winding Resistance Measurement:** All machines have windings made of copper wire. In a three phase machine, all the three phases have identical windings and hence almost same winding resistance. Any marked deviation of resistance value of anyone phase indicates a problem area requiring further investigation.

Winding resistance is measured by ohm meters directly. The measured resistance values are corrected to a common ambient temperature. The resistance so measured is known as cold d.c. resistance. Resistance is measured by isolating the winding under test from other windings.

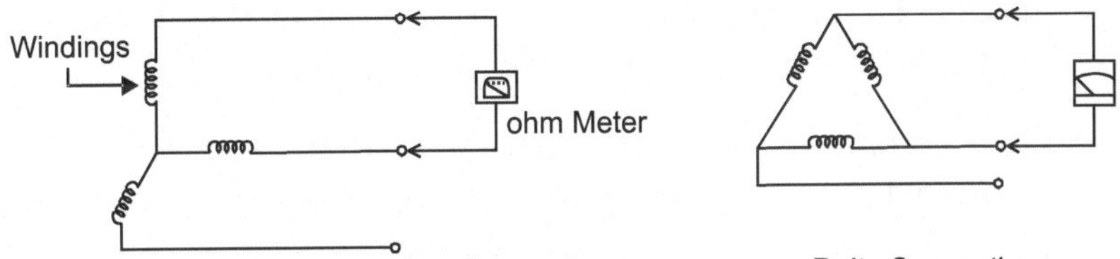

Fig. 24.3 Star and Delta Connected Winding.

(*iv*) **Infra Red Thermography:** Measurement of temperature of machine hot spots, joints, winding *etc.* provides very useful clues. Non-contact type Infrared thermography is extensively used for electrical machines, transmission lines, substation equipments and accessories *etc.*

It is based on the fact that all electrical components heat up as their physical condition or electrical properties deteriorate. All heated objects emit infrared radiation and it is captured by infrared scanner cum recorder. When object has higher temperature than adjoining parts, it appear as very bright. So all loose connections, hot spots, *etc.* can be easily identified and corrective action taken before actual breakdown takes place.

This recording is carried out when the equipments are in operation.

(*v*) **Tan Delta measurement:** Tan-Delta or dissipation factor indicates power dissipation in the insulating material. High voltage winding of electrical machine has capacitance and resistance.

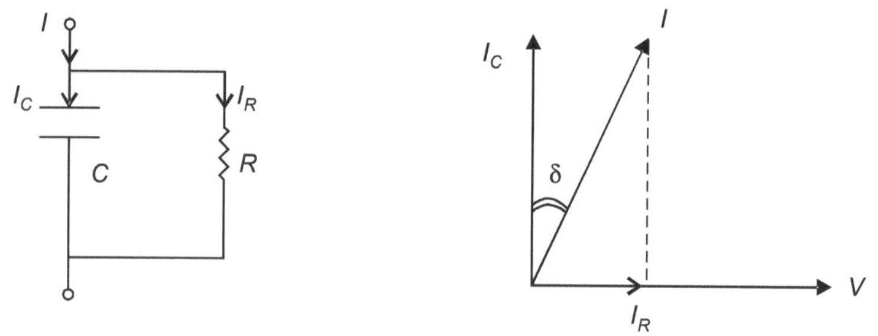

Where I = Total Current
V = Voltage across Winding

Fig. 24.4 Phasor Diagram.

With deterioration in insulation, Tan-delta value increases with increase in applied voltage. So a high value of Tan-delta is indicative of presence of moisture content, voids, contaminants, aging *etc.*

A high voltage winding acts as a capacitor with insulation as the dielectric. An increase in capacitance value with increase in voltage indicates presence of voids or impurity in the insulation. Tan-delta value can be directly measured by using Tan-delta meters.

(*vi*) **High Voltage DC Leakage Current Test:** Measurement of leakage current taken after periodic intervals provides clue about health of insulation. The measured values are plotted and trend analysis is done. A rising pattern whether slow or abrupt indicates deteriorating state of insulation.

(*vii*) **High Voltage Withstand Test:** This test is used to test insulation strength for surges and over voltages under fault condition. AC high voltage at 50 Hz is used for testing purpose. The machine is allowed to withstand this test for one minute.

(*viii*) **Partial Discharge Measurement:** Partial discharge is a localized breakdown of insulation in an electrical machine under high voltage stress. Partial discharge is initiated by presence of void or crack in the insulation.

It can be used in cables, switch gears, transformers, motors, generators, surge arrestors, capacitors *etc.*

(*ix*) **Vibration Analysis:** This test is used for rotating machines such as motors, synchronous generators, synchronous condenses *etc.* It indicates health of mechanical system.

Vibration readings are taken at various points on the machine and trend analysis is carried out over a period of time. A rising trend indicates various mechanical problems. The test readings are taken with a hand held vibration meter.

(*x*) **Measurement of Breakdown Value of Transformer Oil:** This test is used for testing the insulation level of transformer oil. Test are done on oil immersed equipments such as transformer, circuit breaks, oil immersed current and potential transformers, oil immersed starters *etc*.

Breakdown value of oil is measured in a standard measuring test equipment (as per IS:335). The value of oil should be minimum 30 KV rms.

(*xi*) **Current Balance Test:** In this test, current is measured with a clip-on-meter in all the three phases (for 3 phase machine). In ideal condition, all the three currents should be equal. Unbalance current indicates inter-turn short, high contact resistance *etc*.

(*xii*) **Dissolved Gas Analysis:** This technique is based on the fact that all electrical or thermal faults in a oil filled transformer results in decomposition of transformer oil leading to formation of various gases such as hydrogen, methane, ethane, acetylene, carbon-dioxide, nitrogen, oxygen *etc*. By measuring rate of gas generation, interpretation can be provided for faults.

Following three methods used for detection of fault.

(*a*) *Ratio Technique:* This method uses five gases and 3 ratios to detect fault (as per IEC-60599). These ratios are CH_4/H_2, C_2H_4/C_2H_6 and C_2H_2/C_2H_4. The value of these ratios provide idea about the fault.

(*b*) *Key Gas Method:* In this method, few "key gases" generated due to a fault are chosen for analysis. These gases are produced due to four types of faults. These are thermal (due to oil decomposition), thermal (due to cellulose decomposition), electrical (due to corona) and electrical (due to arcing).

(*c*) *IEEE Method:* In this method, interpretation of result is based on concentration limits of CH_4/H_2 and C_2H_2/C_2H_4 gas ratios.

(*d*) *Frequency Response Analysis:* This method is applicable to transformers. Winding is supplied with sinusoidal input signal of 2 V rms at frequency of 2 Khz, 20 Khz and 2 Mhz through co-axial cable leads. Output is measured at other end. Voltage transfer function is measured for three standard frequencies using a Network/spectrum analysis.

Each winding design has a characteristic frequency response and any shift or movement due to fault changes inductance and capacitance value of windings. These charges are reflected in frequency response curve.

24.7 BATTERIES

A battery is an electro-chemical energy storage device. It stores electrical energy as d.c. energy. If an electrical load is connected across the terminals of battery, energy flows from battery to the load.

Batteries are principally of two types: (i) primary and (ii) secondary. A primary battery is one in which chemical reaction taking place inside the battery is irreversible *i.e.* battery cannot re-charged for reuse. Examples are dry cell, Lechlanche cell *etc.* A secondary battery is one in which the chemical reaction taking place inside the battery is reversible. Hence, such batteries can be reused by recharging. Examples of secondary or rechargeable batteries are—lead acid and alkaline batteries. The majority of industrial applications use either lead acid or alkaline batteries. Hence, focus of discussion is confined to these batteries.

Batteries are primarily used to provide a secure source of power supply. Such power source is needed for emergency lighting (at the time of power failure), d.c. power supply for operation of protective devices such as circuit breakers, relays *etc*, for indication such as fire escape, control panel for operation *etc*.

24.8 TYPES OF BATTERIES

Following are various types of batteries in use:

(*i*) Lead acid.

(*ii*) Nical-cadmium (Ni-cd).

(*iii*) Nical-Iron (Ni-Fe).

(*iv*) Nical metal hydride.

(*v*) Nical zinc (Ni-Zn).

(*vi*) Lithium iron.

(*vii*) Lithium sulphur.

(*viii*) Silver oxide.

(*ix*) Mercury battery.

(*x*) Lithium iron phosphate.

(*xi*) Nical Hydrogen.

(*xii*) Zinc-carbon.

(*xiii*) Zinc-chloride.

(*xiv*) Lithium-Thiongle chloride.

Out of all batteries mentioned above, rechargeable ones are lead acid, nical-cadmium, nical-hydrogen, nical metal hydride, lithium-iron, lithium-sulphur and lithium-iron phosphate.

However, lead acid and nical-cadmium batteries, find extensive use in industries.

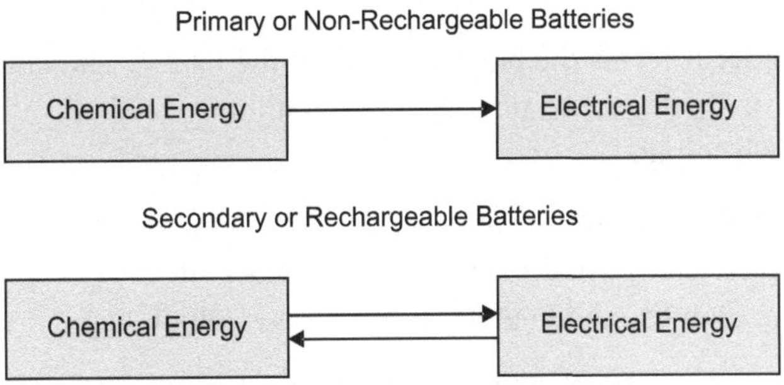

Fig. 24.5 Primary and Secondary Battery.

24.9 APPLICATIONS OF BATTERIES

(*i*) In electric sub station for emergency lightening.

(*ii*) In electric sub station for providing reliable *d.c.* supply to various relays and circuit breakers.

(*iii*) In battery operated vehicles *e.g.*, fork-lift trucks, Tow trucks etc.

(*iv*) As reliable supply to telephone exchanges, broadcast stations, wireless transmitting towers *etc.*

(*v*) In automobiles for starting purpose *e.g.* cars, buses, trucks *etc.*

(*vi*) In trains for lighting and air-conditioning.

(*vii*) In inverters for domestic, commercial and industrial use.

24.10 CONSTRUCTION AND OPERATION OF LEAD ACID BATTERIES

A lead acid battery is major source of reliable d.c. supply in industry. It is used extensively due to low cost compared to Nical-cadmium batteries. Although life of lead acid batteries are much less (3–4 years) compared to Nical-cadmium batteries (10 years).

A lead acid batteries consists of following components:

(*i*) Anode.
(*ii*) Cathode.
(*iii*) Electrolyte.
(*iv*) Separator.
(*v*) Vent plugs.
(*vi*) Battery terminals.
(*vii*) Cell connectors.
(*viii*) Container.

(*i*) **Anode:** It forms the positive plate of the battery and consists of lead per oxide deposited on a frame of antimony-lead alloy. It has a dark brown colour.

(*ii*) **Cathode:** It forms negative plate of the battery and consists of porous spongy lead pasted on a frame. Number of cathode plates are one more than the anode plate. It has a gray colour.

(*iii*) **Electrolyte:** Electrolyte is the chemical in which both cathode and anode are immersed so that chemical reaction can take place. Sulphuric acid is the electrolyte which has a specific gravity of 1.230 at 27°C when fully charged.

(*iv*) **Separator:** Purpose of separator is prevent short circuit between anode and cathode plates. These consists of thin, porous sheets suspended in between cathode and anode plates. These are generally of porous rubber or PVC.

(*v*) **Vent Plugs:** Vent plugs allows escape of gases produced due to chemical reaction. It also allows pouring of water into the electrolyte. Plugs are made of rubber or polystyrene.

(*vi*) **Battery Terminals:** These are made of nical plated copper alloy with nut and bolt arrangement so as to secure cable connections.

(*vii*) **Cell Connectors:** These are used to connect cells of similar type that is all positive cells are connected together and all negative cells are connected together. Each cell has a voltage of 2V.

(*viii*) **Container:** It houses anode, cathode, electrolyte, separator, terminals and vent plugs. It is made of hard rubber, ebonite or moulded plastic.

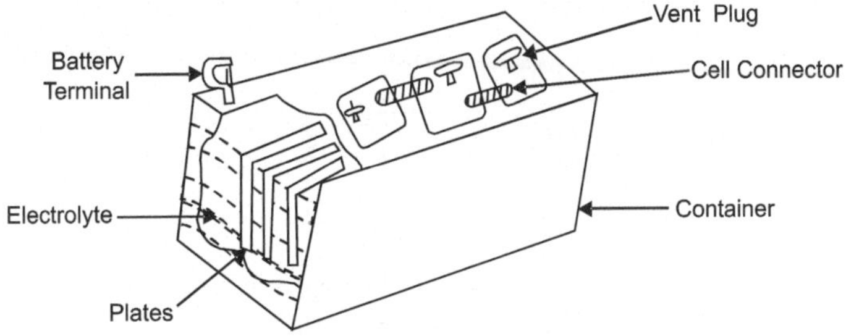

Fig. 24.6 Construction of Lead Acid Battery.

24.11 CHARGING AND DISCHARGING OF LEAD ACID BATTERY

(*a*) **Discharging Process:** An external load is connected across anode and cathode so that electric current flows from battery to the load. This is known as discharge of battery.

Discharging process results in lowering of cell voltage and specific gravity of electrolyte. A fully discharged cell has a cell voltage of 1.8 V and specific gravity of 1.180.

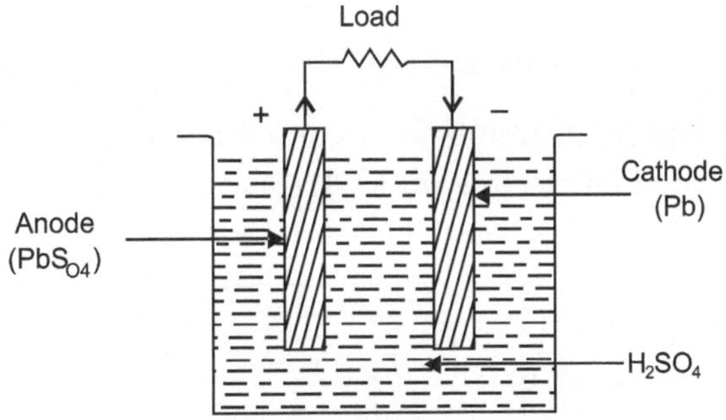

Fig. 24.7 Discharge of Battery.

Chemical reactions taking place

At anode:

(*i*) $PbO_2 + H_2 + H_2SO_4 \rightarrow PbSO_4 + H_2O$

At Cathod

(*ii*) $Pb + SO_4 \rightarrow PbSO_4$

(*b*) **Charging process:** D.C. supply is fed to the battery so as to charge it. Positive of DC supply is connected to anode and negative of DC supply is connected to cathode. During charging process, cell voltage rises to 2V and specific gravity of electrolyte rises to 1.230.

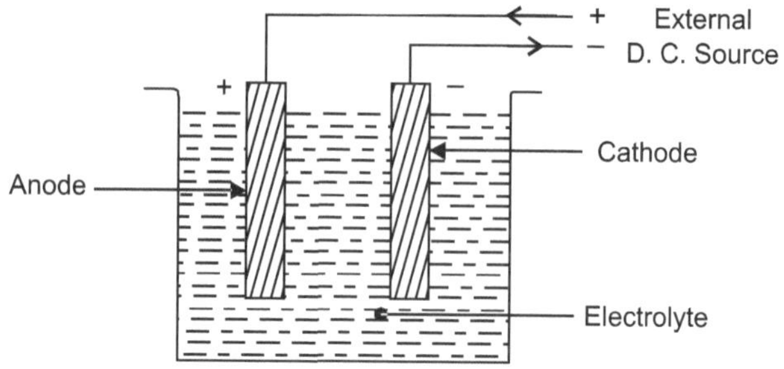

Fig. 24.8 Charging of Battery

Chemical reactions taking place during charging process are:

At anode

(*i*) $PbSO_4 + SO_4 + H_2O \rightarrow PbO_2 + 2H_2SO_4$

At cathode

(*ii*) $PbSO_4 + H_2 \rightarrow Pb + H_2SO_4$

Overall reaction

$$Pb + PbO_2 + 2H_2SO_4 \underset{\text{Charging}}{\overset{\text{Discharging}}{\rightleftharpoons}} 2PbSO_4 + 2H_2O$$

24.12 NICAL-CADMIUM BATTERY

Nical-cadmium and Nical-Iron batteries are also known as alkaline batteries due to alkaline nature of electrolyte (KOH).

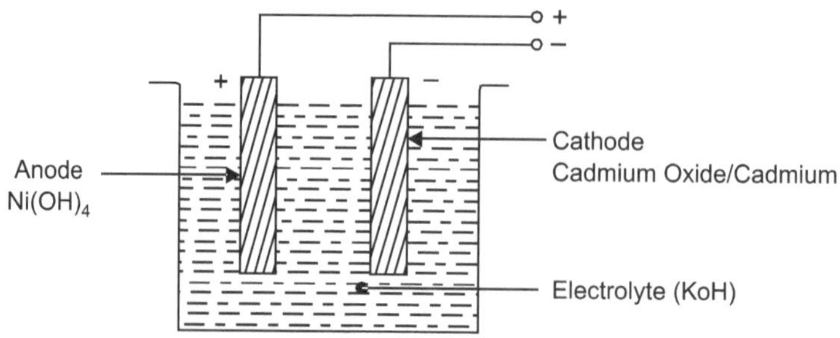

Fig. 24.9 Ni-Cd Battery

Nical-Cadmium battery consists of the following major components:

(*i*) **Anode:** $Ni(OH)_4$ (Nical-hydroxide) forms positive plate.

(*ii*) **Cathode:** Cadmium or cadmium oxide forms negative plate.

(*iii*) **Electrolyte:** KOH or potassium hydroxide forms electrolyte. Specific gravity of electrolyte is 1.200 at 16°C. Each cell has a voltage of 1.2 V at discharge and 1.3 V after charge.

Advantage of Nical-Cadmium battery is that it is not damaged by over-charging and has a long life of 10 years.

Comparison of Lead-acid and Nical-cadmium batteries is given in table 24.4

Table 24.4 Comparison of Batteries

S. No	Parameter	Lead-acid	Nical-cadmium
1.	Anode	Lead per oxide	Nickel hydroxide
2.	Cathode	Spongy lead	Cadmium or Cadmium oxide
3.	Electrolyte	Sulphuric acid	Potassium hydroxide
4.	Voltage/Cell	2 Volts cell	1.3 volts cell
5.	Specific gravity of electrolyte (fully charged)	1.230	1.200
6.	Cost	Cheap	Costly
7.	Life	Short	long
8.	Amper hour capacity	90–95%	60–80%

24.13 MAINTENANCE OF LEAD-ACID BATTERIES

1. Regular cleaning of battery terminals and vent plug is required.
2. White petroleum jelly is applied to terminals to protect them from sulphation.
3. Regular topping up of distilled water is required.
4. Periodic tightening of all battery connections is required.
5. Specific gravity and voltage of each cell is recorded in a log book. Any deviation noticed is investigated and set right.

24.14 MAINTENANCE OF NICAL-CADMIUM BATTERIES

1. Regular topping up of distilled water.
2. Tightening of all battery connections.
3. Recording of specific gravity and cell voltage. If specific gravity falls below 1.170 at 16°C then complete electrolyte is replaced.

24.15 FACTORS AFFECTING LIFE OF BATTERIES

Following factors affect the life of batteries:

(*i*) Keeping discharged battery for a long time.

(*ii*) Frequent boost charging.

(*iii*) Higher rate of charging and discharging.

(*iv*) Low level of electrolyte.

(*v*) Higher temperature of electrolyte during charging.

(*vi*) Poor maintenance.

(*vii*) Repeated partial discharge.

24.16 EARTHING

Earthing means any connection to general mass of earth. So all electrical appliances, equipments and machinery are properly earthed.

The purpose is to save human beings from electric shock.

24.17 NEED FOR EARTHING

Earthing is done due to following reasons:

(*i*) To save human being from electric shock.

(*ii*) To provide discharge path to lightening and over voltages.

(*iii*) To improve protective relaying.

(*iv*) To keep system neutral at ground potential.

(*v*) To improve life of insulation of motors, generators, transformers *etc.*

(*vi*) To provide path to stored energy in high voltage/extra high voltage lines before undertaking maintenance.

24.18 TYPES OF EARTHING

Earthing is of two types:

(*i*) **Neutral Earthing:** It involves connection of neutral point of power system such as generator or transformer neutral point to earth. Purpose is to keep neutral at ground or zero potential.

(*ii*) **Equipment Earthing:** It involves connection of outer body or non-current carrying part of any equipment to earth. Purpose is to save human beings from electric shock.

24.19 TYPES OF NEUTRAL EARTHING

(*i*) Solid earthing.
(*ii*) Resistance earthing.
(*iii*) Reactance earthing.
(*iv*) Resonant earthing.

(*i*) **Solid Earthing:** In this method, neutral point of transformer or generator is connected to earth using earth electrodes, pipes or rods.

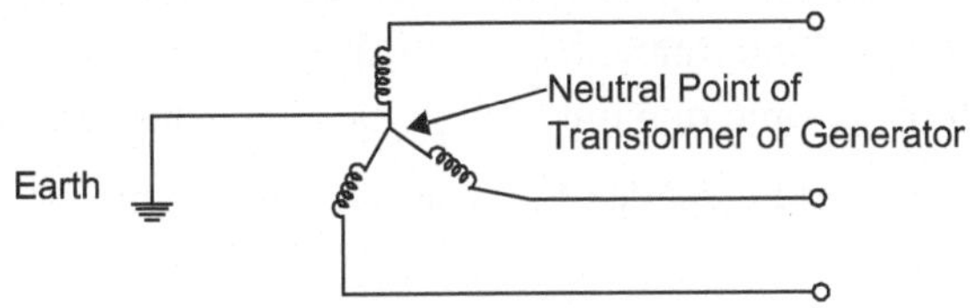

Fig. 24.10 Solid Earthing.

(*ii*) **Resistance Earthing:** With higher system voltage, earth fault current also increases. So, in order to limit earth fault current, a resistance is introduced in the earth path. This method is used in 3.3 KV to 11 KV voltage systems.

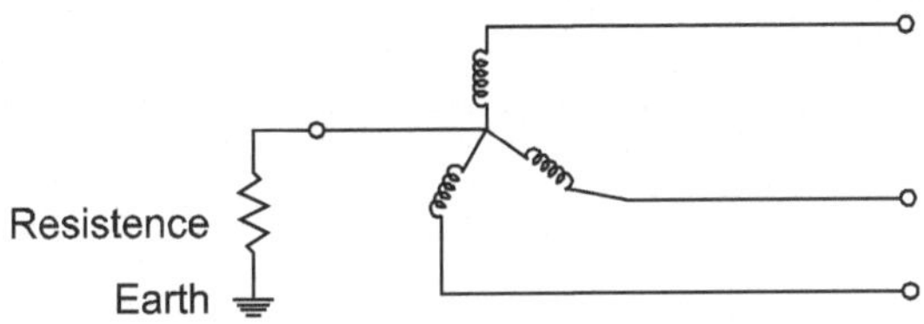

Fig. 24.11 Resistance Earthing.

(*iii*) **Reactance Earthing:** In this method, an inductive reactance is connected between neutral point and earth. Major advantage over resistance earthing is that energy dissipated in reactance is much less and also size is much smaller for same earth fault current.

(*iv*) **Resonant Earthing:** This method utilizes arc suppression coil called Peterson coil. The value of inductive reactance is such that the reactance current in coil neutralised network capacitive current. The resultant current being zero and hence unable to maintain an arc.

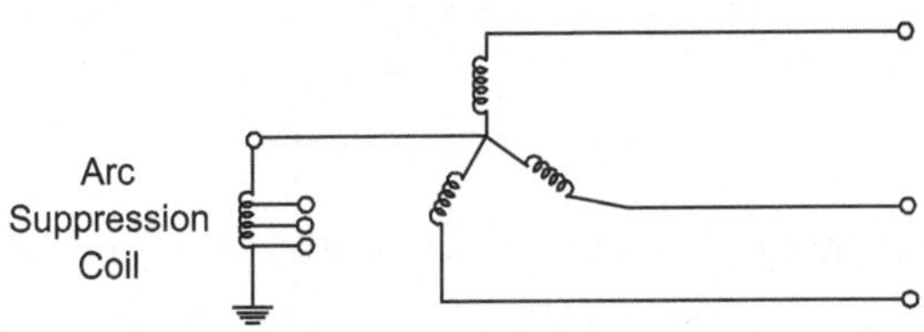

Fig. 24.12 Resonant Earthing.

24.20 VARIOUS METHODS OF EQUIPMENT EARTHING

Purpose of equipment earthing is to prevent electric shock to personnel. So external body of all electrical equipment and machinery are connected to earth.

The various methods are following:

(*i*) **Pipe Earthing:** A galvanized iron pipe of minimum 40 mm diameter and minimum length of 2.75 m is buried in the earth. Alternate layer of salt and charcoal are placed around the pipe upto 2 m from the bottom. Height of each alternate layers of salt and charcoal is 15 cm. The salt being hygroscopic in nature keeps earth resistance low.

An arrangement is made to pour water periodically so as to keep earth resistance low.

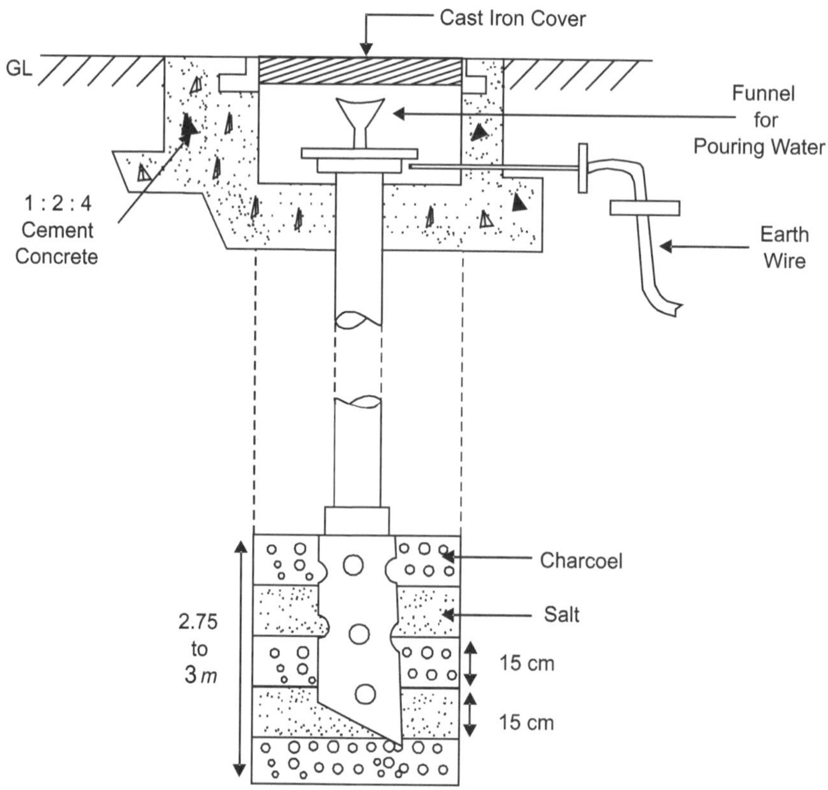

Fig. 24.13 Pipe Earthing

(*ii*) **Plate Earthing:** A G-I Plate (600 × 600 × 6.36 mm) or copper plate of (600 × 600 × 3.18 mm) is buried upto 3 to 4 m (depending upon moisture in the soil) depth in soil. Alternate layers of salt and charcoal are spread around the plate. Plate is connected to a earth wire. A GI pipe of 12.7 mm diameter is also laid so as to pour water.

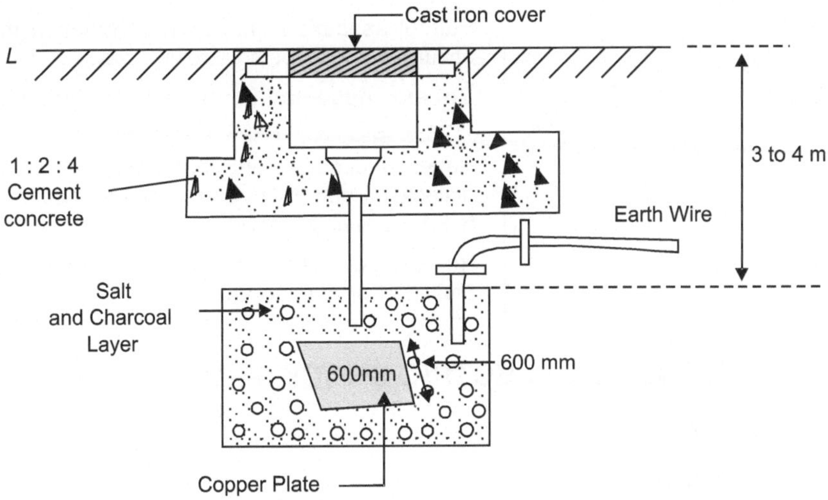

Fig 24.14 Plate Earthing

(*iii*) **Rod Earthing:** A rod of 12.5 mm diameter (copper) or 19 mm diameter (GI) is driven into ground upto a depth of 5 to 6 m. Earth wire is carried along with the rod.

This type of earthing is used in sandy soil.

(*iv*) **Strip Earthing:** A copper strip (minimum 25 mm × 1.6 mm) is buried horizontally in trenches of sufficient depth (minimum 8 m or higher). A grid of such strips is laid to lower earth resistance. This type of earthing is used in rocky area.

24.21 ELECTRICAL SAFETY AND FIRE FIGHTING

Precious human lives are lost due to electrical accidents. A good understanding of handling of electricity goes a long way in avoiding accidents.

Hazards associated with electricity are following:

(*i*) Electric shock.

(*ii*) Electric burn.

(*iii*) Electric fire and

(*iv*) Explosion due to electrical spark or leakage.

Table 24.5 Physiological Effect of Electric Current

Current (ma)	Physiological Effect
0.5–2	Threshold of perception.
2–10	Painful sensation, muscular contraction may occur leading to being thrown away.
10–25	Severe muscular contraction, difficulty in breathing.
25–80	Severe muscular contraction leading to bone dislocation and fracture, chances of respiratory failure, possible cardiac arrest.
More than 80	Burns at point of contact, death from ventricular fibrilation or cardiac arrest.

24.22 FIRST AID TO ELECTRIC SHOCK VICTIM

(*i*) Isolate power source.

(*ii*) Extinguish flame.

(*iii*) Cool the burn with water until skin temperature returns to normal.

(*iv*) Elevate burned limbs to reduce swelling.

(*v*) Treat for shock.

(*vi*) Do not feed anything through mouth.

(*vii*) Give CPR (Cardiac Pulmonary Resuscitation) if victim is not breathing.

(*viii*) Keep victim warm during transport to medical centre.

24.23 CAUSES OF ELECTRIC FIRE

Following are the causes of electric fires:

(*i*) Short circuit.

(*ii*) Overloading of electrical circuit.

(*iii*) Leakage of electric current.

(*iv*) Presence of inflammable material near electrical distribution system.

(*v*) Use of poor quality electrical wires/cables, switches *etc.*

24.24 STEPS TO AVOID ELECTRIC SHOCK

(*i*) All appliances, apparatus *etc.* must be properly earthed.

(*ii*) All installation must have proper earthing and earth resistance must be measured periodically.

(*iii*) All repair work must be carried out by authorized persons only.

(*iv*) While working on electrical system, proper insulated tools must be used.

(*v*) In case of electric shock to a person, the live wire must be removed with a dry wooden stick, broom *etc*. (any material that does not conduct electricity). Thereafter, power must be switched off.

24.25 EXTINGUISHING ELECTRIC FIRE

Electric fires are special type of fires and care must be taken to extinguish it. Only two types of fire extinguishers are suitable for electric fires carbon dioxide type and dry chemical powder type.

Summary

Electrical machines are used for power generation, for providing motive power, for protection and control of electricity *etc.* Accordingly, machines are classified as AC or DC machine, single phase or three phase (for AC machines), static or rotating.

Ventilation of electrical machines is most important as the life of insulation is dependent upon temperature rise. Generally three methods of ventilation are used *viz*-radial, axial and combined radial and axial ventilation.

Principally there are three categories of electrical faults—open circuit, short circuit and earth fault. There are numerous reasons for faults as discussed in the chapter. Recent advances in condition monitoring techniques have made it possible to detect incipient faults, or deteriorating condition. These are infrared thermography, Tan-delta measurement, High Voltage d.c. leakage current test, partial discharge measurement *etc.*

Next, batteries are briefly discussed. Batteries provide source of stored energy. These are used for emergency lighting, power to operate control and protection system during fault, as a source of motive power for fork-lift trucks, in telephone exchanges *etc.* A wide range of batteries are available. However, for electrical engineering applications, mainly lead acid and Nical-cadmium batteries are used.

Earthing of electrical machines is important from safety point of view. Generally, pipe, rod and strip earthing are used depending upon the type of soil and terrain.

Electrical safety aspect covers two areas—fire due to electrical reasons and electric shocks. So causes and prevention methods of electric fires and electric shocks are discussed including first aid. In short, chapter describes basics of electrical machines.

Review Questions

1. What are different types of ventilation used in electrical machines?
2. What are various types of faults that take place in electrical machine?
3. Discuss various causes of faults in electrical machine.
4. Discuss role of Megger as a condition monitoring instrument.
5. Define term "Tan-Delta". How can this parameter be used for condition monitoring?
6. What do you mean by Partial Discharge? How measurement of partial discharge helps to monitor health of insulation of windings?
7. Compare dissolved Gas Analysis and Frequency Response Analysis methods.
8. Discuss maintenance procedure for lead-acid batteries.
9. What is earthing? Why is earthing required?
10. Write a brief note on electrical fire safety.

Discussion Questions

1. "Detection of deterioration of condition of electrical machine is more complex than mechanical machines". Comment.
2. Discuss "Ventilation plays a major role in the life of windings".
3. Justify "Earthing of equipments saves human life due to accidents".
4. Comment "Main enemies of electrical insulation are temperature, humidity and contamination".

Multiple Choice Questions

1. Insulation resistance is not affected by

   (*a*) Temperature. (*b*) Humidity.
   (*c*) Dust. (*d*) Air-pressure.

2. Earthing resistance depends on

   (*a*) Soil temperature. (*b*) Moisture in soil.
   (*c*) Type of soil. (*d*) None of these.

3. Solid earthing is preferable for
   (*a*) High voltage installation. (*b*) Low voltage installation.
   (*c*) High power installation. (*d*) Low power installation.

4. Alkaline batteries are better than lead acid batteries due to
   (*a*) Low maintenance. (*b*) Low cost.
   (*c*) Longer life. (*d*) None of these.

5. A polarization index of 2.5 is considered
   (*a*) Poor. (*b*) Very good.
   (*c*) Good. (*d*) Doubtful.

6. Infrared thermography is useful in detecting
   (*a*) Over load. (*b*) Short circuit.
   (*c*) Hot spots. (*d*) None of these.

7. High voltage withstand test is carried out for
   (*a*) One second. (*b*) One minute.
   (*c*) Five minutes. (*d*) None of these.

8. The minimum breakdown value of transformer oil should be
   (*a*) 10 KV rms. (*b*) 20 KV rms.
   (*c*) 30 KV rms. (*d*) 40 KV rms.

9. Electrical fires can be extinguished with
   (*a*) Soda-acid extinguisher. (*b*) Carbon dioxide extinguisher.
   (*c*) Water. (*d*) None of these.

10. Electrical fires are due to
    (*a*) Over load. (*b*) High voltage.
    (*c*) High temperature. (*d*) None of these.

True/False Questions

1. Condition of insulation decides the life of electrical machine.
2. Ingress of moisture improves insulation resistance.
3. Partial discharge increases deterioration of electrical insulation.
4. Polarisation index is a true measure of health of insulation.
5. Dissolved gas analysis method can be used on generators.
6. Rapid charging and discharging of batteries lead to reduced life span of batteries.

ANSWERS

Multiple Choice Questions

1. (*d*)	**2.** (*b*)	**3.** (*b*)	**4.** (*c*)
5. (*b*)	**6.** (*c*)	**7.** (*b*)	**8.** (*c*)
9. (*b*)	**10.** (*a*).		

True/False Questions

1. True	**2.** False	**3.** True	**4.** True
5. False	**6.** True		

❑❑❑

25

Thermal Insulation and Refractory

Learning Objectives

After reading this chapter, you will be able to:

- Understand classification and properties of insulating material.
- Appreciate importance of optimum thickness of insulating material.
- Understand characteristics, properties and selection of refractory.
- Develop understanding of application of various refractory.

25.1 INTRODUCTION

A thermal insulation is a poor conductor of heat. It is used to prevent heat-leakage from hot surfaces. It is also used to prevent heat-ingress in low temperature applications. Thermal insulation is used in wide variety of applications such as boilers, furnaces, cold boxes, steam pipelines, hot water pipelines, chilled water pipelines, gas turbine exhaust ducts, etc. Accordingly, a large number of insulating materials are available to cater to variety of applications.

This chapter covers basics in medium temperature range *i.e.* 50°C to 600°C in thermal insulation section. Beyond 600°C temperature, refractory materials are used and are discussed in subsequent section.

25.2 FUNCTIONS OF THERMAL INSULATION

A thermal insulation provides following function:

(*i*) Maintenance of temperature in a process.

(*ii*) Fire protection to equipment.

(*iii*) Energy saving by preventing heat loss or heat gain.

(*iv*) Protection to personnel by keeping surface temperature to a safe level.

(*v*) Better working condition by containing heat dissipation to atmosphere.

(*vi*) Prevention of corrosion in refrigerated system exposed to atmosphere.

(*vii*) Prevention of condensation in cold surfaces.

(*viii*) Conservation of products at desired temperature.

25.3 PRINCIPLE OF INSULATING MATERIALS

Insulating materials are poor conductor of heat having low thermal conductivity. These materials are porous containing large number of dormant air cells (diameter of air less than 0.09 micron). A large air pocket cell is not dormant because convection currents are set-up which ultimately increases rate of heat flow.

The heat transfer through insulation takes place through various modes such as gaseous conduction, radiation across gas/air gaps, convection across gas/air space, solid conduction, etc. Thermal conductivity of insulating material varies with its density, porosity, pore size and fibre diameter.

25.4 CLASSIFICATION OF INSULATING MATERIAL

Insulating materials are classified according to temperature ranges. These temperature ranges relate to usage and are following:

(*i*) **Low Temperature Insulation (upto 90°C):** Material in this range are used for cold and hot water systems, refrigerators, storage tanks *etc*. Materials used are expanded polystyrene, polyurethane, cork, mineral fibre *etc*.

(*ii*) **Medium Temperature Insulation (90°C-325°C):** Material in this range are used in steam pipelines, flue gas ducts *etc*. Materials used are Asbestos, Calcium silicate, Mineral fibre *etc*.

(*iii*) **High Temperature Insulation (above 325°C):** Material in this range are used in super-heated steam system, furnaces, ovens *etc*. Materials used are Asbestos, Calcium silicate, Mineral fibres, Mica, Fire clay, Ceramic fibres *etc*.

25.5 PROPERTIES AND USES OF INSULATING MATERIALS

(*i*) **Cork and Felt:** These are natural and cheapest available heat insulating or lagging materials. These materials are used for surfaces having temperature less than 200°C. These are inflammable above this temperature.

(*ii*) **Asbestos:** It is a natural occurring material which can be used upto 250°C. It is a fireproof material. It is available in opened fibre, sheet and mattresses. It is used in boilers, piping, furnaces *etc.*

(*iii*) **Glass Wool:** It is used upto a maximum temperature of 550°C in loose form and upto 250°C in bonded form. It causes skin irritation.

(*iv*) **Slag Wool and Rock Wool:** These are made by centrifuging molten slag or mineral rocks. These can be used upto 850°C temperature. These are available in moulded sections, boards, blankets *etc.* It is used in heat exchangers, driers, boilers and high temperature pipings.

(*v*) **Ceramic Fibres:** This can be used upto 1700°C and are available in blanket form. It is used as furnace backup refractory, fire boxes, induction coil insulation *etc.*

(*vi*) **Polyurethane:** It can be used from –178°C to 4°C. It is inflammable and produces toxic vapours. It is used in refrigeration pipings, deep freezing cabinets, cold rooms *etc.*

(*vii*) **Polystyrene:** It can be used from –167°C to 82°C. it is inflammable and UV light degradable. It is used in refrigeration pipings.

(*viii*) **Calcium Silicates:** It can be used upto 1050°C. it is light weight, moisture resistant, non-corrosive and does not burn. It is used to insulate boilers, furnace walls, flue gas linings *etc.*

(*ix*) **Fibre Glass:** It can be used upto 540°C. It used as insulation for boilers, heat exchangers, ovens, driers *etc.*

25.6 FORMS OF INSULATING MATERIALS

Insulating materials are available in four forms *viz.* granular, fibrous, cellular and reflective. Granular materials are composed of skeletons of di-atoms having air entrained in the material (*e.g.* calcium silicate). Fibrous materials contain air between fibres (*e.g.* mineral wool). Cellular materials have small air or gas cells sealed from each other (*e.g.* foamed plastic). Reflective materials consists of numerous layers of low emissivily foils. However, any combination of above four types can be used.

25.7 SELECTION OF INSULATING MATERIAL

Choice of an insulating material depends upon numerous factors given below:

(*i*) Total cost.

(*ii*) Operating temperature of system.

(*iii*) Resistance to weather.

(*iv*) Ability to withstand vibration and accidental mechanical damage.

(*v*) Resistance to chemicals.

(*vi*) Resistance to fire.

(*vii*) Thermal conductivity of insulating material.

(*viii*) No cracking or shrinkage during use.

25.8 OPTIMUM ECONOMIC THICKNESS OF INSULATION

Cost of insulation plays an important factor in selection of type and thickness of insulation. Higher the thickness of insulation, lesser is the heat loss. However, higher thickness increases cost of insulation. Hence, there is an economic limit to thickness of insulation.

The above graph in fig. 25.1 shows, cost of insulation increases with increase in insulation thickness but cost of heat loss also decreases. Hence, the total cost *i.e.* sum of insulation cost and cost of heat loss is plotted and is minimum at point '*A*'. Hence, the shaded portion upto point '*A*' is the economic thickness.

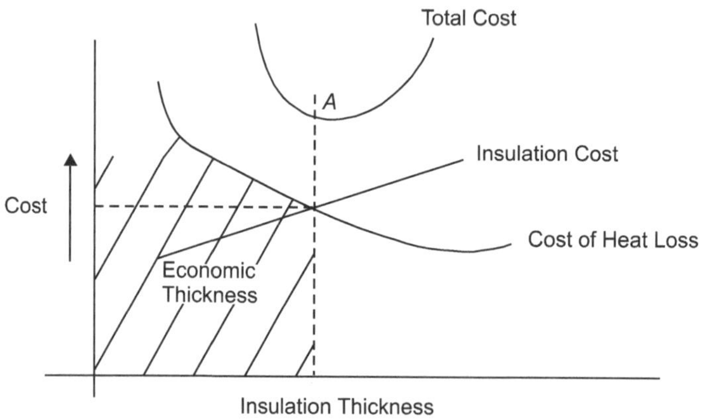

Fig. 25.1 Effect of insulation thickness on cost.

Optimum economic thickness depends upon the following factors:

(*i*) Cost of fuel.

(*ii*) Annual operating hours.

(*iii*) Calorific value of fuel.

(*iv*) Efficiency of combustion.

(*v*) Operating, surface temperature.

(*vi*) Pipe diameter.

(*vii*) Estimated cost of insulation.

(*viii*) Heat loss per linear meter or square meter (depending upon whether surface is flat or not).

25.9 EXAMPLE OF OPTIMAL INSULATION

Consider a black steel pipe of diameter 89 mm at a surface temperature of 90°C. Let the pipe be covered with 50 mm insulation and 100 mm insulation.

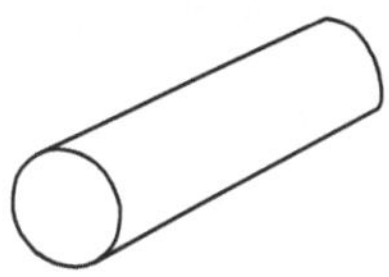

Fig. 25.2 50 mm Uninsulated pipe

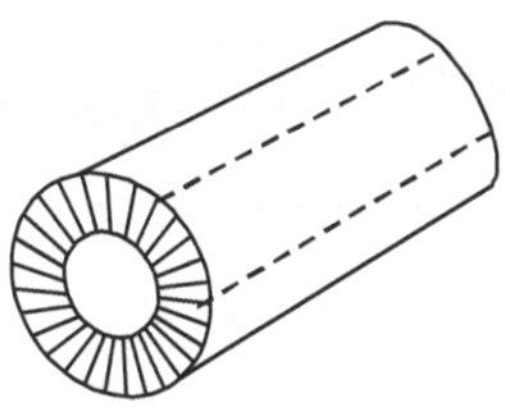

Fig. 25.3 50 mm insulation

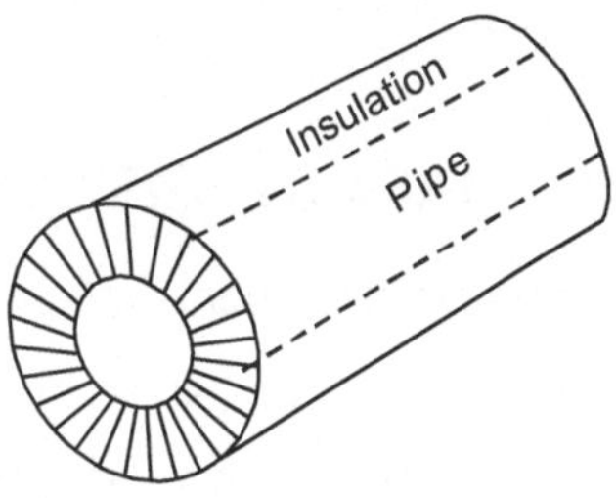

Fig. 25.4 100 mm insulation

Without insulation, the steel pipe dissipates 275 kcal/m of heat. With 50 mm insulation on pipe, the heat dissipated is 25 kcal/m. With 100 mm insulation, the heat dissipation is reduced to 15 kcal/m.

Now, energy saved by 50 mm insulated pipe = 275 – 25
= 250 kcal/m
(equivalent to 260 litres of oil/year)

Energy saved by 100 mm insulated pipe = 25 – 15 = 10 kcal/m
(equivalent to 9 litres of oil/year)

It shows that with 50 mm thick insulation, the fuel saved compared to uninsulated pipe is 260 litre/year. While with 100 mm thick insulation, fuel saved is 9 litre/year more. Hence, as the thickness of insulation is increased, there is little incremental increase in fuel saving. Hence, an optimum thickness of insulation gives not only fuel saving but also lower cost of insulation.

Cost of heat loss/year can be estimated from following formula

$$= \frac{Q \times N}{n + H} \times P \text{ Rs/year}$$

where

P = Price of fuel (Rs./kg)
N = Operating hours/year
Q = Heat (Kcal/hr)
N = Efficiency of energy conversion process
H = Gross calorific value of fuel (kcal/kg)

25.10 REFRACTORY

Refractory is a material that can withstand action of abrasive or corrosive solids, liquids or gases at high temperature without suffering deformation in shape. Depending upon operating conditions, refractories are available in wide range and shapes.

25.11 CHARACTERISTICS OF A REFRACTORY MATERIAL

The general requirements of refractory materials are given below:

(*i*) Ability to withstand high temperature.
(*ii*) Ability to withstand abrasive and corrosive conditions.
(*iii*) Ability to withstand sudden change in temperature.
(*iv*) Low coefficient of thermal expansion.
(*v*) Low thermal conductivity.
(*vi*) Should not react/contaminate materials it comes in contact with.

25.12 CLASSIFICATION OF REFRACTORY MATERIAL

(*i*) **Classification based on chemical composition:**

(*a*) *Neutral Refractory:* It does not combine with acid or base *e.g.* Graphite, pure-Alumina, Chrome, Zirconia *etc.*

(*b*) *Acid Refractory:* It reacts with base and is resistant to acid slags *e.g.* Silica, Alumino-Silicate *etc.*

(*c*) *Basic Refractory:* It is resistant to basic slags *e.g.* Magnetite, Dolomite, Chrome-Magnesite *etc.*

(*ii*) **Classification based on method of manufacture:** Depending upon method of manufacture, these are classified as hand moulded, dry press process, fused cast, formed and unformed.

25.13 PROPERTIES OF REFRACTORY

Various properties of refractory are listed below:

(*i*) **Bulk Density:** It is the amount of material present per unit volume. A high bulk density means high volume stability, high heat capacity and high resistance to slag penetration.

(*ii*) **Porosity:** It is the ratio of pore volume to the bulk volume. It affects penetration of gases, slag *etc*. to greater depth and results in reduction of life.

(*iii*) **Cold Crushing Strength:** It is the measure of ability to withstand rigours of transport, application and usage.

(*iv*) **Refractoriness:** It is the ability to withstand heat without appreciable deformation.

(*v*) **Chemical Inertness:** It is the ability of refractory to remain inert in the presence of slags, gases, fuel ashes and products used in furnaces.

(*vi*) **Thermal Conductivity:** It indicates amount of heat loss. It should be low.

(*vii*) **Thermal Spalling:** Sudden change in temperature results in uneven expansion and contraction of refractory material. This leads to stresses causing cracks in the material.

(*viii*) **Electrical Conductivity:** Refractory used in electrical furnaces should have low electrical conductivity.

(*ix*) **Thermal Expansion:** Refractory material should have low coefficient of thermal expansion.

(*x*) **Mechanical Resilience:** This property allows refractory material to be fabricated and assembled without getting damaged.

(*xi*) **Thermal Shock Resistance:** It is the ability of material to withstand rapid changes in temperature. This property allows faster heating and cooling cycle.

25.14 FACTORS AFFECTING HEAT LOSSES

Following factors affect heat losses from the walls of boilers, furnaces, ovens etc.

(*i*) **Inside temperature:** Higher the inside temperature, more the heat loss.

(*ii*) **Ambient air temperature:** Lesser the ambient temperature, more the heat loss.

(*iii*) **Ambient air velocity:** Higher the air velocity, more the heat loss.

(*iv*) **Emissivity of walls:** Lesser the emissivity, lesser the heat loss.

(*v*) **Thickness of walls:** Higher the wall thickness, lesser the heat loss.

(*vi*) **Thermal conductivity of walls:** Higher the thermal conductivity, more the heat loss.

(*vii*) **Nature of operation whether continuous or intermittent:** Heat loss is more with intermittent operations.

25.15 SELECTION OF REFRACTORY

Refractory for a particular application is selected with the aim to get best performance. Some of the refractory materials widely used are fire bricks, high aluminium bricks, silica bricks, magnesite bricks, chromite bricks, zirconia bricks *etc*.

Following are factors affecting selection of a particular refractory:

(*i*) Working temperature.

(*ii*) Stress due to temperature gradient and temperature fluctuations.

(*iii*) Heat transfer.

(*iv*) Area of application.

(*v*) Extent of abrasion, corrosion and impact.

(*vi*) Chemical inertness to products inside furnace.

(*vii*) Structural load of furnace.

(*viii*) Capital cost.

25.16 SOME REFRACTORIES AND THEIR APPLICATIONS

(*i*) **Fire Clay Refractory:** Fire clay refractory consists of Aluminium Silicates and various amounts of Silica upto 78% and Al_20_3 upto 44%. This material is cheaper and hence widely used in iron and steel industry, glass industry, cement industry, pottery kilns *etc*.

(*ii*) **High Alumina Refractory:** High Alumina refractory contains more than 45% Alumina. High Alumina content increases refractoriness. It is used in hearth and shaft of blast furnaces, ceramic kilns, cement kilns and metal melting crucibles.

(*iii*) **Silica Bricks:** It contains 93% or more Silica. It does not soften under high load and is flux and slag resistant. It has high volume stability and high thermal shock resistance. It is extensively used in iron and steel melting furnaces, glass furnaces *etc.*

(*iv*) **Magnesite:** It contains more than 85% Magnesium oxide. It is made from naturally occurring Magnesite. It is used in furnaces operating under oxidizing and reducing conditions.

(*v*) **Chromite Refractory:** Chrome Magnesite refractory contains 15–35% Cr_2O_3 and 42–50% MgO and Magnesite-Chromite refractory contains at least 60% MgO and 8–18% Cr_2O_3. It can withstand high temperature and corrosive slags and gases. It is used for building high temperature area of furnaces.

(*vi*) **Zirconia Refractory:** It consists of zirconium dioxide with small quantities of Calcium, Magnesium and Cerium oxide as stabilizing agent. It can withstand very high temperature typically 1500°C. It has very low thermal conductivity and does not react with liquid metal. It is used in glass furnaces and crucibles for metallurgical purposes.

(*vii*) **Alumina Oxide Refractory:** It consists of Aluminium oxide with traces of pure Alumina. It is mechanically strong and insoluble in water, superheated steam and inorganic oxide and alkalies. It is used for fusing of Sodium Carbonate, Sodium Hydroxide and Sodium Peroxide.

(*viii*) **Monolithic:** Monolithic refractory are replacing conventional fixed refractory due to numerous advantages such as easy transportation, handling and application, easy repair, better thermal resistance and volume stability. It is applied by ramming, spraying and gunniting the monolithic. It is available in both air setting and heat setting variety.

(*ix*) **Ceramic Fibres:** It is made by blowing high velocity air onto a molten stream of Alumino-Silicate material with a controlled composition. This results in a fluffy, white cotton like fibres that can be spun and fabricated into blankets, felts, boards, flocks, wood, ropes *etc*. It has low thermal conductivity, light weight, immunity to thermal shocks and chemically stable. It can be used upto 1325°C temperature.

Summary

Thermal insulation and refractories act as heat shield for boilers, steam and gas turbines, furnaces, steam pipelines, chilled water pipelines, cold boxes, exhaust ducts of D.G. sets and gas turbines *etc.* Thermal insulation is used for both hot as well as cold surfaces.

Insulating materials are classified into three categories viz. low temperature insulation (upto 90°C), medium temperature insulation (90°C to 325°C) and high temperature insulation (above 325°C).

Insulating materials are available in four forms viz. granular, fibrous, cellular and reflective.

Thickness of insulating material plays an important role towards cost of heat loss and cost of insulation.

Refractory material is available as bricks. These are used for high temperature applications such as boiler, furnace, reformer *etc.* Refractory material withstand high temperature as well as abrasive and corrosive action of solids, liquids and gases. All refractory materials are classified into three categories *viz.* neutral refractory, acid refractory and basic refractory.

Heat loss from refractory surface depends mainly on inside temperature, ambient air temperature and velocity, wall emissivity, thickness and thermal conductivity.

A wide variety of thermal insulation are available such as cork, felt, expanded polystyrene, polyurethane (cold insulation), ceramic fibres, fibre glass, calcium silicate (hot insulation). Some of the refractory materials are silica brick, chromite refractory, zirconia refractory, high alumina refractory *etc.*

Review Questions

1. Discuss functions of thermal insulation.
2. Explain in detail with example the classification of thermal insulation.
3. What are various factors affecting choice of insulating material.
4. Discuss factors affecting optimum economic thickness of insulation.
5. What is a refractory? How is it different from other types of thermal insulation?
6. Discuss various characteristics of refractory material.
7. Discuss factors affecting choice of right type of refractory.

8. What are the various classification of refractories?
9. Why low-thermal conductivity is an important property of a refractory?
10. Explain terms (a) thermal spalling (b) thermal shock resistance (c) thermal expansion.

Discussion Questions

1. "Optimum thickness of insulation not only saves cost of insulation but also fuel cost". Elaborate.
2. "Refractories are given least attention although maximum energy saving potential lies there". Justify.
3. "Porosity of insulating material helps to prevent heat loss or gain by forming an envelope of air". Support your answer with example.
4. "Porosity of refractory material causes reduction of life due to penetration". Comment.

Multiple Choice Questions

1. Refractory material is used for
   (*a*) High temperature. (*b*) Low temperature.
   (*c*) Medium temperature. (*d*) None of these.
2. Low temperature insulating materials are:
   (*a*) Calcium silicate. (*b*) Asbestos.
   (*c*) Cork. (*d*) Mica.
3. High temperature insulating material are:
   (*a*) Calcium silicate. (*b*) Mica.
   (*c*) Asbestos. (*d*) All of these.
4. Examples of neutral refractory are:
   (*a*) Zirconia. (*b*) Silica.
   (*c*) Magnetite. (*d*) Dolomite.
5. A good refractory material should not possess
   (*a*) High thermal conductivity.
   (*b*) Low coefficient of thermal expansion.
   (*c*) Ability to withstand high temperature.
   (*d*) Ability to withstand thermal shock.

6. Heat loss from a hot body is affected by
   (*a*) Emissivity of walls.
   (*b*) Thickness of walls.
   (*c*) Ambient air velocity.
   (*d*) All of these.
7. Ceramic fibre can be used upto a temperature of
   (*a*) 1000°C.
   (*b*) 500°C.
   (*c*) 1300°C.
   (*d*) 50°C.
8. Cost of insulation increases with
   (*a*) Increase in thickness of insulation.
   (*b*) Decrease in thickness.
   (*c*) Thickness has no effect.
   (*d*) None of these.
9. Sudden change in temperature of refractory can cause
   (*a*) Thermal spalling.
   (*b*) Porosity.
   (*c*) Deformation.
   (*d*) None of these.
10. For oxidizing and reducing environment, refractory used are:
   (*a*) Zirconia.
   (*b*) Aluminium oxide.
   (*c*) Magnetite.
   (*d*) Silica.

True/False Questions

1. A good insulating material should be fire resistant.
2. A good refractory should possess high bulk density.
3. Heat loss can be reduced by increasing emissivity of walls.
4. Intermittent nature of operation of furnace causes more heat loss.
5. A higher ambient air velocity can cause more heat loss from a hot body.

ANSWERS

Multiple Choice Questions

1. (*a*)	**2.** (*c*)	**3.** (*d*)	**4.** (*a*)	**5.** (*a*)
6. (*d*)	**7.** (*c*)	**8.** (*a*)	**9.** (*a*)	**10.** (*c*)

True/False Questions

1. True	**2.** True	**3.** False	**4.** True	**5.** True

❑❑❑

26

Energy Conservation and Energy Audit

Learning Objectives

After reading this chapter, you will be able to:

- Understand meaning of energy conservation and energy efficiency improvement.
- Examine possible opportunities in energy conservation in industries.
- Develop keen understanding of energy efficient technologies.
- Understand energy audit process.

26.1 OVERVIEW OF ENERGY SCENARIO

Energy is one of the major inputs for economic development of a country. Economic growth and energy demand are closely linked albeit in a non-linear manner. For example, 6% increase in India's Gross Domestic Product (GDP) would impose an increased demand of 9% on its energy resources. Ratio of energy demand to GDP is a useful indicator. A high ratio indicates energy dependence and a strong influence of energy on GDP growth. Developed countries focus on energy efficiency and follow lower energy intensive routes to maintain energy to GDP ratios less than one.

With fast depleting conventional energy resources such as coal, oil, gas *etc.* there is urgent need to conserve these resources and also explore alternate sources of energy such as wind, solar, bio-gas, tidal *etc*. The economic progress leads to widening of gap between demand and supply of energy. Apart from this, consumption of fossil fuels adds to global warming and environmental pollution. The contribution of non-conventional sources of energy in the total energy mix is marginal in India though rising. Hence, need of the hour is to find ways and means to conserve energy through reduction of wastage, efficient usage, curtailing peak demand, by shifting of load to non-peak hours *etc*.

Examination of energy consumption pattern by different sectors of economy *viz*. industry, agriculture, transport *etc*. shows that almost 50% of total commercial energy is consumed by industries. Hence, needless to say, scope for energy saving is very large in industrial sector. However, agricultural and commercial sectors also provide good scope for saving. In this chapter, we shall concentrate on ways and means to conserve energy.

26.2 NEED FOR ENERGY CONSERVATION

Eighty-five percent of primary energy comes from non-renewable energy sources and we have consumed 60% of all resources in the last 200 years. Hence, energy conservation is need of the hour. The need for energy conservation are due to following:

(*i*) Conservation of limited energy resources for future use.

(*ii*) Improved energy security.

(*iii*) Reduced cost of energy consumption.

(*iv*) Maintaining sustainable environment by reduced emission of greenhouse gases.

(*v*) Increased productivity.

(*vi*) Improved quality.

(*vii*) Longer equipment life due to lesser heat generation (because of reduced losses).

(*viii*) Higher equipment efficiency.

(*ix*) Reduced import bill (for petroleum products).

26.3 ENVIRONMENTAL ASPECTS OF ENERGY CONSERVATION

Usage of conventional fossile fuel leads to environmental damage by polluting the atmosphere. Examples of air pollution are carbon dioxide, sulphur dioxide, nitrogen oxide, carbon mono oxide and particulate matters from boilers, furnaces, transport vehicles *etc.*

Increased fossil fuel use leads to increased carbon dioxide emission leading to greenhouse effect and global warming. The heating-up of earth's atmosphere due to trapping of long wavelength infrared rays which are reflected from earth's surface by the carbon dioxide present in atmosphere is called greenhouse effect. Due to this greenhouse effect, there is rise in global mean temperature. This increase in temperature will cause melting of arctic ice leading to rise in sea level. This will cause sub-mergence of low-lying areas.

Release of sulphur dioxide and nitrogen di-oxide mix with water vapour in atmosphere and forms sulphuric and Nitric acid respectively. This causes acid rain which affects plant, wild life and human beings.

Hence, environmental impact of using fossil fuel is enormous. So, it is imperative to reduce consumption of fossil fuels to save environment from impending disaster.

26.4 ALTERNATE SOURCES OF ENERGY

To reduce dependence on fossil fuel and to find environmentally clean substitute, alternate sources of energy need to be explored. The various such sources are solar, wind, bio-mass, tidal, geothermal *etc.* These are environmental friendly and sustainable sources of energy.

Although the cost of power produced from non-conventional sources of energy are higher compared to those obtained from fossil fuels but the cost per unit will go down due to mass production and improvement in technology.

Due to these reasons, India has developed wind power stations on a large scale and has given thrust to solar and bio-mass power generation. These power sources operate as standalone unit or feed power to the grid.

India having blessed with sunshine in large parts of the country almost throughout the year, provides ideal ground for power generation through solar means. This can be solar photovoltaic cells that convert solar energy directly into electrical power or through solar thermal route which uses solar heat energy.

Given below is the potential and utilization of renewable energy sources in India (2004–05):

Table 26.1 Renewable Energy Potential and Achievement in India

	Potential	Cumulative Achievement
Biogas plant	120 Lakh	36.71 lakh
Wind	45000 MW	2980 MW
Small hydro	15000 MW	1693 MW
Biomass power/ co-generation	19,500 MW	727 MW
Solar PV	20 MW/sq km	191 MW
Solar water heating	1400 lakh sq m collector area	10 lakh m collector area
Waste to Energy	1700 MW	46.5 MW

Source: MNES Annual Report 2004-05

The above table amply demonstrates that there is large untapped potential for the renewable energy development and usage in India.

26.5 ENERGY EFFICIENCY—SIGNIFICANCE

Energy conservation and energy efficiency are two separate things. Energy conservation is achieved if the growth of energy consumption is reduced. Energy conservation may be due to several reasons such as technological improvement, productivity increase, waste reduction *etc.* While energy efficiency is achieved when energy intensity in a specific product, process or area of production or consumption is reduced without affecting output, consumption or comfort level. Hence, promotion of energy efficiency will contribute to energy conservation.

This can be illustrated with the help of comparison between incandescent bulb (ordinary household bulb) and CFL (compact fluorescent lamp). A 60 watt incandescent lamp produces same amount of light as is produced by a 15 watt CFL. But the energy consumed by CFL is $1/4^{th}$ of that of incandescent bulb. Hence, energy intensity of CFL is lesser than that of incandescent bulb thereby promoting energy conservation.

26.6 ENERGY CONSERVATION IN INDUSTRIES

Industry consumes about 50% of total commercial energy among various other sectors such as transport, commercial, agricultural sectors *etc.* Hence, the potential for saving energy is highest in industrial sector. Significant savings can be realized with minimum investment to the tune of 15% to 20% of industry's present level of consumption.

In this context, we shall examine energy saving opportunities in electrical area, boilers, furnaces, steam system, cooling towers, refrigeration and air conditioning systems, co-generation *etc*. This will provide a broad perspective towards areas of improvement.

26.6.1 Energy Conservation Opportunities in Boilers

Various energy conservation opportunities in boilers are given below:

(*i*) **Low stack temperature:** Stack temperature should be low. But it should not be so low that water vapours present in flue gases condense and react with sulphur present in flue gases. This produces sulphuric acid leading to corrosion. Higher stack temperature indicates need to clean water side/fire side surface.

(*ii*) **Feed water pre-heating:** Feed water can be preheated through economiser which uses heat of flue gases thus saving energy.

(*iii*) **Combustion air pre-heating:** A 20°C rise in temperature of combustion air, increases thermal efficiency by 1%. So pre-heating of combustion air is another energy saving option.

(*iv*) **Excess air control:** Excess air is required to complete combustion. However, control on excess air is required to reduce flue gas losses. For example, 1% reduction in excess air, there is gain of 0.6% in efficiency. Excess air can be controlled by using continuous oxygen analyser and draft gauge.

(*v*) **Automatic blowdown control:** Continuous blowdown should be controlled by sensing water conductivity and PH.

(*vi*) **Reduction of boiler steam pressure:** Fuel consumption can be reduced by decreasing boiler steam pressure.

(*vii*) **Variable speed drive for fans, pumps and blowers:** Instead of throttling dampers to reduce flow, a variable speed drive gives energy savings.

(*viii*) **Optimum boiler loading:** Optimum efficiency of boilers occur at 65–85% of full load. Hence, it is better to run number of smaller capacity boilers rather than one large boiler at low level.

26.6.2 Energy Conservation Opportunity in Furnaces

Following are energy conservation opportunities in furnaces:

(*i*) **Optimum operating temperature:** Operating furnaces at high temperature causes heat loss, oxidation and over stressing of refractories. Hence, furnace should be operated at optimum temperature.

(*ii*) **Control of furnace draft:** A negative pressure in furnace causes infiltration of air thereby reducing operating temperature inside furnace. For optimum fuel consumption, slight positive pressure should be maintained in the furnace. Also all openings must be properly sealed and insulated.

(*iii*) **Furnace loading:** Thermal efficiency of furnace is maximum at a particular loading. Hence, furnace should be operated near to its rated capacity.

(*iv*) **Waste heat recovery:** Flue gases leave the furnace at a higher temperature. Hence, this heat can be utilized for charge preheating, combustion air pre-heating or to generate steam or hot water.

(*v*) **Proper combustion:** Amount of heat loss in flue gases increases with increase in excess air. Hence, it is necessary to control excess air for complete combustion.

(*vi*) **Use of special coating:** Use of ceramic coating in furnace chamber promotes uniform heat transfer thereby extending life of refractories. These coatings applied on refractory can withstand temperature upto 1350°C.

26.6.3 Energy Conservation Opportunities in Steam Systems

Following are energy conservation opportunities in steam system:

(*i*) **Stopping steam leakages:** Steam leak at high pressure is very expensive. Hence, a regular surveillance should arrest steam leakages from pipelines, valves, flanges, joints, and steam traps.

(*ii*) **Using steam at low acceptable level:** A low pressure steam possesses high latent heat required for process heating. However, with reduction in steam pressure, steam consumption goes up with greater processing time. So pressure should be so selected as to avoid longer processing time but simultaneously keeping pressure low.

(*iii*) **Condensate recovery:** Steam condenses after giving off latent heat and the condensate still contains a sizable portion (about 25%) of heat as hot water. If this hot water is returned to boiler, the fuel consumption will be reduced.

(*iv*) **Air venting:** Air acts as heat insulation barrier. Presence of air in the equipment reduces partial pressure of steam thereby reducing overall steam-air mixture temperature. Hence, adequate air vent should be available in the pipeline to purge the air quickly.

(*v*) **Flash steam recovery:** Flash steam is produced when condensate at high pressure is released to low pressure. This flash steam can be used for heating purpose.

(*vi*) **Proper functioning steam traps:** Steam traps remove condensate and incondensable gases from steam pipelines. An improperly working steam trap retains condensate and incondensable gases leading to less heat transfer.

(*vii*) **Insulation of steam pipelines:** All steam pipelines including flanges must be adequately insulated to prevent heat loss.

26.6.4 Energy Conservation Opportunities in Co-generation System

In a conventional power plant with condensing type steam turbine, the efficiency is about 30%. However, co-generation system produces both steam and power and the co-generation efficiency can go upto 90%. Such system is extensively used in sugar, pulp and paper industries, heavy chemical, cement *etc*.

26.6.5 Energy Conservation Opportunities in Cooling Tower

Following are various energy saving opportunities:

(*i*) **Improvement of cycle of concentration (COC):** Concentration of dissolved solids increases due to evaporation. Hence, make-up water need to be added to maintain desired level of total dissolved solids (TDS). By careful, chemical analysis of water, amount of make-up water used can be reduced. This reduces load on water softner and pumping energy apart from saving water and water treatment chemicals.

$$\text{Cycle of Concentration (COC)} = \frac{\text{Dissolved solids in cooling water}}{\text{Dissolved solids in make-up water}}$$

(*ii*) **Use of water treatment chemicals:** Certain water treatment chemicals such as biocides, fungicides, corrosion inhibitors, PH treatment, *etc*. are used. These chemicals prevent formation of algae, slimes *etc.* in cooling tower which floats and passes through water pipelines and causes fouling leading to inefficient heat transfer. Similarly corrosion inhibitors and pH treatment chemicals reduce the rate of corrosion in pipelines and equipments thereby enhancing heat transfer capacity.

(*iii*) **Use of FRP blade in cooling tower fan:** A cooling tower fan consumes good amount of energy. This can be reduced by using lighter, fibre re-inforced plastic (FRP) fan. This is not only lighter, but also corrosion proof. This gives good energy saving.

(*iv*) **Use of two speed motor for cooling tower fan:** Depending upon ambient temperature, the speed of cooling tower fan can be reduced thereby reducing the energy consumption.

(*v*) **Efficiency improvement of cooling tower pumps:** Any improvement in efficiency of cooling tower pumps gives significant saving in energy.

26.6.6 Energy Conservation Opportunities in Electrical System

Following are the energy conservation opportunities in electrical system:

(*i*) **Control of maximum demand charges:** There is large penalty for exceeding contracted maximum demand. This can be reduced by rescheduling of load *e.g.* to night time, shedding of non-essential loads, storage of products (chilled water, brine, hot water using electricity).

(*ii*) **Improving power factor:** There is a penalty for power factor going below 0.9 in most electricity board supply. Hence, use of automatic power factor correction using capacitors can maintain high power factor. This not only saves from penalty of poor power factor but also reduces energy consumption.

(*iii*) **Use of energy efficient motor:** High energy efficiency motors have efficiency 3 to 4 % higher than standard efficiency motors.

(*iv*) **Variable frequency drives:** Devices requiring variable speed can be provided with variable frequency drives rather than controlling flow by throttling. Even large motors can be started using soft starters which reduces energy consumption during starting.

(*v*) **Changing star connection to delta:** Motors operating at less than half load can have their connection changed to star connection from delta. This will reduce energy consumption.

(*vi*) **Voltage optimization:** Maintaining correct voltage, improves power factor, reduces current drawn by motor, improves motor efficiency and reduces system loss.

(*vii*) **Optimum operation of transformer:** Transformers supplying power to the plant should operate 70–80% of load for better performance. Should not operate idle because it reduces power factor drastically and voltage setting should be at optimum position for better performance.

(*viii*) **Use of amorphous metal distribution transformer:** Amorphous metal transformer reduces no load losses upto 80% compared to conventional transformer. Following table gives comparison of amorphous metal and conventional transformer.

Table 26.2 Comparison of no load loss for 11000/433 V transformer

Transformer Rating (kva)	No load losses (watt)	
	Conventional	Amorphous Metal
500	990	250
1000	1450	350
1500	2100	490
2000	2500	500
2500	2800	550

26.6.7 Energy Saving Opportunities in Compressed Air System

Following are the energy saving opportunities in compressed air system:

(*i*) **Generation of compressed air close to process equipment:** This way pressure drop and leakage losses are reduced.

(*ii*) **Use of correct capacity of air compressor:** This minimizes no-load operation time.

(*iii*) **Segregation of high pressure and low pressure users:** High pressure and low pressures users should be supplied from separate lines.

(*iv*) **Operate screw compressor as base load:** Among variety of compressors such as reciprocating, screw, centrifugal *etc.*, it is economical to run screw compressor as base load.

(*v*) **Minimization of air leakages:** Air leakage must be plugged because it causes power loss.

(*vi*) **Installation of inter and after coolers:** Installation of inter and after coolers in air compressors reduces power consumption.

(*vii*) **Use of multi-stage compressor:** If the compression ratio is more than 4 then multi-stage compressors are economical.

(*viii*) **Replacement of compressed air for agitation:** Compressed air must not be used for agitation *e.g.* for stirring liquid in a container or mixing resins in a demineralised water plant mixed bed. Rather an agitator operated by motor and roots blower must be used respectively.

(*ix*) **Reduced discharge pressure:** A higher discharge pressure reduces compressor capacity, increases energy consumption, increase leakage loss and increases loading on piston rods. Reduction in delivery pressure by 1 kg/cm^2 reduces power consumption by 6 to 10%.

26.7 ENERGY EFFICIENT DEVICES AND TECHNOLOGIES

Due to technological development, energy efficient devices are available such as high efficiency motors, amorphous core transformer, fluidized bed combustion boilers *etc*. The devices consume much less power compared to their conventional counterparts. However, the initial cost of these devices are higher compared to conventional devices. But, depending upon usage, the additional cost is recovered within a short period and is a profitable proposition in the long run. Hence, it pays to replace existing energy inefficient devices with these latest devices.

Following are various energy efficient devices:

(*i*) **Energy Efficient Motors:** These motors have high efficiency and consume less power for any given load. Efficiency of these motors are 1 to 4 % higher than standard efficiency motors. Also power factor of these motors is higher and remains fairly constant over wide range (50–100% of load). However, cost of these motors are 15 to 30% higher.

High performance is obtained by using low loss steel, thinner lamination, reduced air gap, higher cross-section of copper for winding and using copper bars in place of aluminium bars in rotor.

Advantages of energy efficient motors are following:

(*a*) Reduced power consumption.
(*b*) Lower temperature rise and hence longer life.
(*c*) Constant efficiency over a broad band.
(*d*) Improved power-factor.

(*ii*) **Amorphous Metal Transformer:** Conventional distribution transformers use cold-rolled grain oriented silicon steel (CRGO) as core. But amorphous metal transformer can reduce no-load losses by

80% compared to CRGO steel transformer. Hence, these transformers are now widely used in distribution system.

(*iii*) **Energy Efficient Lighting:** 40 watt conventional fluorescent lamp can now be replaced with energy efficient 36 watt fluorescent lamp with electronic ballast.

Also incandescent lamps can be replaced with more energy efficient compact fluorescent lamps (CFL). Next on the horizon is LED or Light emitting diode lamps which are not only energy efficient but also have longer life and are free from mercury.

Lighting energy consumption can also be reduced by using light dimming devices and occupancy sensors. Occupancy sensors switch off lights in selected area when no motion or movement is detected by the sensor for a given time.

(*iv*) **Fluidized Bed Combustion Boilers:** FBC boilers can burn high ash content coal with a combustion efficiency of over 95%. The overall efficiency is 84% ± 2%. FBC when combined with conventional steam turbine improves efficiency of electricity generation. The potential fuel savings are at least 4%.

Apart from this, sulphur dioxide and NOX formation is greatly reduced.

(*v*) **Use of Vapour Absorption Machines with Waste Heat:** Vapour absorption machines are used to provide air conditioning. The heat required to run vapour absorption machines can be provided by recovering waste heat of gas turbines, DG-sets, furnaces *etc*. This proves to be an extremely energy efficient method.

(*vi*) **Heat Wheel:** A heat wheel is used to recover waste heat. It consists of a porous disc of a material having high heat capacity and rotates inside two ducts connected side by side. Hot gases enters and leaves from one duct while the cold gas enters from opposite direction so that it forms a counter flow system. The disc rotates slowly thereby transferring heat from hot gas to cold gas. The efficiency of such system is generally 85%.

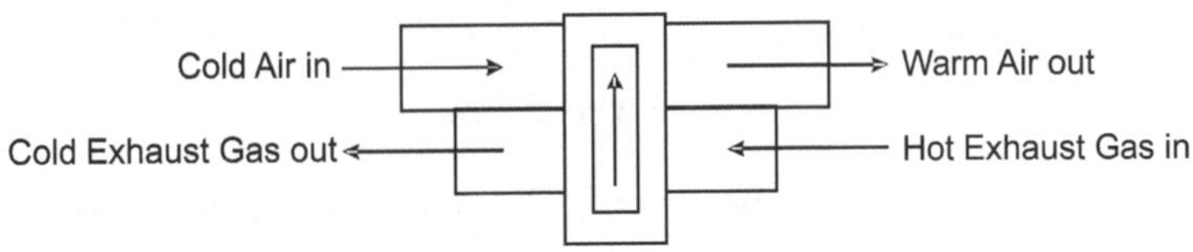

Fig. 26.1 Heat Wheel.

(*vii*) **Thermo-compressor:** This device converts very low pressure steam into medium pressure steam using very high pressure steam.

It consists of a device through which high pressure steam is accelerated through a nozzle. Due to rise in velocity of high pressure steam, low pressure steam is sucked in. The low pressure steam is compressed in the convergent section of venturi tube and then discharges through a divergent section.

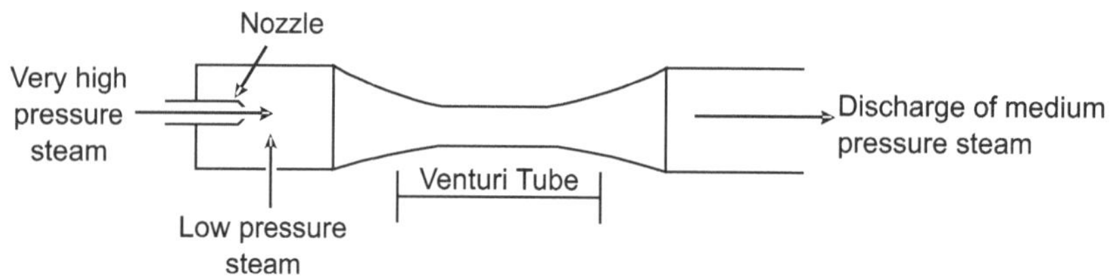

Fig. 26.2 Thermo-Compressor.

(*viii*) **Heat Pipe:** A heat pipe is a thermal energy absorption and transferring device. It consists of a long pipe filled with a working fluid which evaporates upon receiving heat. The inside of the pipe is lined with metal mesh wick which acts a return path for working fluid. The pipe is a sealed tube under vacuum.

The heat from external source causes the working fluid to vapourise. These vapours absorb latent heat of vaporization and travel to other end of pipe. Thermal energy is removed at this end thereby giving up latent heat of condensation. So, one end of pipe acts as evaporator section and the other end as condenser section.

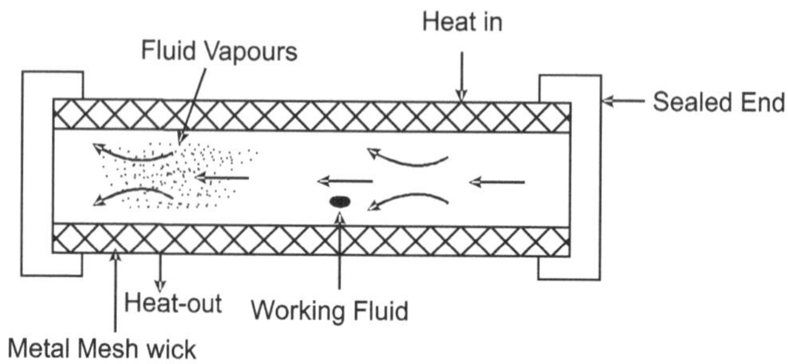

Fig. 26.3 Heat Pipe.

(*ix*) **LED as a Lighting Source:** LED or light emitting diodes is semi-conductor device that emits light due to movement of electrons. Advantage of LEDs are numerous as these are extremely energy efficient, do not emit heat radiation, have very long life (upto 1,00,000 hours), does not use mercury, available in different colours (red, blue, green, orange, amber etc.) Major advantage is that these do not fail suddenly rather light output gradually reduces with time. These are widely used for traffic light, vehicle tail light and dash board illumination, torch and advertising and architectural purpose. LEDs for general lighting purpose have started appearing in the market.

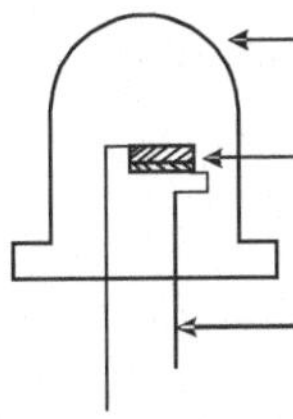

Fig. 26.4 Light Emitting Diode.

26.8 CONCEPT OF GREEN BUILDING

A green building is a building that uses minimum natural resources for construction and operation and uses energy efficient devices to minimize energy consumption. It uses renewable energy sources, water and waste management and uses materials with high recycle content for construction.

Benefits of green building are following:

(*i*) 30% to 40% reduction in energy cost.

(*ii*) 20% to 30% reduction in water consumption.

(*iii*) Recycling of waste (*e.g.* use of flyash bricks).

(*iv*) Environmental friendly construction and operation.

Some features of green building are following:

(*i*) Use of renewable energy sources to meet energy requirement of the building.

(*ii*) Use of energy efficient air conditioning and lighting system.

(*iii*) Use of BMS (Building Management System) for efficient operation and control.

(*iv*) Segregation, collection and disposal of waste stream at source.

(*v*) Use of salvaged or refurbished material.

(*vi*) Use of low volatile organic compound (VOC) paints and adhesives.

(*vii*) Use of daylight to the maximum possible extent.

(*viii*) Rainwater harvesting.

(*ix*) Recycling of 100% grey water and treatment.

(*x*) Use of energy efficient motors and variable frequency devices for AHU (Air handling units of air conditioning system).

(*xi*) Use of energy efficient motors and variable frequency devices for AHU (Air handling units of air conditioning system).

(*xii*) Use of fly ash based cement.

(*xiii*) Use of autoclave aerated concrete blocks.

(*xiv*) Installation of carbon dioxide sensors to monitor indoor air quality.

(*xv*) Use of composite wood.

Green buildings are rated through leadership in energy and environmental design (LEED), a green building rating system by US green building council (USGBC). There are three ratings silver, gold and platinum. LEED GBC is a voluntary, consensus based US national standard for developing high performance and sustainable buildings. The first building in India to get platinum rating is CII Godrej green business centre in Hyderabad. Some of the green features incorporated in CII Godrej Green Business centre, Hyderabad, are listed below:

(*i*) Facility for charging electric automobile.

(*ii*) All paved parking and path ways are constructed with porous material so that rain water seeps into the ground and recharges underground aquifiers.

(*iii*) Rainwater recycling.

(*iv*) Recycling of 100% grey water.

(*v*) Solar orientation of building so as to receive maximum daylight.

(*vi*) Reduced lighting power density (in Watt/sq.ft) so as to achieve higher energy efficiency in lighting.

(*vii*) Daylight lighting dimmer control.

(*viii*) Recycle of construction waste material.

Concept of green building is catching up very fast as it not only gives economic benefits but also addresses ecological and environmental issues.

26.9 BUILDING MANAGEMENT SYSTEM (BMS)

All large buildings such as corporate office, mall, hospital, hotel, airport, appartment block *etc*. provide a variety of services like heating, ventilation and air-conditioning (HVAC), voice and data communication, lighting, power (normal and emergency), fire detection, monitoring and fire fighting systems, security and access control *etc*. Generally, all such services are controlled manually by specific personnel including maintenance. However, in a BMS all such operation and control functions are synchronized in a automated manner resulting in efficient utilization of available resources, reduced operation and maintenance cost and improved performance. So a BMS is an automated centralized control of all building services with a view to maximize performance, investment and operating cost.

Salient features of such buildings are following:

(*i*) Automatic control of lighting, ventilation and temperature inside building by tracking movement of human beings.

(*ii*) Distribution of electric power to loads as per demand and activation of emergency generators in case of power failure or excess demand.

(*iii*) Sensing, monitoring and control of fire.

(*iv*) Automatic lifts with sensors and memory.

(*v*) Safety and energy management systems.

(*vi*) Logging of data from various sub-systems so as to provide a base for detection of impending failure and for the purpose of scheduling of maintenance activities.

(*vii*) Reduced dependencey on human beings.

(*viii*) Simplified administration of operation and maintenance.

26.10 ENERGY AUDIT

As per Energy Conservation Act 2001, energy audit is defined as "the verification, monitoring and analysis of use of energy including submission of technical report containing recommendations for improving energy efficiency with cost benefit analysis and an action plan to reduce energy consumption."

Hence, energy audit is an effective tool to control energy wastages and ways and means to improve energy efficiency.

26.11 NEED FOR ENERGY AUDIT

The need for energy audit are following:

(*i*) To reduce energy consumption.

(*ii*) To improve energy efficiency.

(*iii*) To reduce operating cost.

(*iv*) To conserve depleting fossil fuel.

(*v*) To improve return on investment.

26.12 TYPES OF AUDIT

These are two types of energy audit viz.

(*i*) Preliminary audit.

(*ii*) Detailed audit.

(*i*) **Preliminary Audit:** Purpose of preliminary audit is to establish energy consumption in the organization and estimate scope for savings. It helps to identify areas which require no cost or very low cost for energy saving projects. This audit forms the basis for detailed audit.

(*ii*) **Detailed Audit:** This audit evaluates the all major energy using systems. It estimates energy savings and cost of implementing energy saving project. This audit consists of three distinct phase *viz.*

(*a*) Pre audit phase.

(*b*) Audit phase.

(*c*) Post-audit phase.

(*a*) *Pre-audit Phase:* This phase of audit involves basic data collection, site visit, familiarization with the plant process and maintenance practices, issue of questionnaire to each department, collection of design and operating data *etc.*

(*b*) *Audit Phase:* This phase may take several weeks to months depending upon the industry. Detailed studies are carried out to establish material and energy balance of each department. Readings are taken with the help of instruments to establish baseline data.

(*c*) *Post Audit Phase:* It involves firming up proposals for energy saving projects and their analysis/justification. All these are mentioned in the final audit report that is submitted to the management.

26.13 ENERGY MANAGEMENT SYSTEM

Purpose of energy management system is to produce goods and provide services at minimum cost and environmental impact. This begins with production and/or procurement of energy at optimum cost and its optimum utilization. This means improving energy efficiency at generation and utilization stage, preventing energy wastage and recycling and reusing energy (heat recovery).

Energy management system in an organization requires technical and managerial expertise. It begins with identification and establishment of strategic approach to energy management. Thereafter, an energy manager is recruited to look after energy management activities throughout the organization. He is the process owner and key driver in all activities related to energy management and is accountable to top management. The energy manager sets up energy monitoring and reporting system so as to collect and analyse data related to energy cost and consumption.

Next step is to conduct energy audit so as to identify scope for energy savings in the existing system. Thereafter, selected energy saving projects are implemented and monitored. The organizational staff are trained in energy management aspects so as to ensure wider participation and success.

All these are reviewed on annual basis and programs are chalked out for next 12 months.

26.14 ENERGY CONSERVATION ACT, 2001

Government of India has enacted the energy conservation Act, 2001. The act provides necessary legal framework for energy efficiency drive. Under the provisions of this Act, Bureau of Energy Efficiency has been established w.e.f 1st March, 2002. The act has given bureau of energy efficiency (BEE) five years before any prosecution can be launched against anyone who contravenes the provisions of the act.

The energy conservation program would target the industries in cement, pulp and paper, textile, fertilizer, chemical, chlor-alkali, aluminium, sugar, steel and iron and other energy intensive industries. The objective is to promote energy efficiency in industrial as well as commercial users by bench-marking their own levels of efficiency with the best in India:

The salient features of the Act are following:

(*i*) To order mandatory energy audits in specified industries and for designated consumer.

(*ii*) To prescribe energy consumption norms.

(*iii*) To specify standards, specify appliances and equipments to be subjected to labeling regime.

(*iv*) To prohibit manufacture, sale, import or deal in appliances and equipments not meeting standards and labeling requirements.

(*v*) To inspect and enforce compliance.

(*vi*) To direct any consumer to furnish energy related information.

(*vii*) To give direction for appointment of energy manager.

(*viii*) To prescribe energy conservation building codes.

(*ix*) To levy penalties.

(*x*) To certify energy auditors and energy managers.

(*xi*) To spread awarness, demonstrate pilot project and provide financing schemes for energy conservation.

(*xii*) To set up system for adjudication and appellate procedures.

List of energy intensive industries and other establishments specified as designated consumers are given below :

1. Aluminium.
2. Fertilizer.
3. Cement.
4. Pulp and paper.
5. Chlor-alkali.
6. Sugar.
7. Textile.
8. Chemicals.
9. Railways.
10. Port trust.
11. Transport sector.
12. Petrochemicals, refineries, gas crackers.
13. Thermal power stations, hydel power stations, electricity transmission and distribution companies.
14. Commercial buildings or establishments.

26.15 ENERGY SERVICE COMPANIES (ESCO)

ESCO are organizations whose business is to make energy efficiency better at customer's facility. They conduct energy audit, provide details of commercially viable efficiency improvement projects at customer's facility, provide innovative financing options to undertake energy efficiency projects, undertake implementation of project and guarantee a fixed return on investment. The financing scheme could be arranging soft loan from various energy agencies, on mutual profit sharing basis or on build, own, operate and transfer basis (BOOT).

This has opened a vast lucrative business opportunities for ESCOS as well as for host organizations. Both parties benefit from such association.

26.16 ENERGY LABELING OF APPLIANCES AND EQUIPMENTS

Under Energy Conservation Act, 2001, Bureau of Energy Efficiency (BEE) has begun the labeling of household appliances such as air conditioners, ceiling fans, refrigerators, desert coolers etc. Such labeling is voluntary as of now but will become mandatory later on.

The appliances are tested for energy efficiency and are labeled in stars. Five star indicates highest energy efficiency while one star indicates lowest energy efficiency. So, a customer can choose energy consumption of an appliance before buying it. This process of labeling will gradually be shifted to other appliances and equipments.

26.17 ROLE OF ENERGY MANAGER

As per Energy Conservation Act, 2001, certain designated consumers (declared by BEE) have to recruit Energy Manager to look after their energy management system and file energy returns (later on).

The role of Energy Manager are following:

(*i*) To plan effective utilization of energy usage.

(*ii*) To minimize energy wastage in any form (electricity, steam, fuel, water etc.)

(*iii*) To continuous improve energy efficiency by bench-marking with best practices.

(*iv*) To reuse and recycle energy.

(*v*) To examine techno-economic feasibility of energy conservation projects.

(*vi*) To implement energy conservation projects.

(*vii*) To raise awareness and educate employees towards energy conservation.

(*viii*) To maintain all relevant data and records.

(*ix*) To apprise and advice management on energy conservation aspects.

26.18 ENERGY AUDIT QUESTIONNAIRE

Energy audit questionnaire collects first hand information about the energy usage in an organization. Before embarking on energy audit, the first task is to prepare a questionnaire to collect as much data as possible about various aspects of energy usage. Although the exact questions will vary from organization to organization but a general form will help to prepare user specific questionnaire.

I. **Organizational Detail**

(*i*) Name of company ....

(*ii*) Location of plant ....

(*iii*) Name and designation of official responsible for energy management ....

(*iv*) Total number of employees ....

(*v*) Brief detail of past energy audit ....

II. **Sources of energy**

(*i*) What are various energy sources used in the plant?

(a) Coal, (b) electricity gas, (c) Liquid fuels, (d) others (Bagasse, agricultural waste, bio-gas *etc.*) (e) solar

(*ii*) What is the cost of power/fuel generated/purchased?

(a) Coal, (b) Gas, (c) Electricity, (d) Liquid fuel, (e) Others

(*iii*) What is the percentage energy mix?

(*iv*) What are high energy consumption areas?

III. **Energy Usage**

(*i*) What is the period of review of energy consumption?

(*ii*) What is procedure of review of energy consumption?

(a) By total cost, (b) Department wise, (c) Product wise, (d) By monthly bill.

(*iii*) What are various metering arrangements?

(*iv*) Is there any standard consumption for each process?

(*v*) Is the consumption compared with

(a) Previous period, (b) Other companies.

(*vi*) Has management set any energy consumption target?

(*vii*) What are the steps taken to recycle, reuse, reclaim energy from waste?

IV. **Consumption Record**

(*i*) Show energy consumption record of past five years.

(*ii*) Show total energy cost breakup and cost per unit of each fuel.

(*iii*) List standard consumption for each process and account for losses.

V. **House Keeping and Maintenance**

(*i*) How often control mechanisms and meters are tested and calibrated?

(*ii*) Do you carry out preventive and predictive maintenance?

(*iii*) If answer to above question is yes then list areas/equipments covered?

(*iv*) When was fuel storage and handling system was reviewed last?

(*v*) When was fuel/electricity tariff contract was reviewed vis-à-vis consumption?

(*vi*) Have you installed sub-meters to breakdown consumption into controllable units?

VI. **Personnel**

(*a*) Who is responsible for energy management in the plant?

(*b*) When was energy saving awareness campaign was launched last?

(*c*) Is there any energy management cell in the plant?

CASE STUDIES

Case Study-1

(A) **Title:** Replacement of conventional 36 watt tubelight ballast with electronic ballast in a Municipal Corporation maintained street lights.

(B) **Existing System:** 36,780 number of 36 watts tubelights with conventional ballast were used for street lighting purpose.

(C) **Proposed System:** Replacement of conventional ballast by electronic ballast.

(D) **Energy Saving Calculations**

(i) Power consumed by conventional ballast = 60.5 W

(ii) Power consumed by electronic ballast = 4.0 W

Contd...

(iii) Difference in power consumption = 20.5 W
(iv) Total no. of street lights = 36,780
(v) Power saving potential = 754 kw

(E) **Cost Benefits Analysis**

(i) Units saved = 30,27,310 kwH
(ii) Amount saved @ Rs. 3.30/kwH = Rs. 100 lakh
(iii) Investment required = Rs. 84.6 lakh
(iv) Payback period = 1 0 months

Case Study-2

(A) **Title:** Variable Frequency Drive for Naptha Pump

(B) **Existing System:** Naptha is fed through pump for process requirement

Rating of pump = 132 Kw
Design Data = 68 m^3/hr, 615 m head
Actual flow = 20 m^3/hr

Discharge valve is throttled to meet process requirement.

(C) **Proposed System:** Installation of variable frequency drive.

(D) **Energy Saving Calculation**

Considering 36% valve opening

(i) Power consumed = 88.4 kw
(ii) Calculated pump efficiency = 37%
(iii) Design pump efficiency = 70%
(iv) Saving in power with variable frequency drive = 43.3 kw

(E) **Cost Benefit Analysis**

(i) Amount saved by using VFD = Rs. 9.3 lakh
(ii) Cost of VFD = Rs. 8 lakh
(iii) Simple payback period = 10 months

Case Study-3

(A) **Title:** plugging air leakages in compressed air system

(B) **Existing system:** Two air compressors of 50 HP each were run to feed compressed air into the system. Loading/unloading test was performed under idle condition of plant.

Loading time.	= 59%
Unloading time.	= 41%

(C) **Proposed system:** Plugging air leakages

(D) Energy Saving Calculations

(i)	Power consumption.	= 38.7 kw
(ii)	Energy saving after switching off one compressor.	= 1,53,000 kwh

(E) **Cost Benefit Analysis**

(i)	Amount saved after switching off one compressor.	= Rs. 8.32 lakh
(ii)	Investment required.	= NIL

Case Study-4

(A) **Title:** Reduction of contract demand of electricity supplied by electricity board.

(B) **Existing System**

(i) Sectional contract demand = 16,500 KVA

(ii) 85% of contract demand = 14,025 KVA (Minimum payable)

(C) **Proposed System:** Reduction of contract demand as the actual demand never crosses existing sanctioned demand.

(D) **Energy Saving Calculations**

(i)	Proposed contract demand	= 15000 KVA
(ii)	Minimum billable demand	= 12,750 KVA
(iii)	Reduction in demand	= 5400 KVA

(E) **Cost Benefit Analysis**

(i)	Savings due to reduction in demand	= Rs. 11.1 lakh
(ii)	Investment required	= NIL

Case Study-5

(A) **Title:** Reduction of energy consumption of cooling water pumps in a fibre manufacturing plant.

(B) **Existing System:** Two cooling water pumps of 50 kW each are run. Differential temperature of cooling water is 1.5°C. The designed temperature difference is 8°C. The heat load is 5.43 x 10^5 kcal/hr. Flow of each Pump is 180 m^3/hr.

(C) **Proposed System:** To switch off one cooling water pump.

(D) **Energy Saving Calculations**

(i) New temperature difference (with one cooling water pump) = 3°C

(ii) Power saved = 49.4 kw

(iii) Units saved/annum = 4,03,000

(iv) Amount saved = Rs. 12.7 lakh

(E) **Cost Benefit Analysis**

(i) Investment required = NIL

Case Study-6

(A) **Title:** Reduction of energy consumption of glycol chiller in a dairy plant.

(B) **Existing System:** It consists of two glycol chillers of 100 TR capacity (110 kW). One brine pump is used for chilling processed milk.

(C) **Proposed System:** To run two brine pumps as the existing running pump is overloaded due to higher flow (106 %).

(D) **Energy Saving Calculation**

(i) Specific energy consumption of chiller = 0.79 kw/TR

(ii) Specific energy consumption of chiller-2 = 1.24 kw/TR

(iii) Designed specific energy consumption = 0.80 kw/TR

(iv) With brine pumps in operation, additional power consumption = 6.4 kW

(v) Improvement in chiller efficiency due to reduction in power consumption (because of improvement in specific energy consumption) = 28.1 kw

(E) **Cost Benefit Analysis**

(i) Saving in power = 21.7 kw

(ii) Investment required = NIL

Summary

Energy is a driver for economic growth of a country. However, due to depleted fossile fuels, oil prices are sky-rocketing. With a growing economy like India, consumption of fossile fuels are on rise. This poses severe economic burden as 80% of oil is purchased from the international market. Hence, energy conservation assumes very high importance.

India has been focusing on development of renewable energy sources notably wind, solar, bio-mass and hydro. However, development of these sources is slow and is not keeping pace with the rising demand.

Out of all sectors of Indian economy, industrial sector consumes almost half of the total commercial energy. Hence, energy saving potential are more in this sector.

Various energy intensive areas of industry are — boilers, furnaces, steam system, co-generation system, cooling tower, compressed air, electrical system *etc.* Various areas of energy conservation have been discussed in detail.

There are new generation of industrial equipments based on energy efficient technology which improves the efficiency of energy usage. Some of these technologies are: Amorphous metal transformer, fluidized bed combustion boilers, heat wheel, thermo-compressor, heat pipe *etc.*

Apart from industrial sector, buildings in domestic as well as in commercial sector consume a large amount of energy. So concept of green buildings is discussed. It not only reduces energy consumption but also minimizes ecological footprint.

For any energy conservation effort, energy audit forms the foundation. Various methods of energy audit are discussed including a questionnaire.

Lastly, Energy Conservation Act, 2001, makes it mandatory for certain category of Industrial consumers and appliance manufacturers to adhere to energy efficiency norms. The act also makes energy audit mandatory.

Review Questions

1. Explain the importance of energy conservation for a growing economy like India.
2. What is the difference between energy conservation and energy efficiency improvement?

3. Discuss current status of use of alternate sources of energy in India.
4. Discuss the scope of energy conservation in industries and its impact on economy.
5. Discuss scope of energy conservation in industrial furnaces. What are the focus areas?
6. Discuss key areas for energy conservation in steam generation and consumption.
7. Explain why conservation of water is necessary? How can you conserve water in boilers and cooling towers?
8. Write a brief note on energy efficient technologies.
9. What is a green building? How does it help in saving various resources?
10. Discuss the need for energy audit. Describe energy management system.

Discussion Questions

1. "Developing economies use energy more judiciously than developed economies". Comment.
2. "Energy intensity is an indicator of economic development of a country". Justify.
3. "Use of judicious mix of alternate source of energy and fossil fuels will help to mitigate energy crisis in the current scenario". Elaborate.
4. "Energy audit is the first step in exploring energy saving opportunities". Discuss.
5. "Ownership cost of green building is much less than the conventional building". Justify.
6. "A lot of energy can be easily saved by just cutting down wastages". Comment.
7. "Cost of energy produced through non-conventional energy sources are more than that produced from fossile fuels". Do you agree?
8. "Due to acute scarcity of water, its conservation is as important as energy." Explain.
9. "Incentives given by government to generation of non-conventional energy sources is grossly misused." Comment.
10. "Demand side energy management is best way to curb rising energy demand." Justify.

Multiple Choice Questions

1. Greenhouse effect refers to
   (*a*) Trapping of heat in earth's surface.
   (*b*) Development of greenery.
   (*c*) Reduction of generation of $CO_2$and SO_2.
   (*d*) None of the above.
2. Which of the following is not alternate source of energy.
   (*a*) Solar. (*b*) Wind.
   (*c*) Tidal. (*d*) Coal based thermal.
3. Energy efficiency means
   (*a*) Energy conservation. (*b*) Efficient energy utilization.
   (*c*) Energy wastage reduction. (*d*) None of these.
4. Building management system helps to
   (*a*) Reduce power consumption. (*b*) Improve automation.
   (*c*) Proper logging of all key parameters.
   (*d*) All of the above.
5. Purpose of energy audit is to
   (*a*) Identify sources of energy wastages.
   (*b*) Reduce energy consumption.
   (*c*) Fix responsibility for energy wastages.
   (*d*) None of the above.

True/False Questions

1. Non-conventional energy is expensive compared to conventional sources.
2. Energy conservation act does not apply to railways.
3. Rating of green buildings indicates its capacity to save energy.
4. Solar energy is cheaper than hydel energy.

ANSWERS

Multiple Choice Questions

1. (*a*) **2.** (*d*) **3.** (*b*) **4.** (*d*) **5.** (*d*)

True/False Questions

1. True **2.** False **3.** True **4.** False

❑❑❑

Glossary of Key Terms

1. **Alternator—**Rotating power source producing A.C. Voltage.
2. **Asset—**Physical resources of an organisation such as buildings, plant, machinery, fixtures, furniture *etc.*
3. **Autonomous Maintenance—**Maintenance performed by operation staff.
4. **Availability—**Proportion of total time an equipment is capable of performing its intended function satisfactorily.
5. **Bar chart—**Also called gantt chart. It is a format of scheduled activities showing duration and sequence of activities.
6. **Benchmarking—**Process of comparing performance with other organisations on select metrics.
7. **Breakdown—**A type of failure when an equipment is out of operation.
8. **Budget—**A proposal that financial allocates resources.
9. **Built-in test equipment—**Diagnostic devices contained in the equipment to assist in troubleshooting.
10. **Bulk-density—**Weight of a substance per unit volume.
11. **Buffer stock—**Also called reserve stock. It is a safety cushion to prevent stock-out situation.
12. **Bid—**An invitation to supply goods or services.
13. **Cannibalisation—**A term used in spare parts management meaning using spare parts from similar non-functional machines to make another machine functional.
14. **Capital equipment—**Items with long life or high value such as Boilers, Turbine, Furnace, Reactor vessel *etc.*
15. **CMMS—**A computerised system to help in effective and efficient operation of maintenance.

16. **Condition monitoring**—Use of specialised instruments to measure condition of equipment *e.g.* vibration measurement, temperature measurement wear-debris analysis *etc.*
17. **Condition based maintenance**—A maintenance strategy based on measuring condition of equipment in order to assess whether it will fail in near future or not.
18. **Component**—A replaceable part of an equipment.
19. **Consumable**—Items that are used during operation and maintenance process such as nut-bolts, lubricants, cotton waste, gasket material, V-belts, Fuel, energy paper *etc.* These are low value items
20. **Commercial energy**—Energy available for sale *e.g.* electricity produced by utilities for sale to industrial, commercial and residential users.
21. **Corrective maintenance**—Any maintenance carried out to rectify a fault.
22. **Depreciation**—A gradual decrease in net worth of an equipment over a period of time.
23. **Derating**—Using a higher rating item for lower rating application.
24. **Downsizing**—Reduction in number of employees by management decision of retrenchment.
25. **Downtime**—Time that an equipment is out of service due to failure to perform.
26. **Economic life**—Expected service life of an equipment during which it is cheaper to operate and maintain it rather than replace it.
27. **Expert system**—A software based system having ability to take decisions based on rules or experience parameters.
28. **Ex-stock**—An item that is immediately available in the market.
29. **Failure**—Inability of an equipment to perform its intended function.
30. **Failure mode**—Event which leads to failure.
31. **Failure pattern**—Relationship between probability of failure and its age.
32. **Fault**—A defect.
33. **Flash steam**—When condensate (very hot water) is released from high pressure to low pressure, it forms low pressure steam.
34. **Flow chart**—A diagram that uses symbols to show stages in a process.
35. **Gemba Kaizon**—A Japanese concept of continuous improvement.

36. **Hazard**—A set of conditions that could lead to an accident with harmful consequences.
37. **Human factor**—Study of human capabilities and limitations in work place.
38. **ISO-9000**—An international quality process auditing program based on a series of standards published by International Organisation, for standardisations (ISO).
39. **ISO-14000**—International Environmental Management system developed by ISO. It is related to environmental policy, operation, audit, training *etc*.
40. **Insurance spare**—A very high value, long lead time, non-moving spares which may or may not be required during life time of an equipment.
41. **Insulation class**—A categorisation of various types of electrical insulation based on temperature rise e.*g*. class-A (105°C), class B (130°C) *etc*.
42. **Key performance indicators**—A select number of key measures that enables measurement of performance against target.
43. **Lead time**—Time elapsed between placement of order for an item and its receipt.
44. **Lean manufacturing**—Systematic identification and elimination of waste to reduce operating cost.
45. **Life cycle lost**—Estimation of total cost of ownership.
46. **Lubricant**—A material that reduces friction such as oil, grease *etc*.
47. **Maintainability**—Characteristics of design or installation that allows maintenance actions to perform easily, in a given time.
48. **Maintenance audit**—It is a process of verifying actual status of maintenance function in order to improve it.
49. **Maintenance engineering**—Engineering techniques used to design and improve performance of equipments or subsystems.
50. **Maintenance policy**—A statement of principles used to guide maintenance management decision.
51. **Maintenance schedule**—A list of planned maintenance activities to be performed during a specified time.
52. **Maintenance strategy**—A plan covering various aspects of maintenance management.
53. **Make or buy**—A decision in which it is decided whether to manufacture an item in-house or to buy from outside.

54. **Modification**—An activity carried out on an equipment to enhance its performance or remove a bottleneck.
55. **Non-destructive testing**—Testing of equipment which does not destroy it.
56. **Obsolescence**—Decrease in value of items due to availability of superior products.
57. **Order quantity**—Number of items demanded from a supplier.
58. **Original equipment manufacturer**—An organisation manufacturing and selling products under its own brand name.
59. **Out sourcing**—Shifting of maintenance job to an outside agency.
60. **Overhaul**—A detailed examination and restoration of an asset.
61. **Primary energy**—Energy produced from fuels (cool, oil, gas etc.).
62. **Purchase order**—A document released to an external organisation for supply of specific material or services.
63. **Quick-Couplers**—A type of device that joins two flexible hoses in a short time.
64. **Raw material**—Input with which a product is manufactured.
65. **Redundancy**—Having one or more standby equipments.
66. **Refurbish**—Clean and replace worn out parts in a machinery to enhance its operational life.
67. **Re-order point**—Ordering of more items when stock reaches a preset low level.
68. **Repair**—Restoration of a machinery/equipment to an acceptable level of performance.
69. **Risk**—Chance of facing an unwanted incident.
70. **Safety Stock**—Additional quantity of an item in additions to normal supplies required to be maintained for continuity of supply.
71. **Secondary energy**—Energy contained by a medium obtained from primary energy source *e.g.* energy possessed by steam produced by burning fossil fuels.
72. **Shutdown**—Planned stoppage of an equipment/sub-system/plant for maintenance.
73. **Spare part**—A component of an equipment/machinery which returns it to functional state.

74. **Standardisation**—Process of reducing variety of items to bring in uniformity.
75. **Standard operating procedure (Sop)**—Prescribed procedures to be followed for performance in a given situation.
76. **Stock-out**—Condition of having zero availability of an item.
77. **Tan-delta**—A parameter used in determining the health of an electrical insulating material.
78. **Tender**—A document issued by an organisation giving details of items to be procured or services required.
79. **Tools and tackles**—It includes a broad category of items used for reducing maintenance effort *e.g.* pulleys, winches, spanners, hammers etc. It could be specially fabricated item as well.
80. **TS-9000**—A quality certification system for automobile ancilliary suppliers.
81. **Useful economic life**—Duration for which an equipment remains in service till it can be cheaply replaced by a new equipment.
82. **Workload**—Number of manhours required to carry out a task.
83. **Workorder**—A document that details type of job to be done.

❑❑❑

Professional Reference Tables

Table A – Characteristics of Oils

Parameter	Unit	Naptha	HSD	LDO	Fuel Oil		
					LSHS	HPS	HFO
Density	gm/ml	0.66–0.75	0.78-0.8	0.85–0.90	0.90–0.96	0.96	0.95
Pour point	°C	—	Max + 6	12–18	57	72	68
Flash point	°C	23	32	66	75	75	66
Kinematic Viscosity (38°C)	Cst	0.5–0.7	2.0-7.5	2.5–15.7	180 at 50°C	500 at 50°C	370 at 50°C
Gross calorific Value	Kcal /Kg	10,200	10.500	10,800	11,000	9,500	11,000
Water content by volume (max)	%age	0.5	0.2	0.25	1.0	1.0	10.

Table B – Characteristics of Thermic Fluids

Properties	Unit	Petroleum oils	Therminol 66	Marlotherm-s	Dowtherm-A
Density	**Kg/m³**				
20°C		880	1005.5	1029	1059
100°C		830	952.7	974	995
200°C		765	884.6	904	907
300°C		700	816.6	834	907
Kirematic Vicosity	Centi-stoke				
20°C		50.17	119.34	39.94	2.83
100°C		5.31	4.02	3.0	1.01
200°C		1.33	0.99	0.95	0.44
300°C		0.6	0.52	0.51	0.27

Contd...

Properties	Unit	Petroleum oils	Therminol 66	Marlotherm-s	Dowtherm-A
Specific Heat	Kcal/ Kg.-C°				
25°C		0.47	0.36	0.38	0.38
100°C		0.53	0.39	0.44	0.43
300°C		0.74	0.49	0.61	0.57
Freezing Point	°C	—	–26.1	–35.0	12.0
Flash point	°C	204.0	177.0	190.0	124.0

Table C – Mineral Wool Insulation Thickness in mm for Different Pipesizes and Different Fluid Temperature

Pipe dia. (NB) MM	Fluid Temperature°C					
	90	150	200	260	315	375
25	25	40	65	65	75	90
40	25	40	65	65	75	90
50	25	40	65	65	75	90
80	25	40	65	65	75	90
100	25	50	75	75	100	100
150	40	50	75	75	100	115
200	40	50	75	75	100	115
250	40	65	90	90	100	125
300	40	65	90	90	100	125

Note:

1. Mineral wool is assumed to have density of 100–200 kg/m^3 and thermal conductivity of 0.036 kcal/m hr°C.
2. Surface temperature after insulation is 55- 60°C.
3. Insulation upto 65 mm is single layer and above this multi layer.

Table D – Glass Wool Insulation Thickness in mm for Different Pipe Sizes and Different Fluid Temperatures

Pipe dia (NB) MM	Fluid Temperature°C					
	90	150	200	260	315	375
25	25	25	25	25	40	40
40	25	25	25	25	50	50
50	25	25	25	25	50	50
80	25	25	40	40	50	65
100	25	25	40	40	50	65
150	25	25	40	40	65	75
200	25	40	40	40	65	75
250	40	40	40	40	65	75
300	40	40	50	50	65	75

Note:

1. Glass wool is assumed to have a density of 80-100 kg/m^3 and thermal conductivity of 0.045 Kcal/mhr°C.
2. Surface temperature after insulation is 55-60°C
3. Insulation upto 65 mm is single layer and above this multi layer.

Table E – Characteristics of Thermal Insulating Material (50 –1000°C)

Insulation	Type	Density (kg/m³)	Thermal conductivity (kcal/ mhr°c)	Limiting temp °C
Asbestos	Fibrous	80–250	0.0 42 at 40 °c	600
Glass fiber	Fibrous	12–80	0.032 at 40°c to 0.051 at 50 °c	400
Rock wool and slag wool	Fibrous	40–250	0.032 at 40 °c to 0.045 at 110°c	800
Calcium silicate	Granular	240	0.051 at 40°c	850
Magnesia	Granular	200	0.05 at 40°c to 0.065 at 200°c	300
Silica	Fibrous	50–150	0.033 at 40°c 0.07 at 250 °c 0.081 at 250 °c	1000
Alumino	Fibrous	50–250	0.028 to 0.071	1200
Silicate	Granular	500–800	–	1200
Vermi culate	Granular	50–500	0.073 at 40°c 0.093 at 250°c	1100

Table F – Types of Thermocouples

Type	+ve wire	–ve wire	Useful range °C	Limit of error for temperature in °C%
J	Iron	Constantan	0 to 750	2.2°c/0.75%
K	Chromel	Alumel	–200 to 1200	2.2°c/0.75%
T	Copper	Constantan	–200 to 250	1.0°c/0.75%
E	Chromel	Constantan	–200 to 900	1.7°c/0.50%
R	87%.Pt+13%.Rh	100%Pt	0 to 1450	1.5°c/0.25%
S	90%.Pt+10%.Rh	100%Pt	0 to 1450	1.5°c/0.25%
B	94%.Pt+6%.Rh	70%.Pt+30%.Rh	0 to 1700	0.5%

Table G – Recommended Water Characteristics for Water Tube Boilers (Feed Water)

Pressure at Boiler outlet (bar)	20	40	60	80	100	120	130
Parameter		Feed	Water				
Total Hardness, mg/lit as $CaCO_3$, Max	10	2	0.5	ND	ND	ND	ND
pH Value	8.5–9.5	8.5–9.5	8.5–9.5	8.5–9.5	8.5–9.5	8.5–9.5	8.5–9.5
Oxygen, mg/lit. as O_2 Max	0.05	0.02	0.01	0.01	0.005	0.005	0.005
(Fe + Cu + Ni), mg/lit Max	–	–	0.02	0.02	0.01	0.01	0.01

Note: ND– Not Detectable

Table H – Recommended Water Characteristics for Water Tube Boilers (Boiler Water)

Pressure at boiler outlet (bar) →	20	40	60	80	100	120	130
Parameter			BOILER	WATER			
Sodium phosphate as Na_3PO_4, mg/lit.	50-100	30-70	20-50	10-40	5-20	3-10	#
Caustic Alkalinity in terms of $CaCO_3$, max in mg/lit	300	150	60	30	10	5	2
Total alkalinity in terms of $CaCO_3$, max mg/lit	700	500	300	200	100	40	10
Silica as SiO_2 max, mg/lit	0.4 of caustic alkalinity	–	20	10	#	#	#
Sodium Sulphite as Na_2SO_2, mg/lit	30–50	20–40	15–30	10–20	#	#	#
Hydrazine as N_2H_4, mg/lit	0.1–1.0	0.1–0.5	0.05–0.3	0.05–0.1	#	#	#
Dissolved Solids, max, mg/lit	3000	2000	1200	700	350	100	15
Chloride as Cl, max, mg/lit	–	–	–	–	10	5	1
Suspended Solids, max, mg/lit	200	50	–	–	–	–	–

Note: # Depends on circumstances.

Table I – Pipeline Capacities at Specific Velocities

Pressure bar	Velocity m/s.	Flow rate – Kg/hr 15 mm	25 mm	40 mm	50 mm	65 mm	80 mm	100 mm	150 mm
0.4	15	7	24	37	99	145	213	394	917
	25	10	40	62	162	265	384	675	1457
	40	17	64	102	265	403	576	1037	2303
1.0	15	8	29	65	112	182	260	470	1020
	25	12	48	100	193	300	445	730	1660
	40	19	71	172	311	465	640	1150	2500
2.0	15	12	45	100	182	280	410	715	1580
	25	19	70	162	295	428	656	1215	2530
	40	30	115	275	475	745	1010	1895	4175
3.0	15	16	60	127	245	385	535	925	2040
	25	26	100	225	425	632	910	1580	3440
	40	41	157	357	595	1025	1460	2540	5940
4.0	15	19	70	156	281	432	635	1166	2460
	25	30	115	270	450	742	1080	1980	4225
	40	49	197	456	796	1247	1825	3120	7050
5.0	15	22	87	187	352	526	770	1295	2833
	25	36	135	308	548	885	1265	2110	5150
	40	59	225	495	855	1350	1890	3510	7870
6.0	15	26	105	225	425	632	925	1555	3400
	25	43	162	370	658	1065	1520	2530	6175
	40	71	270	595	1025	1620	2270	4210	9445
7.0	15	29	110	260	445	705	952	1815	3990
	25	49	190	450	785	1205	1750	3025	6900
	40	76	303	690	1210	1865	2520	4585	10880
8.0	15	32	126	285	475	800	1125	1990	4540
	25	54	205	465	810	1260	1870	3240	7120
	40	84	327	730	1370	2065	3120	5135	12470
10.0	15	41	155	372	626	1012	1465	2595	5860
	25	66	257	562	990	1530	2205	3825	8995
	40	104	408	910	1635	2545	3600	6230	14390

Table J – World Health Organisation (WHO) Drinking Water Standard

Parameter	Limit (mg/lit)
Total dissolved solids	500
Colour	5 units platinum – cobalt scale
Turbidity	5 units
Iron	0.3
Managanese	0.1
Copper	1.0
Zinc	5.0
Calcium	75
Magnesium	50
Sulphate (SO_4)	200
Chloride (Cl)	200
PH range	7 to 8.5
$Mg+Na_2SO_4$	500

Table K – Characteristics of Lamps

Lamp type	Efficacy (lm/watt)	Life (hrs)	Wattage (w)	Color Rendering Index
GLS	12-18	1000	25 – 1000	100
Fluorescent tube	60	5000	20/40	67–77
CFL	55–65	8000	7–36	85
High Pressure Mercury Vapour lamp	58	5000	80–1000	45
High Pressure Sodium Vapour lamp	145	15000	50–1000	25-60
Low Pressure Sodium Vapour lamp	200	Sensitive	18–180	0
Metal Halide	85	7500	Upto 1800	70
Halogen	22	2000	300/500/1000	100

Table L – Selection Table for Fuses/Switchs/ Cables for DOL Motors (3 Phase)

HP at 415 V, 50 hz	Full load Current (A)	HRC Fuses (A)	Switch (A)	Cable Size (mm^2) Aluminium
0.5	1.2	6	25	2.5
0.75	1.6	6	25	2.5
1	1.8	6	25	2.5
1.5	2.6	10	25	2.5
2	3.5	15	25	2.5
3	5	15	25	2.5
5	7.5	20	25	4
7.5	11	25	25	4
10	14	35	63	6
12.5	18	50	63	6
15	21	50	63	10
20	28	63	63	16
25	35	80	100	25
30	40	100	100	25
40	55	120	100	35
50	65	160	250	50
60	80	200	250	70
75	100	200	250	95
100	135	320	400	185
150	200	350	400	400

Table M – Selection Table for Fuses/ Switches/ Cables for Star Delta Motors (3 Phase)

HP 415V, 50 hz'	Full load line current (A)	Phase current (A)	HRC fuses (A)	Switch (A)	Supply side cable (mm^2)	Motor side cable (mm^2)
3	5	2.9	16	25	2.5	2.5
5	7.5	4.3	20	25	2.5	2.5
7.5	11	6.3	25	25	2.5	2.5
10	14	8.1	25	25	4	2.5
12.5	18	10.4	35	63	4	2.5
15	21	12.1	35	63	6	2.5
20	28	16.2	63	63	10	4
25	35	20.2	63	63	16	6
30	40	23	63	100	16	6
40	53	30.6	100	100	25	16
50	63	38.2	100	100	35	16
60	80	46.2	120	250	50	25
75	100	57.5	160	250	70	35
100	135	77.9	200	250	95	50
125	165	95	250	250	120	70
150	200	115	250	250	185	70
175	230	132.8	320	400	225	120
200	275	159	350	400	400	150
250	325	187	350	400	400	185
275	360	208	400	400	400	185
300	385	222	400	400	500	225
400	500	289	500	630	625	400

Table N – Current Rating of PVC Cables (Aluminium)

Cable size mm^2	Current Rating (A)							
	3&3½ core, 1.1kV grade		3 & 3½ core 3.3 kV		3 & 3½ core, 6.6 kV		3 & 3½ core, 11 kV	
	Ground	Air	Ground	Air	Ground	Air	Ground	Air
2.5	21	18	-	-	-	-	-	-
4	28	23	-	-	-	-	-	-
6	35	30	33	29	33	29	-	-
10	46	40	43	39	43	39	-	-
16	60	51	56	52	56	52	-	-
25	76	70	95	90	93	106	93	106
35	92	86	115	110	111	123	111	123
50	110	105	140	135	130	153	130	153
70	135	130	170	170	160	182	160	182
95	165	155	200	205	191	221	191	221
120	185	180	225	240	217	254	217	254
150	210	205	255	275	243	291	243	291
185	235	240	285	315	273	330	273	330
240	275	280	325	370	317	390	317	390
300	355	365	370	430	357	450	357	450
400	335	375	435	490	408	525	408	525

Standard Conditions of Cable Lying

(*i*) Maximum Conductor Temperature → 90 °C

(*ii*) Ambient Air Temperature → 40°C

(*iii*) Ground Temperature → 30°C

(*iv*) Thermal Resistivity of Soil → 150°C cm/watt

(*v*) Depth of Laying (for 1.1kV) → 75 cm

(For 3.3 kV to 11kV) → 90 cm

Table O – Derating Factor for Variation in Temperature

Temperature	Ground	Air
15°c	1.12	–
20°c	1.08	–
25°c	1.04	1.14
30°c	1.00	1.10
35°c	0.96	1.05
40°c	0.91	1.00
45°c	0.87	0.95
50°c	0.82	0.89

Table P – Output of Common Thermo-couple types in Micro volts (Assuming 0°C reference temperature)

Temp (°C)→ Type of thermo-couple	–50	–25	0	25	50	100	150	200	400	500
(I) K (Nicr/NiAl)	-1889	-968	0	1000	2022	4095	6137	8137	16395	20640
(ii) T (Cu/con)	-1819	-940	0	992	2035	4277	6702	9286	20869	—
(iii) J (Fe/con)	-2431	-1239	0	1277	2585	5268	8008	10777	21846	27388
(iv) R (Pt/Pt-13% Rh)	-226	-123	0	141	296	647	1041	1468	3407	4471

❑❑❑

References

1. Apte, S.S., "*Failure and Hazard Analysis*," 8th Advanced Maintenance Management Program for Senior Executives, 24-28 June, 1995, IIPE, N.Delhi.
2. Apte, S.S., "*Spare Parts Inventory Control Models*," Workshop proceedings on "Spare Parts Management – A Tool for Cost Control," 5-8 Nov. 2003, IIPE, N. Delhi.
3. Apte, S.S., "*Subversive Analysis Applied to Design for Maintainability*", 21st Advanced Maintenance Management Program, 2-4 July 2009, Hyderabad, IIPE.
4. Apte, S.S, "*Economics of Overhaul*", 18th Advanced Maintenance Management Program, 3-6 Nov. 2004, IIPE, N. Delhi.
5. Anderson R.T. and Lewis, N., "*Reliability Centered Maintenance Management and Engineering Method*, McGraw Hill, 1990.
6. Anthony, K., "*Maintenance Planning and Control*", East West Press Pvt. Ltd, New Delhi, 1991.
7. Anthony Kelly, "*Maintenance Planning and Control*" Butterworth, 1984.
8. Alexandrov, "*Material Handling Equipment*," MIR Publications, 1981.
9. Bagadia, Krishan, "*Micro Computer Aided Maintenance Management*" Marcel Dekker Inc., New York, 1987.
10. Balaguruswamy, E., "*Reliability Engineering*, "Tata McGrawHill, New Delhi, 1984.
11. Bhadury, Bikash, "*Principle and Practice of Total Productive Maintenance*", Allied Publishers Ltd, India.
12. Bhadury, Bikash, "*Total Productive Maintenance*, 'Allied Publishes Ltd. New Delhi 1998.
13. Bhadury, B. and Basu, S.K., "*Terrotechnology, Reliability Engineering and Maintenance Management*, "Asian Books Pvt., New Delhi.

14. Baily P. and Farmer D., "*Purchasing Principles and Management*, "Pitman Publishing, 7th edition.
15. Berk, J. and Berk, S., "*Total Quality Management: Implementing Continuous Improvement*," Excel Books, New Delhi.
16. Bhatia, Manish, "*Dense Phase Pneumatic Conveying*," International Seminar on Material Handling System, 16- 18 Nov 2000, IIPE, New Delhi.
17. Blanchard, B.S., Verma, D. and Patterson, E.L., "*Maintainability*," John Wiley and Sons, New York, U.S.A.
18. Black, Rolland P. "*Industrial Safety*," Prentice Hall of India Pvt. Ltd., New Delhi, 1973.
19. Blanchart, B.S., Verma D. and Patterson, E.L., "*Maintainability: A Key to Effective Serviceability and Maintenance Management*," John Wiley and Sons Inc.
20. Briggs, M. and Atkinson, C., "*Strategies for Effective Maintenance*", Cromwell Press, U.K, 2000.
21. Broadcast Engg. Society (India), "*Financial Rules and Operational Procedures* (FR and OP), 2008.
22. Buffa, E.S., "*Production/Operation Management*", John Wiley, 1984.
23. Brian, Rothery, "ISO 14000 and ISO 9000", Gower Publication, 1995.
24. Campbell, J.D., "*Strategies for Excellence in Maintenance Management*", Productivity Press.
25. Collacott, R.A., "*Mechanical Fault Diagnosis and Condition Monitoring*," Chapman and Hall, London, 1977.
26. Condra, L.W., "*Reliability Improvement with Design of Experiments*," Marcel Dekker, New York, 1993.
27. Corder, A.S., "*Maintenance Management Techniques*," McGraw, 1976.
28. Das, S.C. "*Total Productive Analysis: An Innovative Approach to Maintenance Management*," 12th Advanced Maintenance Management Program, 19-23 Jan 1999, IIPE, New Delhi.
29. Deb, Tanmoy, "*A case Study of Abnormal Vibration of* 13.5 *MW Alternator*," Journal of Plant Engg., Jan-Mar, 2002, IIPE, New Delhi.
30. Deb, Tanmoy, "*A Case Study of Frequent Burning of Dense Ash Bagging Motor*," Journal of Plant Engg., July-Sept, 2005, IIPE, New Delhi.

31. Deb Tanmoy, "*A Textbook of Installation, Testing, Commissioning and Maintenance of Electrical Apparatus*," Dhanpat Rai and Co., 2007.

32. Deb, Tapomoy, "*Human Resource Development*", Ane Books Pvt Ltd; 2006.

33. Deb, Tapomoy, "*Training and Development—Concepts and Applications*", Ane Books Pvt. Ltd., 2008.

34. Detoro, I.J. and Tanner, A.R., "*Total Quality Management: Three Steps to Continuous Improvement*," Addition Wesley Publishing Co.

35. Davies, A. (editor), "*Handbook of Condition Monitoring Techniques and Method*", Chapman and Hall, 1998.

36. Dhillon, "*Engineering Maintainability: How to Design for Reliability and Easy Maintenance*," Prentice Hall of India.

37. Duffupa, S.O., Rouf, A. and Campbell J.D., "*Planning and Control of Maintenance Systems: Modeling and Analysis*", John Wiley, 1999.

38. Ebeling, C.E. "*An Introduction to Reliability and Maintainability Engineering*," McGraw Hill Company Inc, New York, 1977.

39. Edward, Hartmann, "*Maintenance Management*," Industrial Engineering and Management Press, 1987.

40. Finely, H.F., *Modern Maintenance Management System*," The Howard Finely Corporation, Houston, 1981.

41. Frank, Herbaty, "*Cost Effective Maintenance Management*," Noyce Publication, U.S.A, 1987.

42. Gopal Krishnan P., and Banerji A.K., "*Maintenance and Spare Parts Management*," Prentice Hall of India, 1991.

43. Gopal Krishnan P., and Sudarshan M., "*Materials Management – An Integrated Approach*", Prentice Hall of India, 1996.

44. Govil O.K., "*Reliability Engineering*," Tata McGraw Hill, New Delhi, 1983.

45. Goto F., "*Autonomous Maintenance*," Chapter 4 in Nakejima S. (Editor), TPM Development Program: Implementing Total Productive Maintenance, Productivity Press, Cambridge, Massachusettes, U.S.A, 1989.

46. Gulati, Vinod and Roy S.K., "*An Automated Material Management System- Experience in NHPC*", Workshop Proceedings on Spare Parts Management a Tool for Cost Control, 5-8 Nov, 2003, IIPE.

47. Hartman, Edward, "*Maintenance Management*," Industrial Engineering and Management Press, 1987.
48. Higgins, L.R., "*Maintenance Engineering Handbook*," McGraw Hill, New York, 1988.
49. Hirani, Mohan, "*Computer Aided Maintenance Management System in Core Sector*", Advanced Maintenance Management Program, IIPE, New Delhi, 1994.
50. Jardine, A.K.S. (Editor), "*Operational Research in Maintenance*", Manchesters University Press, Manchester, England/New York, 1970, Pp. 72-79.
51. Jardine, A.K.S., "*Maintenance Replacement and Reliability*," Pitman Publishing, London, 1973, Pp 31-34.
52. Joel. L., "*The Handbook of Maintenance Management*," Industrial Press Inc., New York 1997.
53. Joel, L., "*Internet Guide for Maintenance Management*," Industrial Press Inc., New York, 1998.
54. Juran, J.M. and Gryna, F.M. Jr., "*Quality Planning and Analysis: from Product Development through Usage*," McGraw Hill Book Co., New York, 1970.
55. Karkada, N.A. and Trasi, G.G., "*Reconditioning Techniques for Reducing Downtime*," 6th Advanced Maintenance Management Program, 1994, IIPE, New Delhi.
56. Khartam, N.A., Al-Diahane, T.G., Al-Bahar, J.F., "*Professional Project Management Practices in Kuwait – Issues, Difficulties and Recommendation*," International Journal of Project Management, 2000, 18(4), Pp 96-281.
57. Khatana, A.A, "*Energy Optimization and Efficiency Improvement in Power Distribution System*," National Seminar on Energy Optimization and Efficiency Improvement in Infrastructural and Industrial Sector, 29-30 Oct, 1998. IIPE, New Delhi.
58. Kegal, J.A, "*Role of Total Productivity Maintenance in Power Plant Performance*," 21st Advanced Maintenance Management Program, 2-4, July 2009, IIPE.
59. Kelly, A., "*Topics in Terotechnology—An Instructional Series-10, Motivation of Maintenance Trade Force*," Maintenance Management International, Vol. 4 No. 2, Jan 1984, pp 71-81.

60. Kog, Y.C., Chua, D.K.D., Loh, P.K., Jaselskies, E.J., "*Key Determinants for Construction Schedule Performance*", International Journal of Project Management, 1999, 17(6).

61. Krishna, Vijay (Maj. Gen.), "*Industrial Accidents: Courses, Cost Analysis and Prevention,*" 12th Advanced Maintenance Management Program, 19-23 Jan, 1999, IIPE, New Delhi.

62. Krishnan, I. (Maj.Gen.), "*Design of Maintenance System,*" 7th Advanced Maintenance Management Program, 1994, IIPE, New Delhi.

63. Krishnan, N.V., "*Safety Management in Industry,*" Jaico. Publishing House, Bombay. 1997.

64. Levitt, J., "*Managing Factory Maintenance,*" Industrial Press Inc., 1996.

65. Lee, L. and Cobble, D.W., "*Purchasing and Materials Management,* McGraw Hill, 1977.

66. Lindley, Higgins, R. and Mobley, R., "*Maintenance Engineers Handbook*", McGraw Hill, 6th edition.

67. Maheshwari, S.N. and Mittal, S.N., "*Cost Accounting – Theory and Problem,*" Mahabir Publication, 1992, New Delhi.

68. Majumdar, Milli, "Green Building Design – An Integrated Design Process to Reduce Energy Demand and Maximize Energy Efficiency," Air Conditioning and Refrigeration Journal, Pp 55-62, Vol. 7, no. 2 Apl-June 2004.

69. Maubray, J., "*Reliability Centered Maintenance,*" Butterworth Heinemann, Oxford, 1991.

70. Mishra and Pathak, "*Maintenance Engineering and Management,*" Prentice Hall of India, 2001.

71. Mobley, R.K., "*An Introduction to Predicative Maintenance,*" Van Nostrand Reinhold, 1990.

72. Modern, J.J., Philips, C.R. and Davis E.W., "*Project Management with CPM and PERT*", Van Nostrand Reinhold, 3rd Edition, New York, 1993.

73. Nagpal, B.S. (GP), "*Accidents, Causes and Prevention*". 12th Advance Maintenance Management Program, 19-23 Jan, 1999, IIPE, New Delhi.

74. Nakajima, S., "*The Organization of Maintenance in Japanese Industry*" Maintenance Management International, Vol.3, No. 2, Aug 1982, Pp. 83-88.

75. Nakajima, S., "*TPM Challenges to the Improvement of Productivity by Small Group Activities*", *Maintenance Management International,* vol-6, 1986, Pp 73-86.

76. Nakra, B.C, Yadava, G.S. and Thuestad, L., "*Vibration Measurement and Analysis*," National Productivity Council, 1989.

77. Nayak, S.K., "*Energy Conservation Techniques*," Workshop on Energy Conservation, 29-30 Jan, 2004, Electrical Research and Development Association, Baroda.

78. Nebhrani, N.C. and Sahoo, P., "*Predictive Maintenance and Residual Life Assessment of Rotating Machinery in Power Plant*," 24-28, Jun. 1995, IIPE, New Delhi.

79. Niebel, B.W., "*Engineering Maintenance Management*," Marcel Dekker, 1994.

80. Nigel, Slack, Stuart and Robert Johnston, "*Operations Management*," Pearson (U.K.), 3rd Edition, 2001.

81. Nikander, I.O., Eloranta, Eero, "*Project Management by Early Warnings,*" International Journal of Project Management, 2001, 19(7), Pp 99-385.

82. Nyman, D. and Levitt, T, "*Maintenance Planning, Scheduling and Co-ordination*," Industrial Press Inc., New York, 2001.

83. Palmer, D., "*Maintenance Planning and Scheduling Handbook*," McGraw Hill, 1999.

84. Pillai, G.M. (Editor), "*The New Energy Economy*," Word Institute of Sustainable Energy, Pune, 2005.

85. Raheja, Ashish, "*Air-conditioning the CII Godrej Green Business Centre* ", Air-conditioning and Refrigeration Journal, Pp 25-30, Vol. 7, no. 2, April-June 2004.

86. Raghupathy, S., "*Green Building – Opportunities and Benefits*," Air-conditioning and Refrigeration Journal, Pp 65-68, Vol. 7, No. 2, April-June 2004.

87. Rao, B.K.N. (Editor), "*Handbook of Condition Monitoring*," Elsevier Science Let, 1996.

88. Rausand, M. "*Reliability Centered Maintainence,*" Reliability Engineering and System Safety, Vol. 60, Pp 121-132, 1998.

89. Ravel, B.N., "*Feedback on Energy Audit Conducted at TPS*", Workshop on Energy Conservation, 29-30 Jan. 2004, ERDA, Baroda.

90. Sasha, B.N., "*Integrated Maintenance Management – Concept to Computerization*," S.B.A. Publication, New Delhi.

91. Sangerud, O.T., "*Practical Condition Monitoring of Rotating Equipment*," Norwegian Maritime Research, Vol. 4, 1975.
92. Shiroz, K., "*Equipment Effectiveness, Chronic Losses and Other TPM Improvement Concepts*," Chapter 2 of TPM Development Program: Implementing Total Productive Maintenance [Nakajima S. (Editor)] Productivity Press, Cambridge, Massachusetts, USA, 1989, Pp 27-84.
93. Shooman, M.L., "*Probabilistic Reliability: An Engineering Approach*," McGraw Hill, New York, 1968.
94. Sinha, S.K., "*Reliability and Life Testing*," Wiley Eastern Ltd., New Delhi, 1986.
95. Smith, G.D., "*Wear Debris Monitoring*."
96. Srinath, L.S., "*Reliability Engineering*," Affiliated East-West Press, 1991.
97. Suzaki, K., "*The Manufacturing Challenge: Techniques for Continuous Improvement the Free Press* (A division of MaC Million Inc.), New York, 1987, Pp 31-41and113-123.
98. Takahashi, Y., "*Maintenance Oriented Management via Total Participation Total Production Maintenance, A New Task for Plant Managers in Japan*," Trotechnica, Vol. 2, No. 2, May 1981, Pp 79-88.
99. Theusan, H.G., Fabrycky, W.J. and Theusen, G.J., "*Engineering Economy*," Prentice Hall of India, 4th Edition, 1975, Pp 168-179, New Delhi.
100. Tiwari, S.K. and Kumar A., "*Reliability and Maintainability – A Diagnostic Approach*," 7th Advanced Maintenance Management Program, 1994, IIPE, New Delhi.
101. Turner, J.R. and Simister, S.J., "*Project Contract Management and A Theory of Organization*," International Journal of Project Management 2001, 19(8), Pp 64-457.
102. Velan, B., "*Options in Selection of Equipments and System for Inplant Bulk Handling*," International Seminar on Bulk Handling, 16-18 Nov., 2000, IIPE, New Delhi.
103. Venkatraman, K., "*Maintenance Engineering and Management*," Prentice Hall of India, 2007.
104. White, E.N., "*Maintenance Planning, Control and Documentation*," Gower Press, U.K.
105. Wiest, J.D., and Levy, F.K., "*Management Guide to PERT/CPM*", Prentice Hall of India, New Delhi, 1978.

106. Wills, A., "*Asset Maintenance Management – A Guide to Developing Strategy and Improving Performance,*" Industrial Press Inc., New York, 2002.

107. Wireman, T. "*Total Productive Maintenance: An American Approach,*" Industrial Press Inc., New York, 1991.

108. Wireman, T., "*World Class Maintenance*", *Industrial Press Inc.*, New York, 1990,

109. Yadava, G.S., "*Maintenance Audit for Improved Plant Efficiency*", National Seminar on Energy Optimization and Efficiency Improvement in Infrastructure and Industrial Sectors, 29-30 Oct. 1998, IIPE, New Delhi.

110. www.maintenance resource.com

111. *elsmar.com/planned – maintenance/*

112. www. idcon. Com/article — revitalizing.htm

113. www.ndmindia.nic.in

114. www.idcon.com

❑❑❑

Index

❑❑❑